AF581644

Urban and Industrial Wastewater Disinfection and Decontamination by Advanced Oxidation Processes (AOPs)

Urban and Industrial Wastewater Disinfection and Decontamination by Advanced Oxidation Processes (AOPs): Current Issues and Future Trends

Editors

Stefanos Giannakis
Ana Rita Lado Ribeiro
Jorge Jesus Rodríguez-Chueca

MDPI • Basel • Beijing • Wuhan • Barcelona • Belgrade • Manchester • Tokyo • Cluj • Tianjin

Editors
Stefanos Giannakis
Universidad Politécnica de Madrid (UPM)
Spain

Ana Rita Lado Ribeiro
University of Porto (FEUP)
Portugal

Jorge Jesus Rodríguez-Chueca
Universidad Politécnica de Madrid (UPM)
Spain

Editorial Office
MDPI
St. Alban-Anlage 66
4052 Basel, Switzerland

This is a reprint of articles from the Special Issue published online in the open access journal *Water* (ISSN 2073-4441) (available at: https://www.mdpi.com/journal/water/special_issues/Disinfection_Decontamination).

For citation purposes, cite each article independently as indicated on the article page online and as indicated below:

LastName, A.A.; LastName, B.B.; LastName, C.C. Article Title. *Journal Name* **Year**, *Volume Number*, Page Range.

ISBN 978-3-0365-1112-2 (Hbk)
ISBN 978-3-0365-1113-9 (PDF)

Contents

About the Editors

Stefanos Giannakis, Doctor of Environmental Engineering from the Aristotle University of Thessaloniki, Greece, and the Polytechnic University of Catalonia, Spain, is currently a "Ramón y Cajal" Fellow at the Polytechnic University of Madrid, Spain. His main research lines include the development of novel, sustainable catalytic (waste)water treatment processes, unraveling the fundamentals governing microorganism disinfection, as well as elucidating the photochemistry and photobiology of natural waters. He has participated in 6 National and International projects on the chemical and microbiological quality of water and wastewater in Europe, Africa, and Latin America. Dr. Giannakis is currently the author of 75 peer-reviewed research articles, has contributed to 35 conference works and has an h-index of 25.

Ana Rita Lado Ribeiro holds a Researcher position at Laboratory of Separation and Reaction Engineering–Laboratory of Catalysis and Materials (LSRE–LCM) hosted by Faculty of Engineering of University of Porto and is Associate Editor of *Chemical Engineering Journal Advances*—Elsevier. Her main research interests include the analysis of contaminants of emerging concern in the environment; novel miniaturized sample preparation methodologies; water/wastewater treatment by advanced oxidation processes; and identification of reaction by-products. She is Principal Investigator of a funded project at LSRE-LCM (SAMPREP) and FEUP team leader of another one (ENANTIOTOX), having participated as team member in >15 research projects. She has co-authored 2 patent requests, 55 publications in JCR indexed journals and 8 chapters in books or in book series with peer review (h-index 25), presenting her work at highly ranked international conferences.

Jorge Jesus Rodríguez-Chueca, Doctor of Chemical Engineering and Environmental Technologies from the University of Zaragoza, Spain, is currently Associate Professor I3 at the Polytechnic University of Madrid, Spain. Jorge has contributed 40 peer-reviewed research articles, 7 technical articles in non-indexed journals, 3 book chapters; in addition, he was the editor of the book of abstracts of IWA Young Water Professionals Conference 2019, with an h-index of 17. He has participated in 7 Regional, National and International projects on water treatment and management, being the Principal Investigator in three of them. His main research interests are the new technologies for environmental remediation, mainly focused on water and wastewater treatment (disinfection, removal of micropollutants, organic matter, etc.) especially through the use of Advanced Oxidation Processes.

Editorial

Urban and Industrial Wastewater Disinfection and Decontamination by Advanced Oxidation Processes (AOPs): Current Issues and Future Trends

Ana Rita Lado Ribeiro [1], Jorge Jesus Rodríguez-Chueca [2] and Stefanos Giannakis [3,*]

1 Laboratory of Separation and Reaction Engineering—Laboratory of Catalysis and Materials (LSRE-LCM), Faculdade de Engenharia, Universidade do Porto, Rua Dr. Roberto Frias s/n, 4200-465 Porto, Portugal; ritalado@fe.up.pt

2 Departamento de Ingeniería Química Industrial y del Medio Ambiente, Universidad Politécnica de Madrid (UPM), E.T.S. de Ingenieros Industriales, c/ José Gutiérrez Abascal 2, 28006 Madrid, Spain; jorge.rodriguez.chueca@upm.es

3 Departamento de Ingeniería Civil: Hidráulica, Energía y Medio Ambiente, Canales y Puertos, Universidad Politécnica de Madrid (UPM), E.T.S. Ingenieros de Caminos, Unidad docente Ingeniería Sanitaria, c/ Profesor Aranguren s/n, 28040 Madrid, Spain

* Correspondence: stefanos.giannakis@upm.es

Citation: Lado Ribeiro, A.R.; Rodríguez-Chueca, J.J.; Giannakis, S. Urban and Industrial Wastewater Disinfection and Decontamination by Advanced Oxidation Processes (AOPs): Current Issues and Future Trends. *Water* **2021**, *13*, 560. https://doi.org/10.3390/w13040560

Academic Editor: Anas Ghadouani

Received: 4 February 2021
Accepted: 19 February 2021
Published: 23 February 2021

Without any doubt, the 21st century has kick-started a great evolution in all aspects of our everyday life. The tremendous advances in medicine, production of goods (such as chemicals, plastics), etc. have all contributed to an increase in the standards and the comforts of the modern society. However, combined with the growing demands in water due to the uneven population increase and product manufacturing needs, the rapid change has impacted the wastewater flows and composition of both urban and industrial origin. Environmental engineers have been facing a constantly changing landscape, where new problems seem to emerge in an alarming, "whack-a-mole"-like situation. From antibiotic resistance to microplastics, from the ever-growing list of contaminants of emerging concern to their transformation products, the need for proper, advanced wastewater treatment methods are dominating environmental engineering research to alleviate the shift of urgent needs of the water sector. Nonetheless, the multidisciplinary nature of this huge environmental problem spans a number of research topics, namely environmental, chemical and process engineering, water and sanitary engineering, environmental science, chemistry, biology and related disciplines, material science, toxicology, risk assessment, economy, social sciences, ecology and environmental policy, among others. Therefore, in order to develop efficient solution tools to tackle the urban and industrial wastewater challenges, it is of utmost importance to promote the knowledge exchange between all expertise fields involved, thus promoting a scientific, technological, and societal cooperation for a more sustainable environment and society.

It has also become clear over the past two decades that the perception of water as an infinite resource is distorted and a dangerous idea to hold on to. As a result, novel and more holistic approaches, such as effluents' reuse, the circular economy of water, the one-health integration in wastewater treatment, have gradually replaced the archaic views on treatment and discharge of wastewater. The (well-deserved) price paid for adopting such viewpoints into policy and legislation has been partially transferred to the end of the line, namely wastewater treatment plants (WWTPs). Lower limits, recovery schemes, safer disposal and handling of wastes illustrate the reality of the treatment schemes nowadays.

In 2020, the Special issue "Urban and industrial wastewater disinfection and decontamination by Advanced Oxidation Processes (AOPs): current issues and future trends" was launched to gather the innovations of the research community on (waste)water treatment worldwide. Following the call, several articles on various aspects of (waste)water treatment have been published, both in form of original research papers and literature reviews. Authors affiliated with institutions and companies from Colombia to Switzerland, from Portugal to India and from Spain to Iran, have contributed by disseminating their research results within this Special Issue; a brief summary of the published papers follows:

Guerra-Rodríguez et al. [1] overviewed the different strategies that can be adopted in the water sector in the context of the circular economy. Wastewater reuse is a mature and well-established strategy in many water-scarce regions of the world, but its reuse poses associated risks to ecosystems and humans due to the presence of pathogens, contaminants of emerging concern, and microbial resistance genes. However, there are other options for applying the circular economy in the water sector that are still largely unexploited, such as the recovery of materials from aqueous streams or energy recovery; Guerra-Rodríguez et al. reviewed all these options, highlighting the challenges and opportunities in their application.

Camargo-Perea et al. [2] reviewed extensively the data on the degradation of emerging pollutants by ultrasound AOP, with a special emphasis on pharmaceuticals, providing a deep understanding on the operation fundamentals and the parameters that most affect its efficiency, namely, the ultrasonic frequency, the electrical power, the pH and temperature of the solution, the nature of the target contaminant, the constituents of the water matrix, and the type and the geometry of the sonochemical reactor. In addition to the applications that were detailed, this review discussed the future perspectives and the cost implications of ultrasound AOP as a possible candidate to tackle water pollution by emerging pollutants.

Guateque-Londoño et al. [3] studied the removal of the antihypertensive losartan as a pharmaceutical model in simulated fresh urine through the application of ultrasound and UVC/H_2O_2. They reported higher selectivity of the sonochemical process traduced by a higher ratio between the degradation rate constants obtained in fresh urine and distilled water for the removal of losartan than that obtained in the UVC/H_2O_2 system. Although neither of the treatments reached the mineralization of the pollutant in distilled water, it was confirmed that the sonochemical system reduced the phytotoxicity of the sample, showing the potential of this technology on the regeneration of wastewater. Finally, the authors studied the generation of transformation products from the degradation of losartan, and the fitting of these results with theoretical methods to predict and rationalize the attack of degrading species. The computational analyses confirmed that the atoms on imidazole moiety of losartan were the most susceptible to transformation by the radical species, in agreement with the experimental results obtained.

The presence of antibiotics in wastewater effluents is a problem of emerging concern threating ecosystems and human health. Moles et al. [4] evaluated the behavior of five antibiotics (amoxicillin, enrofloxacin, sulfadiazine, trimethoprim, and azithromycin) in the influent and effluent of different WWTPs located in the north of Spain. Moles et al. determined that azithromycin was the antibiotic with the highest mass load followed by enrofloxacin, trimethoprim, sulfadiazine, and amoxicillin. Regarding the behavior of these substances in the WWTPs, they observed that biological treatments have a significant influence on removal, especially the use of trickling filters that showed the highest degree of removal. Besides, the authors studied the performance of a TiO_2 photocatalytic treatment plant installed as tertiary treatment that minimized the Ti release into the environment and allowed catalyst reuse. This technology totally removed sulfadiazine, amoxicillin and azithromycin, while 80% of trimethoprim and enrofloxacin was removed.

Hospital effluents are the main emitters of pollutants of emerging concern such as pharmaceuticals and antibiotics. Rosero Parra et al. [5] described the treatment of hospital wastewater using two novel catalysts supported on graphene (SnO-GO and TiO_2-GO). Both heterogeneous photocatalysts were demonstrated as alternatives for abatement of pollution in this kind of effluents, reducing chemical oxygen demand by 85%, the dissolved organic carbon by 94%, and 80% of phenols, allowing the discharge of the effluents in compliance with current legislation.

Karbasi et al. [6] studied a semiconductor photocatalytic material based on Bi and W oxide, as an alternative to the traditional TiO_2 photocatalytic disinfection process. Owing to its large band gap and exceptional structure (micro/nanohierarchical, flower-like morphology) it absorbs more photons and further into the visible range, while it reduces the kinetic limitations associated with semiconductors. After the complete characterization of the newly synthesized material, *E. coli* bacteria were used as a model microorganism and the disinfection capabilities of Bi_2WO_6 were found to match the benchmark TiO_2 P25 under solar light but had superior bactericidal efficacy under visible light. This opens the possibilities of further studying this material and advance towards indoor applications, where sterility is a prerequisite and plain fluorescent lighting is provided.

The photocatalytic degradation of potassium hexacyanoferrate (III), as a model cyanocomplex of gold mining wastewaters, was studied by Arce-Sarria et al. [7] in a bench-scale compound parabolic collector (CPC) reactor assisted with a light-emitting diode (UV-A/LED) and a hydrothermally treated TiO_2 P25, which showed a specific surface area 2.5-fold superior to the original catalyst, a slightly higher band gap energy, and a mesoporous structure. Although the TiO_2 P25 performed better, the higher free cyanide release achieved with the modified material in this specific case may be beneficial for its reuse in the gold extraction process. This report opens new prospects for future research on the structural changes of the catalyst and their potentialities for photocatalytic applications.

Graça et al. [8] proposed a combined technology (ozonation followed by ultrafiltration) to simultaneously reduce organic contaminants and microbial load of secondary urban wastewater for its further reuse, complying with the legislation for water quality for irrigation. The estrogenic activity, cell viability and cellular metabolic activity were also monitored in order to assess the impact of the treatment on the biological effects. Although most organic microcontaminants were removed (except citalopram and isoproturon), the biological effects did not suggest the production of toxic by-products, and the reduction of the bacterial loads targeting the water quality standards for irrigation was effective, the authors highlighted that microbial regrowth was observed after storage, with the concomitant increase in the genes 16S rRNA and *intI*1.

Finally, Acosta-Herazo et al. [9] showed the main features of the PHOTOREAC tool developed, as well as the results in different simulations, thanks to the experience gathered by their research groups at Cartagena University (Cartagena de Indias, Colombia) and the Universidad del Valle (Cali, Colombia) during the last twenty years of research in heterogeneous solar photocatalysis, and on extensive literature research in photoreactor engineering. PHOTOREAC is an open-access application developed in the graphical user interface of MATLAB® that allows a user-friendly evaluation of the solar photoreactors operation. Furthermore, they showed the potential of the tool in several case studies such as the removal of dichloroacetic acid and methylene blue in different types of photoreactors, as well as the modeling of radiation in a Flat Plate Photoreactor.

Considering the above contributions and the current issues dominating literature, the recent challenges faced by WWTPs exceed their regular capacity, and the need for appropriate wastewater treatment calls for novel, sophisticated methods of decontamination prior to its discharge or reuse. Emerging threats such as antibiotic-resistant bacteria, antibiotic-resistance genes, and the contaminants of emerging concern (chemicals, microplastics) demand efficient, end-of-pipe solutions before their discharge into the environment or reclamation for reuse purposes. Under this scenario, AOPs have been procured as effective methods for wastewater disinfection and decontamination. Despite the huge amount of works developed in the last decade in this area, some research opportunities and future directions can be highlighted.

One important feature in any comparative analysis between different treatments or studies is the examination of the same endpoints for a safe effluent discharge or reuse, namely the abatement of chemical contaminants, mitigation of by-products, reduction of the toxicity of the effluent, bacterial inactivation, minimization of regrowth, control of antibiotic resistance determinants, and cost of treatment. This comparison is often difficult when different studies are focused in dissimilar endpoints, different matrices, and distinct operational conditions (typical at small-scale tests). It is important to highlight that larger-scale studies than some published in this Special Issue and a further cost effectiveness analysis of scaled-up processes are valuable for a comprehensive analysis of any given treatment option. In this regard, solar-driven AOPs are very promising when compared with other advanced technologies, being a research topic that has plenty room for further advances, for instance by developing and tailoring new photocatalysts aimed to be active under sunlight.

Regarding the endpoints, the development of tools to predict the fate and behavior of emerging contaminants during wastewater treatment would be a great asset. Given that the elimination of chemical and biological contaminants of emerging concern from wastewater by AOPs largely depends on the composition of the water matrix, studies conducted under environmental-like conditions are crucial to advance our knowledge on AOPs. In this sense, environmental relevant concentrations of mixtures of contaminants and realistic water matrices should be used instead of purely artificial conditions because the efficiency of each process results from a complex balance between inhibitory and promoting effects originated by the water components. The available data about such matrix effects on the elimination efficiency of contaminants in WWTPs is still limited and research towards the impact of the main water/wastewater constituents and the mechanisms governing it are required. These environmental-like conditions are also imperative for a better understanding of the competition effects between different contaminants; thus, the approach of studying spiked synthetic solutions spiked with one or few compounds at concentrations orders of magnitude above than those found in the environment tends to become obsolete, except for some specific applications such as performance studies of new catalysts, elucidation of degradation pathways, and characterization of chemical structures of transformation products.

As far as the fate of contaminants during treatment is concerned, the structural elucidation of the transformation products is pivotal for an accurate evaluation of the degradation pathways and for a deep understanding of their possible side-effects, like toxicity. The ecotoxicological studies provide a useful tool to clarify the harmful effects of both parent contaminants and their transformation products on the ecosystems and human health, as well as to understand the possible impact of the reagents and catalysts used. Combining other technologies can be an interesting option to deal with toxic transformation products, for instance by coupling an adsorptive post-treatment. On the other hand, AOPs are very promising as post-treatment to treat the resulting concentrate of membrane-driven processes.

Concluding, the water/wastewater treatment is a hot topic of research that has been challenged in the last decade by newly identified threats (e.g., antibiotic resistance, microplastics) that have provided new perspectives and recent opportunities for investigation. Besides the importance of bearing in mind the multidisciplinary of this topic and the need for knowledge exchange between different subjects in order to generate innovation and scientific advances, raising awareness within society and promoting individuals' involvement in the mitigation of wastewater problems is a prerequisite for a more proficient development and implementation of possible solutions.

Author Contributions: The authors made equal contributions to this editorial. All authors have read and agreed to the published version of the editorial.

Funding: Ana Rita Lado Ribeiro would like to acknowledge the support of Base Funding—UIDP/50020/2020 of the Associate Laboratory LSRE-LCM—funded by national funds through FCT/MCTES (PIDDAC); and FCT funding under DL57/2016 Transitory Norm Programme. Jorge Rodríguez-Chueca acknowledges the Community of Madrid (Comunidad de Madrid) for funding the research project IN_REUSE (APOYO-JOVENES-X5PKL6-88-KZ46KU) within the framework of the multi-year agreement with the Universidad Politécnica de Madrid. Stefanos Giannakis would like to acknowledge the Spanish Ministry of Science, Innovation and Universities (MICIU) for the Ramón y Cajal Fellowship (RYC2018-024033-I) and the "NAVIA" Project (PID2019-110441RB-C32).

Institutional Review Board Statement: Not applicable.

Informed Consent Statement: Not applicable.

Data Availability Statement: Not applicable.

Acknowledgments: The editors would like to gratefully acknowledge the reviewers who kindly reviewed the submissions to this Special Issue, dedicating their time to provide precious scientific input towards the improvement of this collection of articles.

Conflicts of Interest: The authors declare no conflict of interest.

References

1. Guerra-Rodríguez, S.; Oulego, P.; Rodríguez, E.; Singh, D.N.; Rodríguez-Chueca, J. Towards the Implementation of Circular Economy in the Wastewater Sector: Challenges and Opportunities. *Water* **2020**, *12*, 1431. [CrossRef]
2. Camargo-Perea, A.L.; Rubio-Clemente, A.; Peñuela, G.A. Use of Ultrasound as an Advanced Oxidation Process for the Degradation of Emerging Pollutants in Water. *Water* **2020**, *12*, 1068. [CrossRef]
3. Guateque-Londoño, J.F.; Serna-Galvis, E.A.; Ávila-Torres, Y.; Torres-Palma, R.A. Degradation of Losartan in Fresh Urine by Sonochemical and Photochemical Advanced Oxidation Processes. *Water* **2020**, *12*, 3398. [CrossRef]
4. Moles, S.; Mosteo, R.; Gómez, J.; Szpunar, J.; Gozzo, S.; Castillo, J.R.; Ormad, M.P. Towards the Removal of Antibiotics Detected in Wastewaters in the POCTEFA Territory: Occurrence and TiO_2 Photocatalytic Pilot-Scale Plant Performance. *Water* **2020**, *12*, 1453. [CrossRef]
5. Rosero Parra, L.; Guerrero Pantoja, L.; Mena, N.L.; Machuca-Martínez, F.; Urresta, J. Evaluation of TiO_2 and SnO Supported on Graphene Oxide (TiO_2-GO and SnO-GO) Photocatalysts for Treatment of Hospital Wastewater. *Water* **2020**, *12*, 1438. [CrossRef]
6. Karbasi, M.; Karimzadeh, F.; Raeissi, K.; Rtimi, S.; Kiwi, J.; Giannakis, S.; Pulgarin, C. Insights into the photocatalytic bacterial inactivation by flower-like Bi_2WO_6 under solar or visible light, through in situ monitoring and determination of reactive oxygen species (ROS). Water (Switzerland) MDPI AG. *Water* **2020**, *12*, 1099. [CrossRef]
7. Arce-Sarria, A.; Aldana-Villegas, K.M.; Betancourt-Buitrago, L.A.; Colina-Márquez, J.Á.; Machuca-Martínez, F.; Mueses, M.A. Degradation of Hexacyanoferrate (III) from Gold Mining Wastewaters via UV-A/LED Photocatalysis Using Modified TiO_2 P25. *Water* **2020**, *12*, 2531. [CrossRef]
8. Graça, C.A.L.; Ribeirinho-Soares, S.; Abreu-Silva, J.; Ramos, I.I.; Ribeiro, A.R.; Castro-Silva, S.M.; Segundo, M.A.; Manaia, C.M.; Nunes, O.C.; Silva, A.M.T. A Pilot Study Combining Ultrafiltration with Ozonation for the Treatment of Secondary Urban Wastewater: Organic Micropollutants, Microbial Load and Biological Effects. *Water* **2020**, *12*, 3458. [CrossRef]
9. Acosta-Herazo, R.; Cañaveral-Velásquez, B.; Pérez-Giraldo, K.; Mueses, M.A.; Pinzón-Cárdenas, M.H.; Machuca-Martínez, F. A MATLAB-Based Application for Modeling and Simulation of Solar Slurry Photocatalytic Reactors for Environmental Applications. *Water* **2020**, *12*, 2196. [CrossRef]

Review

Towards the Implementation of Circular Economy in the Wastewater Sector: Challenges and Opportunities

Sonia Guerra-Rodríguez [1], Paula Oulego [2], Encarnación Rodríguez [1], Devendra Narain Singh [3] and Jorge Rodríguez-Chueca [1,*]

1 Department of Industrial Chemical Engineering & Environment, Escuela Técnica Superior de Ingenieros Industriales, Universidad Politécnica de Madrid, Calle José Gutiérrez Abascal 2, 28006 Madrid, Spain; sonia.guerra@upm.com (S.G.-R.); encarnacion.rodriguez@upm.es (E.R.)

2 Department of Chemical and Environmental Engineering, University of Oviedo, c/Julián Clavería s/n, E-33071 Oviedo, Spain; oulegopaula@uniovi.es

3 Department of Civil Engineering, Indian Institute of Technology Bombay, Powai, Mumbai 400076, India; dns@civil.iitb.ac.in

* Correspondence: jorge.rodriguez.chueca@upm.es; Tel.: +34-910-677-334

Received: 11 March 2020; Accepted: 13 May 2020; Published: 18 May 2020

Abstract: The advancement of science has facilitated increase in the human lifespan, reflected in economic and population growth, which unfortunately leads to increased exploitation of resources. This situation entails not only depletion of resources, but also increases environmental pollution, mainly due to atmospheric emissions, wastewater effluents, and solid wastes. In this scenario, it is compulsory to adopt a paradigm change, as far as the consumption of resources by the population is concerned, to achieve a circular economy. The recovery and reuse of resources are key points, leading to a decrease in the consumption of raw materials, waste reduction, and improvement of energy efficiency. This is the reason why the concept of the circular economy can be applied in any industrial activity, including the wastewater treatment sector. With this in view, this review manuscript focuses on demonstrating the challenges and opportunities in applying a circular economy in the water sector. For example, reclamation and reuse of wastewater to increase water resources, by paying particular attention to the risks for human health, recovery of nutrients, or highly added-value products (e.g., metals and biomolecules among others), valorisation of sewage sludge, and/or recovery of energy. Being aware of this situation, in the European, Union 18 out of 27 countries are already reusing reclaimed wastewater at some level. Moreover, many wastewater treatment plants have reached energy self-sufficiency, producing up to 150% of their energy requirements. Unfortunately, many of the opportunities presented in this work are far from becoming a reality. Still, the first step is always to become aware of the problem and work on optimizing the solution to make it possible.

Keywords: circular economy; wastewater reuse; zero waste; zero energy; substances recovery; sustainable development goals

---

1. Introduction

It is evident that without water, there is no life. Water is not only essential for human survival, but also for all living organisms. On 28 July, 2010, water and sanitation were recognised as human rights by the United Nations General Assembly [1]. Moreover, water is essential for human activities in many sectors of the economy—and not only from a biological point of view. In the previous century, the increase of human activities in all economic sectors has favoured the development of society (e.g., increasing life expectancy). However, this was at the expense of the over-exploitation of planet resources, accompanied by the emission of greenhouse gases into the atmosphere, organic and biological pollutants into water bodies, and the generation of tonnes of solid waste [2]. In terms of water, not only the quality of water bodies has worsened due to contamination from human activities, but there has also been increased withdrawal and consumption of water resources [3]. For these reasons, the number of regions in the world with hydric stress has grown considerably—even regions with a regular precipitation regime [4].

Under this paradigm, humanity must make decisions to increase the quality of water bodies and to increase sustainability in the management of the resources. The 2030 Agenda for Sustainable Development of the United Nations established a specific Sustainable Development Goal (SDG) to ensure availability and sustainable management of water and sanitation for all. This is the SDG 6, and it presents a specific target, 6.4 for instance, aimed to increase water-use efficiency in all of the sectors, ensuring sustainable withdrawal and supply of freshwater to address water scarcity. There are other SDGs not directly focused on water, but with specific targets involving it. This is the case of SDG 11 (to make cities and human settlements inclusive, safe, resilient, and sustainable) and 12 (to ensure sustainable consumption and production patterns). In December 2015, the European Commission adopted a Circular Economy Plan called "Closing the loop—a European Union (EU) action plan for the Circular Economy" [5]. It is a strategy to change the consumption model of European citizens towards a climate neutral, circular economy, which would minimise the impact on the environment. This plan is mainly focused on reducing, recycling, and recovering of waste, focusing on paper, ferrous metals, aluminium, glass, plastic, and wood. However, in context to the idea of "closing the loop", the circular economy can go beyond any sector, for instance, the wastewater treatment sector.

The circular economy can also be applied in the wastewater treatment sector. For example, thinking about the reclamation and reuse of wastewater seems to be an excellent option to increase water resources by reducing the environmental impacts [6]. The technology can treat wastewater to the point of being fit for use even by humans. As this idea is still unacceptable for most people, and is known as the "yuck factor" [7], the most common uses of reclaimed wastewater are agricultural, industrial, urban, environmental, or recreational purposes [6,8–10]. However, if we think not only about the value of water as a resource, but also about the substances contained in these effluents, we can observe a wide range of possibilities to apply the circular economy. For example, recovery of nutrients [11–14] and energy [15–18] from wastewater is well known, but there are more possibilities, such as the recovery of biomolecules [19–23], metals [24–27], organic and inorganic compounds [28,29], etc. On the other hand, there are other options to apply a circular economy in water, for example, the valorisation of sewage sludge. The traditional valorisation of sludge is through its reuse with agricultural purposes [30,31]. However, they can also be valorised for the preparation of carbonaceous materials with adsorptive [32–35] or catalytic abilities [36–39]. These adsorbents or catalyst can be applied at some stage of the same water or wastewater treatment, to remove organic pollutants by adsorption [40], or for the application of Advanced Oxidation Processes (AOPs) to remove organic or biological pollution [41].

The main goal of this review manuscript is to go deep in the state-of-the-art of the introduction of circular economy in the wastewater treatment sector, understood as the activities related to treatment, and valorisation and reuse of wastewater of any origin (municipal, industrial, or agricultural).

Throughout this manuscript, the degree of application of strategies to close the loop in the wastewater sector will be addressed. Actions with a high degree of maturation will be described, as well as the reclamation and reuse of wastewater, or the nutrients and energy recovery and others in a nascent level of development, in the form of recovery of biomolecules or valorisation of sewage sludge.

2. Wastewater Reclamation and Reuse

2.1. Definition and Overview of Reclamation around the World

The reclaimed wastewater is treated when the residual effluents are subjected to an additional or complementary treatment to increase or make adequate its quality to the final reuse purposes. Additionally, it is considered as a non-conventional water resource.

The reuse of these reclaimed effluents lies in their application in different purposes, instead of discharging them to water bodies after conventional treatment. According to current national legislations around the world, the reuse purposes are varied, such as agricultural, industrial, urban, recreational, environmental, or even human consumption [6]. The reclamation wastewater treatment plant (RWWTP) is normally a wastewater treatment plant equipped with a tertiary treatment composed of different technologies, depending on the quality requirements of the final effluent regarding the final use.

Besides, there must be a distinction between the indirect and direct reuse of treated wastewater, also called as unplanned or planned reuse respectively [42]. The indirect reuse is defined as the discharge of treated effluents to the watercourses, and after passing through the public hydraulic domain, is subject to its subsequent use. The direct reuse of treated wastewater is that in which the second use occurs after the first, without water being incorporated into the public hydraulic domain between them [43]. This review mainly describes the experiences in direct reuse of the treated effluents.

It is difficult to talk about reclamation and reuse of wastewater when approximately 80% of wastewater is discharged into the world's waterways without any kind of treatment [44]. The construction of RWWTPs would solve the problems associated to the uncontrolled discharge of wastewaters, as well it would help closing the loop reducing the withdrawal of water resources (e.g., with agricultural purposes), thereby, doing more sustainable management of this resource.

Until now, the reclamation and reuse of wastewater has been a priority exclusively for countries with significant problems of hydric stress or water scarcity. Figure 1 shows the top countries reusing wastewater, dividing the figures regarding the total use (m^3/d; Figure 1A), total reuse per capita (m^3/d; Figure 1B), and the ratio reuse:water extraction (Figure 1C) in 2008. Despite the outdated data shown in Figure 1, due to the absence of a current ranking, and the difficulty of individual search for accurate data from each country, these can be used as a basis for understanding which countries have a need to reuse wastewater. In this paradigm, during 2018, China and Mexico were at the top of the list of the countries reusing wastewater, with almost 15 hm^3/d; however, in both cases, this regeneration is happening without additional treatment. If this point is considered, the top country to reuse the reclaimed wastewater was the United States of America, with almost 8 hm^3/d. However, this ranking changes if the volume per capita is considered, or the percentage of reuse (considering the extraction of water resources). Under this classification, Qatar, Kuwait, and Israel would be the top countries, Kuwait being the country with the largest ratio of reuse regarding the withdrawal of water resources (35.2%).

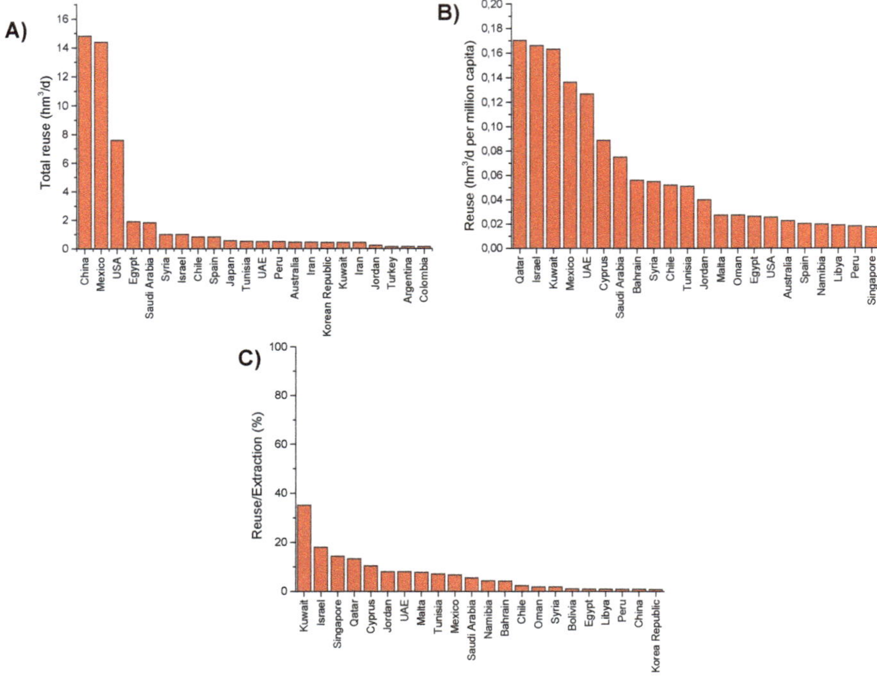

Figure 1. Global situation of wastewater reuse around the world. Adaptation from [45]. (**A**) Total reuse of wastewater (hm^3/d); (**B**) reuse (hm^3/d per million capita); (**C**) percentage of reuse/extraction.

2.2. Legislation and Guidelines Around the World

It is quite difficult to compile all the legislation and guidelines around the world about water reclamation and reuse, for this reason, this section tries to summarize the most important guidelines, as well as the legislation of some leading and emerging countries in this field.

In 2004, the Environmental Protection Agency of the United States of America (US EPA) published *"Guidelines for water reuse"*. This guideline was revised later, in 2012. In this document, different physico-chemical and biological parameters are established regarding the final use of the reclaimed water. In this way, for instance, reuse of reclaimed water is recommended for agricultural purposes only when the concentration of *Escherichia coli* is below 1 CFU/100 mL. This recommendation contrasts with the ones given by the World Health Organization (WHO) in 2006. Thus, the WHO published the *"Guidelines for the safe use of wastewater, excreta and greywater"*. This document tried to establish the maximum admissible values for different physico-chemical and biological parameters posing a threat to human health, considering the reuse of reclaimed wastewater in agriculture and aquaculture. In this case, the WHO considered that reclaimed water can be reused as long as the concentration of *E. coli* is below 1000 CFU/100 mL. This difference between both guidelines is evidently the reason why WHO guidelines are applied in developing countries. Usually, the populations in these countries have limitations (economical, lack of government support, lack of qualified personnel, etc.) in terms of access to technologies for water treatment.

In the European Union, the reclamation and reuse of water encounter numerous barriers in most of the member countries, for example, a supportive and coherent framework in terms of legislation. This is the reason why the Commission is working on common legislation for all of the members in the frame of the Circular Economy Plan. Nowadays, some European countries have the legislation to regenerate and reuse wastewater (e.g., Spain, Portugal, France, etc.); others are preparing legislation, while others have none at all. Table 1 summarizes the sectors in which reclaimed water is currently applied on by country, and the regulations or guidelines of each country (European Union countries and outside EU countries).

Spain was ranked as the country in Europe with the highest rate of wastewater reuse, and, as can be observed in Figure 1, is in the top ten worldwide [46]. The Royal Decree 1620/2007 (RD 1620/2007) establishes the legal regime for the reuse of reclaimed water in Spain, thereby setting up different maximum admissible values for different physico-chemical and microbiological parameters, depending on the final use for the reclaimed water. In relation to the microbiological parameters, *Escherichia coli* and intestinal Nematodes (*Ancylostoma*, *Trichuris*, and *Ascaris* genera) must be controlled for all established uses. *Legionella* analysis is also mandatory for all of the established water uses, except for environmental use, which does not contemplate its control. *Salmonella* spp. should be controlled when the uses are agricultural and industrial. Finally, the control of the parasitic plathyhelminthes *Taenia saginata* and *Taenia solium* is mandatory if the water is used to irrigate pastures for consumption by meat-producing animals, since they are intermediate hosts of these parasites.

In addition to the existing quality criteria in RD 1620/2007, depending on the use to which the reclaimed water is going to be destined, others included in specific matters should be considered. For example, it needs to be in commensuration with the parameters established in Directive 2006/118/CEE, regarding the protection of groundwater against pollution and deterioration, when the reclaimed water is used for environmental purposes. This standard includes Environmental Quality Standards (NCAs) related to nitrates, salinity, pesticides, metals or metalloids, trichloroethylene, tetrachloroethylene, and other hazardous substances. Likewise, Directive 80/68/CEE establishes that, to guarantee effective protection of groundwater, it is necessary to prevent the discharge of List I Hazardous substances and limit the discharge of List II Hazardous substances. Along the same lines, RD 60/2011 on environmental quality standards in the field of water policy establishes NCAs for the so-called preferred, priority, and other pollutants [47].

Regarding outside European Union countries, Table 1 summarizes some information related to the reuse of treated wastewater in India, Mexico, Australia, Jordan, Singapore, South Africa, China, and Namibia. Some of them allow the reuse of treated wastewater as drinking water, as is the case of Singapore or Namibia. However, there are important differences between both countries in terms of legislation and technological infrastructure. In both cases, water is a national priority and both governments have solid strategies to promote water reuse [48,49]. Other countries are promoting water reuse in many purposes except for drinking water; this is the case of Mexico, China, Jordan, Australia, India, and South Africa. Almost all of them have solid legislation or guidelines with limits or restrictions in the use, according to quality water parameters.

Table 1. Legislation regarding the reuse of reclaimed water for various purposes in the different countries of the European Union (EU) (Adaptation from [50]) and outside of the European Union. X: destination for reclaimed wastewater regulated in the legislation.

	Country	Agriculture	Municipal	Potable Unplanned Indirect Reuse	Groundwater Recharge	Industrial	Environment	Drinking Water	Regulations/Guidelines
European Union countries	Austria					X			No
	Belgium	X		X	X	X			Under preparation.
	Bulgaria					X			Under preparation.
	Cyprus	X	X		X	X	X		D 296/03.06.2005
	Czech Republic								No
	Denmark					X			No
	Estonia					X			No
	Finland					X			No
	France	X	X	X	X	X			D 94/463.3.1994 DGS/SD1.D91 Guidelines 1996
	Germany	X	X	X	X	X	X		Under preparation.
	Greece	X				X			JMD 145116/11 GG B′ 192/1997
	Hungary								96/2009 (XII.9)
	Italy	X	X		X	X			D152/2006
	Ireland								No
	Latvia								No
	Lithuania								No
	Luxembourg					X			No
	Malta	X	X			X			Under preparation.
	Netherlands		X			X	X		No
	Poland								Under preparation.
	Portugal	X	X	X	X	X	X		RecIRAR 2/2007 ERDAR Guidelines
	Romania								No
	Slovakia								No
	Slovenia								No
	Spain	X	X		X	X	X		RD 1620/2007 Guidelines from Regional Health Authorities
	Sweden	X			X	X			No

Table 1. *Cont.*

	Country	Agriculture	Municipal	Potable Unplanned Indirect Reuse	Groundwater Recharge	Industrial	Environment	Drinking Water	Regulations/Guidelines
Countries outside European Union	United Kingdom India		X	X	X	X	X		Under preparation. Not yet
	Mexico	X	X			X			CONAGUA (Water National Commission) NOM-003-SEMARNAT-1997; NOM-004-SEMARNAT-2002
	Australia	X	X		X	X	X		Water Quality Australia. Different guidelines for water recycling.
	Jordan	X							Based on the World Health Organization (WHO) guidelines for water reuse in agriculture (2006)
	Singapore	X	X	X	X	X	X	X	Singapore's national water agency. Reclaimed water obeys the World Health Organization's drinking water guidelines
	South Africa	X	X	X	X	X	X		Not yet, just some text addressing reuse briefly: Water Services Act of 1997; National Water Act of 1998, 37(1); DNHPD. Guide: permissible utilization and disposal of treated sewage effluent. Report No. 11/2/5/3; 1978.
	China	X	X	X	X	X	X		GB/T 18920-2002; GB/T 18921-2002; GB/T 19923-2005; GB/T 19772-2005; GB 20922-2007
	Namibia	X	X	X	X	X	X	X	Water supply and sanitation policy, 2008

2.3. Risks of Reclaimed Wastewater

Wastewater discharges, even after treatment, contains a large number of pollutants potentially dangerous to human health and ecosystems. Some of these pollutants, which are being increasingly analysed and found in reclaimed wastewater [51,52], are pathogens [53], micropollutants [54], antibiotic resistant genes [55], nanomaterials, by-products of disinfection, personal care products, and pharmaceuticals.

In terms of reuse of water, the presence of pathogen germs, such as bacteria, virus, or protozoa, represents the most severe threat for human health [56], as they are responsible for hydric waterborne diseases. However, chemical pollutants, such as medicines, heavy metals, hydrocarbons, etc. could also pose a risk to human health. The population could be exposed to these pollutants directly by consumption of polluted water, inhalation, or by skin contact, or indirectly by the consumption of food dropped or produced with polluted water effluents.

Studies are reporting the emission of different pathogens after the discharge of treated effluents. López et al. [57] and Mosteo et al. [58] reported the discharge of different pathogen bacteria, such as *E. coli*, *Enterococcus sp.*, *Pseudomonas sp.*, *Clostridium perfringens*, or *Staphylococcus aureus* in different wastewater treatment plants (WWTPs) located in Spain, with different secondary treatments, and even with, for some of them, tertiary treatment based on ponds. In the research by Mosteo et al. [58] pathogen parasites, such as *Giardia* and *Cryptosporidium* were not detected, but free-living amoebae, which are potentially pathogenic, were observed. However, other authors, such as Ajonina et al. [59], have reported the presence of *Cryptosporidium parvum* in the effluent of WWTP. As the presence of any of these pathogens would be responsible for possible waterborne diseases, the total or partial disinfection of wastewater becomes mandatory in order to be reused. Chlorination is the most common disinfectant agent because it satisfies most requirements as an ideal disinfectant. However, its use as a disinfectant to treat wastewater has the main disadvantage of generating disinfection by-products, such as trihalomethanes (THMs), among others, a consequence of the reaction between organic matter and chlorine [60]. THMs are carcinogenic, and their presence in water bodies represent a new threat for humans and ecosystems. For this reason, in the previous decade, new disinfection treatments have been developed in order to prevent the generation of these disinfection by-products, as well as to increase disinfection efficiency compared with chlorination or traditional technologies. Table 2 shows a summary of the recent studies related to the removal of pathogens in water reclamation. As can be observed in Table 2, the application of AOPs based on the generation of free radicals with a high oxidation potential has increased the efficiency of disinfection of wastewater and, hence, it has become a credible alternative to chlorination. Technologies, such as heterogeneous photocatalysis, Fenton, and photo-Fenton reaction, oxidation by persulfate salts, combination of ozone with UV-C radiation, or H_2O_2, etc., have been widely studied and reported as efficient in the removal of different microorganisms. Besides AOPs, there are other efficient technologies to treat wastewater, such as nanofiltration, reverse osmosis, electrochemical technologies, and coupling between different systems.

Table 2. Recent studies on the detection and/or removal of pathogens in treated effluents.

Pathogens	Technology Applied	Efficiency	Reference
E. coli *Enterococcus sp.*	H_2O_2/UV-vis	4–5 log	[61]
E. coli *Salmonella spp.* *Enterococcus spp.*	Photo-fenton H_2O_2/UV-vis	Above detection limit	[62]
E. coli O157:H7 *Salmonella Enteritidis*	H_2O_2/solar	>6 log	[63]
E. coli O157:H7 *Salmonella Enteritidis*	Fe^{3+}-EDDHA/H_2O_2/UV-vis	5 log	[64]
E. coli	TiO_2/UV	3.30 log	[65]
Staphylococcus aureus	Photo-electro-Fenton	5 log	[66]
Enterococcus sp. *E. coli*	PS/UV-A/Iron	3.5 log 5 log	[67]
Enterococcus faecalis *E. coli*	PMS/UV-C/Fe(II)	5.2 log 5.7 log	[68]
E. coli *Enterococcus faecalis*	PS/solar	6 log	[69]
Total heterotrophic bacteria	Combined photo-Fenton and aerobic biological treatment	2 log	[70]
E. coli *Enterococcus sp.* *Candida Albicans*	UV-C/microfiltration	4.2 log 4.7 log 3.1 log	[71]
Enteric Virus	Wastewater treatment pond	3 log	[72]
E. coli *C. perfringens*	Sequencing batch biofilter granular reactors	4 log 1 log	[73]
Total coliform	Constructed wetlands	5.1 log	[74]
E. coli *MS2 bacteriophage*	Solar Disinfection	6 log 3 log	[75]

PS: persulfate; PMS: peroxymonosulfate; UV: Ultraviolet radiation; EDDHA: ethylenediamine-N,N′-bis(2-hydroxyphenylacetic acid).

The increase of human activities poses a growth in the synthesis and consumption of new chemical substances [76]. These substances are released into the municipal sewer system via human excretion and direct/indirect disposal of medicine and chemical, and without the appropriate treatment, they reach the water bodies at trace level (ppb or ppt). In most cases, the lack of specific regulations and methodologies to detect them at trace level, as well as the largely unknown long-term effects on aquatic ecosystems and human health, making them the contaminants of emerging concern [77]. As previously mentioned, though there are various sources of contaminants of emerging concern, most of them belong to hospitals and veterinary, industrial, and domestic activities. They can be classified in different groups [78]: brominated flame retardants, chlorinated paraffins, pesticides, organofluorine compounds, personal hygiene products, pharmaceuticals, antibiotics, and drugs. These substances are characterized by high chemical stability and a low biodegradability. For this reason, conventional WWTP are not designed to remove them, the occurrence of these substances in the treated effluents being an usual affair [79]. Gros et al. [80] and Ibáñez et al. [81] reported the occurrence of medicines/antibiotics and other micropollutants in wastewater effluents in Greece and Sweden, respectively. However, further research is being done to know if the presence of these substances pose a risk if the reclaimed water is used. Some authors have reported the effects of micropollutants on aquatic ecosystems [82,83], and the World Health Organization (WHO) reported the possible impacts from the consumption of low concentrations of pharmaceuticals present in drinking-water [84,85]. Some authors have verified the WHO conclusions with further research [86,87]. Table 3 shows recent literature references about the occurrence and reduction of micropollutants in wastewater. Between the degradation strategies, AOPs are feasible treatments, because of their capacity to degrade any kind of chemical, without forgetting their combination with other remediation treatments. Table 3 also summarizes other techniques, such as adsorption, biological treatments, filtration technologies, etc. Their inclusion, as a tertiary treatment in WWTP, will solve the threat.

Table 3. Recent studies on the detection and/or removal of organic micropollutants in treated effluents.

Micropollutant	Technology Applied	Removal Efficiency	Reference
DCF, SMX, CBZ, ATN, TCS, SCL	PMS or PS/Fe(II)/UV-C	100%	[68]
ATN, BPA, CBZ, CFN, DCF, IBP, SMX	PMS/Fe(II)/UV-C	40–100%	[88]
25 MPs	H_2O_2/UV-C PMS/UV-C	55% 48%	[89]
Chloroform	Fenton/Radiofrequency	69% DOC	[90]
Metoprolol Metoprolol acid	H_2O_2/UV	71.6 ± 0.8% 88.7 ± 1.1%	[91]
8 MPs	Nanofiltration	98%	[92]
7 MPs	Sorption	>80%	[93]
Metoprolol Benzotriazole, DCF, BPA CBZ, SMX	Constructed wetlands	70% 30–40% 0%	[94]
12 MPs	PS-assisted Membrane distillation	>99%	[95]
ACMP, ATZ, DDVP	Ozonation	80–100%	[96]
DCF, IBP, SMX	Membrane bioreactor	100%	[97]
13 antibiotics	Bioreactors	33–88%	[98]
CBZ, DCF, Iopromide, Venlafaxine	Fungal treatment	55–96%	[99]
Cytostatic compounds	Ozonation	100%	[100]
7 MPs	Activated Sludge	75–93%	[101]
Four benzotriazoles	Moving bed biofilm reactor	31–97%	[102]
DEET, SCL, primidone, TCEP, meprobamate	O_3/granulated AC	45–90%	[103]
48 MPs	Activated Carbon (AC) Adsorption	20–99%	[104]
ACMP	Photocatalytic ozonation	89–100%	[105]
ATZ, IBP	Ozonation	100%	[106]

MPs: micropollutants; DCF: Diclofenac; SMX: sulfamethoxazole; CBZ: carbamazepine; ATN: atenolol; TCS: Triclosan; SCL: sucralose; BPA: Bisphenol A; IBP: Ibuprofen; ACMP: acetamiprid; CFN: caffeine; DDVP: dichlorvos; TCEP: tris (2-carboxylethyl) phosphine; ATZ: atrazine; DEET:N,N-diethyl- m-toluamide.

2.4. Tertiary Treatment in WWTP: Technologies for Wastewater Reclamation

As reported in Section 2.3, the removal of the main risk associated with the reuse of reclaimed wastewater is well known, it is important to understand that there is not just a single technology to regenerate wastewater, and depending on the final quality required for the reuse, different possible configurations of treatment can be adopted. If the reclamation is considered as a tertiary stage in a WWTP, omitting the explanation of primary and secondary treatment, the regeneration line can be divided into four kinds of treatments with different efficiency in terms of microbiological pollution removal. Figure 2 shows the different configuration of tertiary treatments [107]. The most demanding treatment (treatment 1), can be composed of a coagulation–flocculation treatment followed by filtration or decantation; the whole stage has the aim of removing colloidal material and, for that, reducing organic matter concentration and turbidity. Then, it can be incorporated—a separation process based on the use of membranes (i.e., ultrafiltration (UF) or reverse osmosis (RO)). It is important to note that reverse osmosis can separate dissolved solids/salts. In this way, this treatment allows obtaining almost ultrapure water free of organic/inorganic and biological substances. Finally, although reverse osmosis can remove biological pollution, a disinfection step is required to introduce a residual amount of disinfectant to guarantee the quality of water until the point of use. Logically, the application of this treatment will have a higher investment, maintenance, and operating cost. When water with ultrapure quality is not required, the configuration of the line can be slightly modified, substituting, in this case, the UF/RO stage for a simple pre-disinfection based on UV-C radiation (treatment 2). Treatment 2 allows to obtain a high quality effluent, with very low content of organic matter, suspended solids, and turbidity thanks to the coagulation–flocculation stage, and with very high quality in terms of biological pollution, but without guaranteeing the absence of microbiological pollution. Finally, when the required quality for reuse is not high, the treatment can be simplified, up to a point, to a simple filtration and disinfection (treatment 3), or even only a filtration (treatment 4). As mentioned earlier, the reuse could be directed from the secondary effluent of the WWTP in the case where the quality of this effluent is within the permissible limits of the physico-chemical and biological parameters established in the legislation.

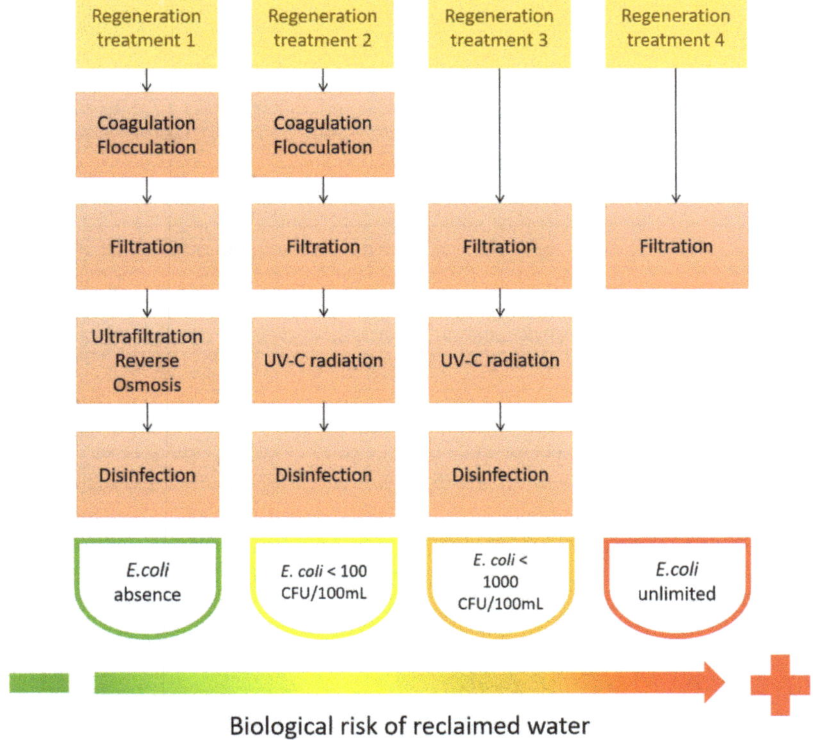

Figure 2. Proposal of wastewater reclamation line and biological risk associated with reclaimed water. Adaptation from [107].

3. Resources Recovery

3.1. Nutrients

Nutrient recovery can be seen as a loop or cycle comprising the following stages: (i) agricultural production; (ii) processing; (iii) consumption; (iv) collection; and (v) treatment of waste and wastewater. Thus, the return of the nutrients reclaimed from wastewater as mineral and organic substances to agriculture closes the cycle. It is for this reason that municipal wastewater is an essential alternative source of nutrients [108].

The development of recovery processes for nutrient recycling is gaining increasing attention due to both economic and environmental reasons, such as improved cost-effectiveness of water treatment plants and reduced drinking water consumption [109].

The modern sanitation approach is focused on the recovery of valuable resources from wastewater by implementing energy recovery processes, maximizing water reuse, and prioritizing the recycling of nitrogen, phosphorus, and organic matter present in sewage [110]. By doing so, wastewater treatment facilities will be converted into resource recovery centres.

Different techniques can be employed for nutrient recovery, including chemical, physical, and biological processes. However, the most widely used method for the simultaneous recovery of nitrogen and phosphorus is the formation of struvite (magnesium ammonium phosphate), through crystallization/precipitation. Struvite is an excellent fertilizer due to its low solubility in water and slow rate of release. Besides, it can be applied directly to soil [111].

Struvite formation requires the presence of ammoniacal nitrogen ($NH_4{}^+$) and phosphate ($PO_4{}^{3-}$), which reacts according to reaction 1 [112]:

$$NH_4^+ + PO_4^{3-} + Mg^{2+} + 6H_2O \rightarrow MgNH_4PO_4{\cdot}6H_2O \tag{1}$$

It should be noted that its formation could be accelerated by the addition of magnesium (Mg^{2+}) in excess [113].

Besides struvite, other products that can be recovered from secondary streams (generated in primary sludge thickening or after dewatering digested sludge) are ammonium sulphate or nitrate and calcium phosphate (see Table 4). To that extent, different technologies, such as Pearl®, PhosNix®, AirPrex®, PHOSPAQ® and Crystalactor® can be used [113,114]. Slow crystallization, which can be found in Ostara® DHV, Crystalactor®, and P-RoC® technologies, and instant precipitation (PRISA) are always required in liquid effluents. This allows recovery efficiencies around 85%–90% of the phosphorus contained in the supernatant [115]. Considering that heavy metals are present in low concentrations in the supernatant of the digester, it is unlikely that they produce soil contamination if the product is applied in agriculture.

The main challenge related to struvite precipitation is the recovery of phosphorus from wastewaters with low content of this element (<50 mg/L) and a high concentration of suspended solids (<2000 mg/L) [116]. For this reason, chemical precipitation of phosphorus with calcium (Ca^{2+}) is also very common. Other reagents, such as aluminium (Al^{3+}) and iron (Fe^{3+}) can also be employed, although its use is not generalised because Al is toxic to most plants and Fe is strongly bound to phosphorus, making it less available [112].

There are also physical processes that can be used for nutrient recovery, such as ion exchange, adsorption, or membrane processes. Thus, Johir et al. [117] coupled a membrane bioreactor (MBR) to an ion-exchange unit, recovering the nutrients contained in the effluent of the MBR effectively. However, the operating cost of this technology is high because chemicals are needed for the regeneration of the spent material.

Membrane processes are adequate to produce a concentrate (including phosphorus, nitrogen, and potassium) that can be used in irrigation or as a raw aqueous stream for nutrient recovery by crystallization or precipitation [112]. The disadvantages associated with these processes are high-energy consumption and membrane fouling due to the accumulation of pollutants or salts, thus, being necessary to be cleaned regularly [116].

Zeolites can also be used as an adsorbent for the treatment of digestate due to their high cation-exchange capability, mainly for potassium and ammonium [118]. Natural zeolites, such as clinoptilolite, can also be employed. In this case, ammonium is adsorbed onto the mineral and when the material is spent, it can be used as a slow-release fertiliser or soil conditioner.

Table 4. Process employed for nutrient recovery from different wastewater streams.

Process	Raw Material	Resource Recovered	Main Results	Advantages	Disadvantages	References
Ostara Pearl®: crystallization	Dewatering liquor, thickening supernatant, digester supernatant	Phosphorus	Recovery around 80%–90% as struvite. Concentration of phosphorus (PO_4-P) in the influent from 100 to 900 mg/L. $MgCl_2$ is added as a source of magnesium.	High degree of recovery. Very high efficiency as fertilizer for acidic soil and high for alkaline soil.	High consumption of $MgCl_2$. High operating costs. Decrease in the acidification potential.	[11,113,115,119]
PhosNix®: crystallization	Sludge dewatering liquor, wastewater after digestion	Phosphorus	Recovery around 80%–90% as struvite. Concentration of phosphorus (PO_4-P) in the influent from 100 to 150 mg/L. $Mg(OH)_2$ is added as a source of magnesium.	High degree of recovery.	High consumption of $Mg(OH)_2$. High operating costs.	[113,120]
AirPrex®: precipitation/ crystallization	Sludge liquor, digested sludge before dewatering	Phosphorus	Recovery around 80% as struvite. Concentration of phosphorus in the influent from 150 to 250 mg/L. $MgCl_2$ is added as a source of magnesium.	High degree of recovery. Very high efficiency as fertilizer for acidic soil and high for alkaline soil.	High consumption of $MgCl_2$. High operating costs. Decrease in the acidification potential.	[11,113,119,120]
PHOSPAQ®: precipitation	Digester supernatant, sludge dewatering liquor,	Phosphorus	Recovery from 70% to 95% as struvite. Concentration of phosphorus (PO_4-P) in the influent between 50 to 65 mg/L. MgO is added as a source of magnesium.	High degree of recovery.	High consumption of MgO. High operating costs.	[113,119,120]

Table 4. *Cont.*

Process	Raw Material	Resource Recovered	Main Results	Advantages	Disadvantages	References
DHV Crystalactor: crystallization	Digester supernatant	Phosphorus	Recovery around 85%–95% as calcium or magnesium phosphate or struvite pellets. Concentration of phosphorus (PO_4-P) in the influent higher than 25 mg/L.	High degree of recovery. High efficiency as fertilizer for acidic soil and moderate for alkaline soil.	High consumption of Mg source. High operating costs. Increase in the acidification potential.	[113,116,119]
P-Roc®: crystallization	Digester supernatant	Phosphorus	Recovery around 85%–95% as calcium phosphate or struvite. Concentration of phosphorus (P) in the influent 25 mg/L.	High degree of recovery. Very high efficiency as fertilizer for acidic soil and high for alkaline soil.	High operating costs. Decrease in the acidification potential.	[113,115,121]
PRISA: precipitation/ crystallization	Digester supernatant	Phosphorus	Recovery around 85%–95% as struvite. Concentration of phosphorus in the influent between 50 and 60 mg/L.	High degree of recovery. Very high efficiency as fertilizer for acidic soil and high for alkaline soil.	High consumption of MgO. High operating costs. Decrease in the acidification potential	[115]
Ion-exchange	WWTP effluent from MBR reactor	Phosphorus, nitrogen	Retention of PO_4^{3-} of 85%, NO_3^- of 95% by two ion- exchange columns. Around 95–98% phosphate and nitrate recovery during regeneration of columns.	High degree of recovery. Simple operation.	Regeneration of the resins. High consumption of NaCl.	[117]
Adsorption with Clinoptilolite (zeolite)	Digestate supernatant	Phosphorus, ammonium, potassium	Removal efficiencies varied from 64% to 80% for orthophosphate, 40% to 89% for ammonium and 37% to 78% for potassium.	High to moderate degree of recovery. Simple operation.	Regeneration or substitution of the zeolite.	[118]

3.2. High Added-Value Products

The food-manufacturing sector produces significant amounts of nutrient-rich effluents, which can be used as a resource to recover high added-value products, thus, developing more environmentally friendly processes. In this sense, during the extraction of olive oil, polyphenols are partitioned between water and oil phases, the main fraction being present in wastewater due to their high polarity and water solubility. The typical concentrations of polyphenols in olive oil wastewaters varied from 5 g/L to 25 g/L [122]. Thus, their recovery is promising since they are natural antioxidants of great interest, for food, pharmaceutical, and cosmetic industries [123]. Different technologies have been proposed for the recovery of olive polyphenols. Yangui and Abderraba [124] evaluated the use of activated carbon coated with a natural agent (milk proteins) for their recovery. An extraction yield of 80% and overall efficiencies for total phenols of 75.4% were obtained. Besides, the recovery of the most valuable compound, hydroxytyrosol, achieved a value of 90.6%. Kiai et al. [125] studied the Cloud Point Extraction (CPE) methodology to pre-concentrate phenolic compounds from olive processing wastewaters. The CPE process is a separation technique with low energy requirement, which is based on the ability of non-ionic surfactants to form micelles in aqueous solutions. Three non-ionic surfactants were analysed: (i) Triton X-100, (ii) Tween 80, and (iii) Genapol X-080. The optimum conditions were found to be a temperature of 70 °C, acidic pH (2.0), 30 min of equilibrium time and 10% (*w/v*) of surfactant. The phenolic compound recoveries achieved under these conditions were 65%, 62%, and 68% for Triton X-100, Tween 80, and Genapol X-080, respectively. Kalogerakis et al. [126] assessed the recovery of total phenols, hydroxytyrosol, and tyrosol by means of liquid–liquid solvent extraction. The effect of different organic solvents, such as diethyl ether, ethyl acetate, and a mixture of isopropyl alcohol and chloroform on efficiency recovery, was studied. The performance of the extraction agents followed this order: ethyl acetate >isopropanol/chloroform/> diethyl ether. It was calculated that around 0.25 kg of hydroxytyrosol, 0.062 kg of tyrosol, and 3.44 kg of total phenols could be attained after treating 1 m^3 of olive wastewater with ethylacetate.

Wastewaters from dairy industries are also characterised by a high content in macronutrients with nutritional value, such as proteins and lipids, which can be employed as animal feedstock. For example, the Gopinatha et al. [127] study focused on the use of a low-cost polymer (lignosulphonate) to recover proteins and lipids from dairy effluent. Under optimum conditions (pH = 3.5, concentration of lignosulphonate = 0.016% (*w/v*), and temperature = 22 °C, a lipid recovery of 96% and a protein recovery (mostly caseins) of 46% was obtained.

Soybean wastewaters contain proteins and anti-nutritional factors (ANF), including the Bowman–Birk protease inhibitor, which showed anti-cancer or cancer-preventive activity [128], Kunitz trypsin inhibitor, and soybean agglutinin. Therefore, the recovery of both proteins and ANF can become an interesting alternative with potential therapeutic value. Li et al. [129] analysed the use of a quick-shearing system to obtain homogeneous electrostatic complexes between soybean whey protein and a high concentration of polysaccharide (4%, *w/v*). Two types of complexing agents, sulphated polysaccharides and carboxylated, were evaluated for protein recovery and trypsin inhibitor activity, paying special attention to the effect of pH. Sulphated polysaccharides were found to be the most suitable complexing agents, achieving a recovery of proteins of 90% and removal of ANF of 80% at pH 2.5.

Coal gasification wastewater is polluted with phenols, including monohydric phenols and dihydric ones, their concentrations ranging from 2900 to 3900 mg/L and 1600 to 3600 mg/L, respectively [130]. Different solvents (diisopropyl ether (DIPE) and methyl isobutyl ketone (MIBK)) can be used for phenol removal, depending on the type of phenols contained in the coal wastewater. Guo et al. [131] assessed the removal of total phenols and hydroquinone (dihydric phenol) using three different solvents: (i) methyl propyl ketone (MPK), (ii) DIPE, and (iii) MIBK. Besides, one-stage, two-stages, and three-stages of countercurrent extraction were also performed. The efficiency of removal using MPK was greater than those achieved using DIPE or MIBK. Besides, the use of three-stage extraction with MPK allows a total phenol recovery of 99.6%, thus, obtaining a final concentration of phenols lower than 100 mg/L. A comparison between MIBK and MPK showed that the decrease in phenol emissions was as high as 52.5% by using MPK. Gui et al. [132] have developed a novel method for phenol and ammonia recovery based on an extraction process using a self-developed reagent, named as IPE. This methodology implies the following stages: (i) pre-extraction with IPE, (ii) stripping out of sour gas and ammonia, and (iii) deep extraction with IPE. Besides, the use of MIBK as an extracting agent was also evaluated for comparison purpose. The removal of total phenols and ammonia was similar with both extraction agents, the values being 90% and 97%, respectively, for IPE and 90.6% and 98%, respectively, for MIBK.

Wastewaters produced from pharmaceutical industries present a variable composition depending on the raw materials, the processes employed in the manufacturing, the season, and the location of the facility, which mainly affects the water quality. Therefore, it is complicated to define a treatment system for such a diverse industry [133]. However, pre-treatment and recovery of various valuable by-products, such as acids, heavy metals, solvents, and active pharmaceutical ingredients (APIs), which are contained in waste streams, is a control strategy of the waste, thus, making the pharmaceutical industry more sustainable. It is worth noting that the recovered waste stream can be used elsewhere in the process and the water can be employed in cooling towers or boiler feed, reducing the consumption of freshwater and, consequently, the operating costs. Taking into account that pharmaceutical wastewaters are composed of substances with molecular weights higher than 250 Da, membrane technologies can be effectively used to recover the product, provided that these are the only substances present in the effluent. In this sense, the application of nanofiltration has rapidly increased in the past decade. It has been commonly employed in aqueous solution with antibiotics [134]. For example, amoxicillin (MW = 365.4 Da), a widely used antibiotic in human and veterinary medicine to prevent respiratory, gastrointestinal, and urinary bacterial infections, can be separated and recovered by nanofiltration. This will mitigate the harm of amoxicillin to the environment as well as enhance the economy of the process.

Shahtalebi et al. [135] observed recoveries of amoxicillin as high as 97% and permeate flux of 1.5 L/min·m^2, using a spiral nanofiltration membrane composed of polypiperazine amide.

Heavy metals, such as Cd and Ni, were also found to be present in pharmaceutical wastewaters, the concentration being around 35 mg/L [136]. Studies at the University of Alicante developed an electrochemical treatment for recycling and recovery metals (Ni, Zn, Pb, among others). This technique allows obtaining metals with high purity, while being less polluting than the classic pyrometallurgy, because it avoids gas emission, sulphur, and metal particles.

Table 5 summarises the high added-value product recovered, the technology employed, and the main results achieved.

Table 5. Process employed for the recovery of high-value added products from industrial wastewaters.

Process	Raw Material	Resource Recovered	Main Results	Advantages	Disadvantages	References
Adsorption	Olive oil wastewater	Polyphenols	Extraction yield 80% and overall efficiencies for total phenols 75.4%. Recovery of hydroxytyrosol 90.6%.	High percentage of recovery. Use of biodegradable and natural coating agent.	Regeneration or substitution of the activated carbon.	[124]
Cloud Point Extraction	Olive oil wastewater	Polyphenols	Recoveries of 65%, 62% and 68% for Triton X-100, Tween 80 and Genapol X-080 with the optimum conditions.	Biodegradable nature of the extractants.	High consumption of chemicals.	[125]
Precipitation	Dairy wastewater	Proteins and lipids	Recovery of proteins and lipids: 46% (mainly caseins) and 96% with lignosulphonate (coagulant).	High percentage of protein recovery. Low temperature.	High consumption of chemicals. Acidic pH.	[127]
Complexation	Soybean wastewater	Proteins	Recovery of proteins 87.5% and 78.5% with ι-carrageenan and dextran sulphate. Recovery of proteins <50% with xanthan gum and sodium alginate.	High percentage of recovery. Use of biodegradable complexing agents	High consumption of chemicals.	[129]
Extraction	Coal gasification wastewater	Phenols	Recovery of total phenol 99.6% with methyl isobutyl ketone (MIBK) after three-stage extraction.	High percentage of recovery.	High consumption of chemicals. Toxic nature of MIBK.	[131]
Nanofiltration	Pharmaceutical wastewater	Active pharmaceutical ingredients (API)	Recovery of amoxicillin 97% with a polypiperazine amide membrane (permeate flux of 1.5 $L/min \cdot m^2$).	High recovery. Simple operation. Low operating costs.	Short membrane lifetime. Membrane fouling and cleaning.	[135]

4. Sewage Sludge Valorisation

The treatment and disposal of sewage sludge is an environmental subject of prime concern due to its high organic load and low dewaterability, which implies great operational costs. Besides, it is a growing problem worldwide since the number of wastewater treatment plants (WWTPs) is rising all over the world, thus, causing a rise in sludge production. For instance, in the EU, more than 10 million tons of dry solids of sewage sludge were produced in 2015 [137]. This fact, together with the tightening of environmental quality standards, makes correct management of this waste essential. Therefore, its comprehensive management is a key point to reduce the operating costs and make wastewater treatment a more environmentally friendly process. Most countries are focused on the recycling of the sludge, for example, using it to produce energy by incineration or anaerobic digestion [138].

The traditional uses of sewage sludge include land application (with or without previous composting), landfilling or energy recovery [31]. Nevertheless, it should be noted that: (i) land application has to compete with other waste streams, the demand is variable and the legislation about this topic is becoming increasingly strict, (ii) landfilling is an unsustainable alternative due to concerns over pollution, loss of recyclable materials and loss of void for waste that cannot be recycled, and (iii) incineration is a high cost/technological option, and is currently only likely to be cost-effective for large cities. Besides, it lacks a high level of public acceptability due to concerns over gas emissions.

In light of the above, the need to develop new sludge management strategies becomes clear [139]. Nowadays, new perspectives in this topic are being opened, especially those focused on recovering valuable compounds, such as nutrients (mainly phosphorus and nitrogen), heavy metals, bioplastics, materials (adsorbents), proteins, and enzymes. In fact, sludge is now recognized as one of the future key raw materials, as stated in the Spanish Bioeconomy Strategy and in the Bio-Based Industries (BBI) European Strategy. Obviously, these new sewage sludge strategies should be designed coherently with the waste hierarchy introduced by the Waste Framework Directive (EU Directive, 2008), which proposes an order of preference for action to reduce and manage waste, i.e., prevention, minimisation, reuse, recycling, energy recovery and, as a final and undesired option, landfilling.

4.1. Nutrients

Sewage sludge is composed of significant amounts of nutrients, such as phosphorus (0.5%–0.7% of total solids) and nitrogen (2.4%–5% of total solids), in the form of protein compounds, which can be employed to produce fertilisers [140–142]. Taking into account that phosphorus is a finite resource extracted mainly from rocks located in a few regions of the world, and the high cost of commercial fertilizers, the recovery of nutrients from sewage systems is gaining attention to improve sustainability in agricultural production [112].

Chemical, physical, and biological processes can also be used to recover nutrients from sewage sludge in the same way as with aqueous effluents (see Table 6). Again, the simultaneous recovery of nitrogen and phosphorus as struvite by means of crystallization/precipitation is the main alternative [111]. Thus, it can be precipitated in the sewage sludge through acidic leaching followed by the chemical precipitation of phosphorus [143]. However, heavy metals can be present in the sludge, thus, requiring a pre-treatment to avoid its presence in the fertilizer. Chemical extraction processes, such as Gifhorn and Stuttgart, in combination with sulphidic precipitation or complexation of interfering ions and precipitation, can also be used to recover phosphorus as struvite and other phosphorus-based compounds (calcium phosphate and iron phosphide) from thickened, dewatered, or digested sludge. A high percentage of recovery from 35% to 60% can be achieved with these processes [11,115,119]. Another chemical process, named ExtraPhos, from the company Budenheim, allows the recovery of phosphorus as dicalcium phosphate, the yielding ranging from 40% to 60%. This is a four-stage technology, which implies: (i) wet chemical leaching, (ii) extraction of CO_2, (iii) solid/liquid separation, and (iv) precipitation of phosphate [115,119,144].

Table 6. Process employed for nutrient recovery from sewage sludge.

Process	Raw Material	Resource Recovered	Main Results	Advantages	Disadvantages	References
Gifhorn: wet chemical and precipitation	Thickened sludge, dewatered sludge, digested sludge	Phosphorus	Recovery around 35%–60% as calcium phosphate, iron phosphate or struvite.	Relatively high recovery. Reduction of the need of flocculating agents in WWTP. High decontamination potential for heavy metals.	High consumption of chemicals. Increase in the Acidification potential.	[11,115,119]
Stuttgart: wet chemical and precipitation	Digested sludge	Phosphorus	Recovery around 35%–60% as calcium phosphate, iron phosphate or struvite.	Relatively high recovery. Reduction of the need of flocculating agents in WWTP. High decontamination potential for heavy metals.	High consumption of chemicals. Increase in the Acidification potential.	[11,115,119]
Budenheim ExtraPhos: wet chemicaland precipitation	Digested sludge	Phosphorus	Recovery around 40%–60% as dicalcium phosphate.	Relatively high recovery. Reduction of the need of flocculating agents in WWTP.	High consumption of chemicals. High operating costs.	[115,119,144]
Aqua Reci: supercritical water oxidation, acidic/alkaline leaching and precipitation	Thickened sludge, dewatered sludge, digested sludge	Phosphorus	Recovery around 40%–60% as calcium phosphate, iron phosphide and aluminium phosphide.	Relatively high recovery. Reduction of the need of flocculating agents in WWTP. High decontamination potential for heavy metals.	High consumption of chemicals. High operating costs. Increase in the Acidification potential.	[11,115,121]
LOPROX PHOXNAM: wet oxidation and precipitation	Thickened sludge, dewatered sludge, digested sludge	Phosphorus	Recovery around 40%–50% as struvite.	Relatively high recovery. Reduction of the need of flocculating agents in WWTP. High decontamination potential for heavy metals.	High consumption of oxygen and chemicals. High operating costs.	[11,115,121]
MEPHREC: Metallurgic melt-gassing	Dewatered Sludge briquettes	Phosphorus	Recovery around 65%–70% as P-rich slag.	Relatively high recovery. Reduction of the need of flocculating agents in WWTP. No need of mono-incineration. No increase in acidification potential.	High consumption of coke, dolomite and oxygen. High operating costs.	[11,115,143]

Oxidative processes, including Aqua Reci and LOPROX/PHOXNAM, can also be employed to recover phosphorus. In the first one, supercritical water oxidation followed by acid/alkaline leaching and precipitation were used, thus, recovering phosphorus as calcium phosphate, iron phosphide, and aluminium phosphide from thickened, dewatered, or digested sludge. In the second technology, a combination of low-pressure wet oxidation, nanofiltration, acid, and hydrothermal digestion for heavy metal separation was employed to obtain struvite and phosphoric acid from thickened sludge [11,115,121]. The percentage of recovery of these techniques varied from 40% to 60% for Aqua Reci and between 40% and 50% for LOPROX/PHOXNAM.

Another technology used for the recovery of phosphorus as P-rich slag is MEPHREC. This technique is based on a metallurgic melt-gassing, and utilizes dewatered sludge briquettes, the yield being around 65%–70% [11,115,144].

Sewage sludge after incineration (sewage sludge ash) can also be employed as a raw material for phosphorus recovery. Two methodologies are mainly employed: (i) wet chemical, and (ii) dry thermal. In the first one, an organic acid or solvent is added in order to extract phosphorus, thus, being recovered from this organic solution. In the second one, phosphorus is recovered by dissolving the ashes [120,145]. The recovery percentages are higher (from 65% to 99%) than those obtained with sewage sludge [11,115].

4.2. Heavy Metals

Various heavy metals, including Cu, Ni, Zn, Cd, Pb, Cr, and Hg are contained in activated sludge, thus, limiting its use for land application due to possible soil and groundwater contamination, which can also have a detrimental effect on human and animal health [140]. It should be noted that the concentration of these elements varies greatly depending on the origin of sewage sludge, the highest values being found for Cu (468 mg/kg) and Zn (803 mg/kg) in biosolids from WWTP of USA [146]. Thus, sludge must be treated for recovering heavy metals prior to its valorisation.

The extraction of heavy metals was analysed by different authors using various extracting agents, such as chelants: EDDS (Ethylenediamine-N,N′-disuccinic acid; biodegradable) and EDTA (Ethylenediaminetetraacetic acid; non-biodegradable), and organic acid (citric acid). It was found that Cu recoveries were approximately 70% at pH values higher than 4.5 with a molar ratio of EDDS to the total heavy metals in sewage sludge of 10:1. Similar Cu recoveries (72%) were obtained with EDTA when the same conditions were used [147]. When citric acid (0.1 M) was used at pH 3–4, Cu and Zn recoveries of 60%–70% and 90%–100% were achieved [148].

The removal of heavy metals from leaching effluents of sewage sludge using supported liquid membranes was studied by Yesil and Tugtas [149]. The highest removal efficiencies were obtained through 20% Aliquat 336-filled PVDF membrane at 35 °C when 1.0 M HNO_3 was used as the permeate solution, the values being 27%, 22%, 30%, and 32% for Cr, Cu, Ni and Zn, respectively.

Medium volatile heavy metals, i.e., Cu and Zn, were removed from sewage sludge by means of calcination in an inert (N_2) or oxidising atmosphere (air) in presence of Cl-donor ($MgCl_2$) as an additive [150]. Thus, a Cl/sewage sludge ratio of 5% was enough to obtain a Cu removal of 80% in air and 88% in N_2. For significant Zn removal (90% in both atmospheres), a higher Cl/sewage sludge ratio (15%) was needed.

Electrokinetic treatment was also studied as a potential technique for heavy metal removal [151]. It was found that the combination of this treatment with the addition of a chelating agent (tetrasodium of N, N-bis (carboxymethyl) glutamic acid), and a biodegradable biosurfactant (rhamnolipid) as electrolyte, results in a high removal efficiencies for Cu, Zn, Cr, Pb, Ni, and Mn, the values being 70.6 ± 3.41%, 82.2 ± 5.21%, 89.0 ± 3.34%, 60.0 ± 4.67%, 88.4 ± 4.43% and 70.0 ± 3.51%, respectively.

Table 7 summarises the recovered heavy metals, the technology employed, and the main results achieved.

Table 7. The Process employed for heavy metal recovery from sewage sludge.

Process	Raw Material	Resource Recovered	Main Results	Advantages	Disadvantages	References
Extraction with chelants: EDDS and EDTA	Sewage sludge from WWTP	Heavy metals	Heavy metals in sewage sludge = 10:1. Higher extraction of Cu than other metals with EDDS. Cu recovery of 70% with EDDS at pH > 4.5. Cu recovery of 72% with EDTA at pH > 4.5.	High degree of recovery. Biodegradable nature of EDDS.	High consumption of chemicals. Non-biodegradable nature of EDTA.	[150]
Extraction with chelants: citric acid	Sewage sludge from WWTP	Heavy metals	Cu and Zn recoveries of 60%–70% and 90%–100% at pH 3–4 with 0.1 M citric acid at 30 °C.	High recovery. Biodegradable nature of citric acid.	High consumption of chemicals. Acidic conditions.	[148]
Supported liquid membranes	Leaching effluents from sewage sludge	Heavy metals	Recovery of 27%, 22%, 30% and 32% for Cr, Cu, Ni and Zn using 20% Aliquat 336-filled PVDF membrane at 35 °C with 1.0 M HNO_3.	Low costs. Simplicity.	Low recovery. Pre-treatment: removal of suspended solids.	[149]
Calcination with air or N_2 using $MgCl_2$ as additive	Secondary sludge from WWTP after dewatering	Heavy metals	Cu removal of 80% in air and 88% in N_2 with Cl/sewage sludge ratio of 5%. Zn removal of 90% in air and N_2 with a higher Cl/sewage sludge ratio (15%).	High recovery.	High costs.	[150]
Electrokinetic and extraction with chelant	Sewage sludge from WWTP	Heavy metals	Recovery of 70.6%, 82.2%, 89%, 60%, 88.4% and 70% for Cu, Zn, Cr, Pb, Ni, and Mn	High recovery. Biodegradable nature of chelant and electrolyte.	High consumption of chemicals. High costs.	[151]

4.3. Adsorbents

Adsorption processes are extensively studied due to its flexibility and ease of operation. Moreover, increasing attention is being given to the use of adsorbents from waste in order to develop low-cost strategies for water treatment [152,153]. In this sense, sewage sludge can be an interesting precursor for the synthesis of adsorbents due to its high content in carbonaceous matter. It should be noted that anaerobically digested sewage sludge, dewatered sewage sludge, and that from municipal/urban or WWTPs, can be used as precursors (Smith et al., 2009). Different methodologies have been proposed for adsorbent preparation, the most widely employed being: (i) carbonisation, (ii) physical activation, (iii) chemical activation, and (iv) combination of physical and chemical activation [154] (see Table 8).

Regarding carbonisation, various conditions were employed (temperatures varying from 650 °C to 1000 °C, heating rates between 5 °C/min and 20 °C/min, and time ranging from 0.5 h to 2 h), with the aim of preparing adsorbents with high BET (Brunauer–Emmett–Teller) surface areas, since this parameter significantly influences the adsorption process. Several authors reported that the increase in the aromatization, which occurred at high temperatures, is responsible for the increase in the BET area [155,156]. Nevertheless, the optimum conditions differed among the authors. In this sense, Zhai et al. [155] reported 850 °C as the most suitable temperature, whereas Bagreev et al. [156] and Inguanzo et al. [157] reported 950 °C and 650 °C, respectively, as the optimum temperatures. This is due to the great variability in the composition of sewage sludge. Another parameter of interest is macroporosity and mesoporosity, which can also be enhanced by raising the carbonisation temperature [158]. Moreover, the use of citric acid for sludge demineralisation also improved the BET area due to the removal of the inorganic content. In this sense, Kong et al. [159] increased the BET area from 6.8 m^2/g to 385.8 m^2/g when citric acid washing was applied.

Regarding physical activation, it implies the gradual burn-off of the carbonaceous matter, thus, leading to a significant increase of the inorganic content, which is mainly non-porous. Therefore, the adsorbents obtained by physical activation generally exhibit low BET surface areas. In order to improve their properties, different activation conditions were used. Various authors reported the use of steam, air, CO_2, O_2/N_2 and air/N_2 as gasifying agents, employing temperatures between 275 °C and 800 °C and times between 0.5 h and 4 h [154]. Among them, the best results in terms of BET area (226 m^2/g) were obtained by Rio et al. [160], who employed a two-stage process, based on a carbonisation at 600 °C for 1 h, followed by activation with steam at 760 °C for 0.5 h.

Concerning chemical activation, several dehydrating compounds can be used, such as KOH, NaOH, $ZnCl_2$, H_2SO_4, K_2S, and H_3PO_4, alone, or in a combination (of two of them) in order to obtain high values of BET surfaces areas. On some occasions, carbonation, followed by chemical activation with KOH or $ZnCl_2$, was employed [154]. Several studies reported the great effectivity of KOH as dehydrating agents in combination with carbonisation. In this sense, BET areas from 1058 to 1882 m^2/g were obtained when different types of sewage sludge were carbonised at 700 °C for 0.5 h and, subsequently, treated with KOH in a ratio of 1:1 at 700 °C for 1 h [161–164].

Table 8. Process employed for adsorbent preparation from sewage sludge.

Process	Raw Material	Material Prepared	Main Results	Advantages	Disadvantages	References
Carbonisation	Sewage sludge from WWTP. Anaerobically digested sludge from WWTP.	Adsorbent	Optimum temperature for sewage sludge 850–950 °C. Optimum temperature for anaerobically digested sewage sludge: 650 °C. Sludge demineralization with citric acid also increased BET area.	High BET area (from 82 to 385.8 m^2/g). Mesoporous or microporous materials.	High operating costs. Release of gases.	[155,156,159]
Physical activation	Sewage sludge from WWTP. Dewatered sewage sludge from WWTP.	Adsorbent	Sewage sludge: BET area from 61 to 70 m^2/g. Dewatered sludge: BET area from 7 to 18 m^2/g.	Non-porous materials.	Low BET areas (from 7 to 70 m^2/g). Release of gases.	[154]
Chemical activation	Dewatered sludge from WWTP.	Adsorbent	BET area = 289 m^2/g (activation with H_3PO_4 at 650 °C). BET area = 472 m^2/g (650 °C activation with $ZnCl_2$). BET area = 658 m^2/g (400 °C carbonisation + 850 °C KOH activation). BET area = 1224 m^2/g (700 °C carbonisation + 700 °C NaOH activation). BET area = 1058 to 1882 m^2/g (700 °C carbonisation + 700 °C KOH activation).	High BET area (from 472 to 1882 m^2/g) when $ZnCl_2$, NaOH and KOH are used.	High consumption of chemicals. High operating costs. Release of gases.	[154,161–163]

4.4. Bioplastics

Polyhydroxyalkanoates (PHAs) are biodegradable polymers with similar characteristics to those of petroleum-based polyolefins and can be considered a sustainable alternative to petroleum plastics. These biopolymers are naturally synthesised by bacterial fermentation of sugar or lipids. Besides, several microorganisms can accumulate PHA; however, its application is limited due to high production costs. In that sense, around 50% of the cost of microbial PHA is related to the cost of substrate or carbon source. This is why the growth of the PHA industry is low [165]. Therefore, it is necessary to find sustainable feedstock. Hence, microorganisms in activated sludge are able to accumulate PHA in the range of 0.3 to 22.7 mg biopolymer/g activated sludge [140]. It should be noted that PHA recovery using sewage sludge depends on various operating parameters. According to Takabatake et al. [166], the capability of sewage sludge to accumulate PHA was higher in the conventional process than in the aerobic–anaerobic ones, achieving an average content of PHA around 19% after the aerobic process. Thus, the addition of small amounts of oxygen in the anaerobic zone can improve the accumulation of PHA. Temperature is another factor that significantly affects the PHA production, which is favoured at low temperatures [167]. Besides, the type of reactor used for its production is also essential. It has been reported that the sequencing batch reactor (SBR) is the most adequate to obtain high production thanks to its flexible operation and simple control [168].

PHA production can be integrated during the treatment of urban water and sludge management in WWTPs [169]. This implies a process with four stages: (i) removal of readily biodegradable carbon from wastewater coupled to the selective biomass growth with capacity for PHA storage, (ii) acidogenic fermentation of sludge in order to produce a liquid stream with high content of volatile fatty acids (VFA), (iii) accumulation of PHA from the liquid stream with high VFA concentration using enriched biomass from stage 1 and, (v) recovery and characterization of PHA (Figure 3).

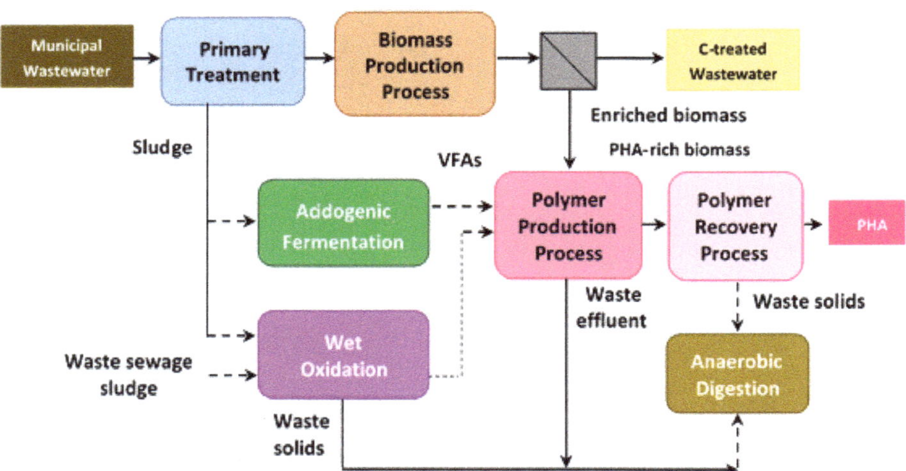

Figure 3. Schematic process flow diagram of municipal wastewater and sludge treatment together with the production of polyhydroxyalkanoates (PHA) (adapted from [168]).

Since PHAs are biodegradable compounds, it is crucial to generalize their use as packaging films or disposable products, such as bottles, cups, cutlery, and diapers, among others, in order to reduce the detrimental effect of petroleum-based plastics in the aquatic environment [170].

4.5. Construction Materials

Taking into account that the content of oxides of sewage sludge (Al_2O_3, CaO, SiO_2, and Fe_2O_3) is similar to that of Portland cement or clay, it has been proposed that it can be used for the production of building or construction materials, including eco-cement, bricks, ceramic materials, supplementary cementitious materials (SCMs), or lightweight aggregates (LWAs) [171]. Nevertheless, a high amount of organic matter is also present in sewage sludge, which can affect the cementitious properties, causing low bonding strength. Therefore, pre-treatment of sewage sludge before being employed in the preparation of cement or concrete is a must.

Considering the studies of various authors [172], the valorisation of sewage sludge as raw material for the preparation of cement-like materials is viable. However, depending on the characteristics and origin of the sludge, different conditioning steps may be needed. In that sense, Xu et al. [172] and Lin et al. [173] reported the use of lime-dried sludge as a substitute of limestone in the production of cement at 1400 °C, the optimum amount of limestone being below 18 wt% in order to facilitate the formation of crystalline phase in the cement clinker. Tay and Show [174] found that a cement-like material, which satisfies the American Society for Testing and Materials (ASTM) requirements for mortars, could be synthesised by incineration at 1000 °C for 4 h using dewatered sewage sludge mixed with lime at a ratio of 1:1. Rezaee et al. [175] reported that dry sewage sludge could be used to partially replace (from 5% to 15%) the traditional raw materials in order to obtain an eco-cement with properties similar to that of Portland cement.

Besides, sewage sludge can also be used to partially replace clinker in cement due to its properties as a self-cementing agent. Thus, Valls and Vazquez [176] partially substitute Portland cement with sewage sludge in a range of 25% to 50%, obtaining different pastes with hydration products, such as those of Portland cement. However, the organic matter contained in the sludge is degraded due to the alkalinity of the cement. Thus, in order to accelerate this process, $CaCl_2$ and $Ca(OH)_2$ can be added as additives. According to Hamood et al. [177], sewage sludge with high moisture content can provide the required water for the preparation of mortars through its conditioning with unprocessed fly ash, thus, avoiding both dewatering and drying processes.

The partial replacement of Portland cement by raw sewage sludge in cement is not a feasible option due to the low pozzolanic activity of the sludge, which needs pre-treatment to improve its activity. Thermal treatments (incineration) have been widely used since the organic matter is decomposed, and high temperature favours the activation of inert minerals, such as kaolin, which is transformed into metakaolin [171].

For the preparation of LWAs, it is necessary to pelletize and thermally treat the raw sewage sludge. Depending on the proportion of sewage sludge, different materials can be obtained. Thus, a mixture of 10% sewage sludge and 90% clay after sintering at 1150 °C for 0.5 h, resulted in an LWA with lower density, higher porosity and reduced compressive strength, in comparison to the traditional one [178]. Blends of waste glass and sewage sludge can also be employed to produce LWA, obtaining higher compressive strength in LWA as the amount of glass increased [179].

4.6. Proteins

Proteins can be considered high-added-value products due to their use as liquid fertilizers, protein foaming agents, or adhesives or animal feed [180]. Based on the high proportion of proteins (up to 61%) in the activated sludge and considering that around 50% of the dry weight of bacteria cells are due to protein content, their recovery is an issue of great interest [181,182]. In this sense, different methodologies have been proposed for protein recovery, the first stage always being the solubilisation of the intracellular material in the sludge [179]. Various treatments, such as physical [183], chemical [184,185], physico–chemical [180,186,187], biochemical [184], and hydrothermal ones [188] have been reported to effectively break the floc structure of the sludge; thus, releasing the intracellular content into the aqueous phase. The physical methods rely on centrifugal, ultrasonic, and membrane processes. The chemical or physico–chemical ones include acid or alkaline treatment alone, or in combination with hydrolysis [184,189]. The biochemical methods consist of enzymatic or enzymatic assisted treatment [184,190]. The hydrothermal methods are based on solubilization using oxidizing (air or oxygen) or inert (nitrogen) atmospheres. It should be noted that after solubilisation, it is necessary to precipitate proteins from the aqueous solution, dried and, on some occasions, purify them, before obtaining the protein product. Table 9 summarises the technology employed, main results achieved, and advantages and disadvantages inherent to the methodology used. It is common that the recovery process causes the solubilisation of heavy metals contained in the sludge, together with the release of intracellular compounds (mainly proteins, carbohydrates, and humic acids). Therefore, it is needed to detoxify the sludge by means of sterilization and remove heavy metals or toxins before its valorisation to obtain proteinaceous products for animal feed to avoid compromise of animal health [191].

Table 9. Process employed for protein recovery from sewage sludge.

Process	Raw Material	Resource Recovered	Main Results	Advantages	Disadvantages	References
Acidic or alkaline hydrolysis	Secondary sludge from WWTP	Protein	Extraction of 84.9% with acidic hydrolysis Extraction of 75% with alkaline hydrolysis.	High recovery	High consumption of chemicals. High-energy consumption in comparison to enzymatic treatments.	[184]
Enzymatic, thermal enzymatic assisted or ultrasonic enzymatic assisted treatment	Secondary sludge from WWTP	Protein	Extraction of 35.8% with the addition of alkaline protease. Extraction of 43.6% with the addition of neutral and alkaline proteases. Extraction of 56.5% with sonication (1 W/mL) and the addition of alkaline protease. Extraction of 54.5% with thermal treatment and the addition of alkaline protease.	Low energy consumption in comparison to acidic or alkaline hydrolysis.	Low-moderate degree of recovery. Enzymatic activity is pH-dependent.	[184]
Alkali treatment and ultrasonication	Sewage sludge from WWTP	Protein	Recovery of 80.5% at pH 3.3, Protein concentration in the supernatant being 3177.5 mg/L.	Similar composition to that of the commercial protein for animal feed. Heavy metals were removed. Toxins and harmful microorganisms not detected.	High operating costs	[180]

Table 9. *Cont.*

Process	Raw Material	Resource Recovered	Main Results	Advantages	Disadvantages	References
Physico-chemical treatment	Sludge from anaerobic ammonium oxidation (anammox)	Protein	Recovery of 10 mg/gVSS with sonication (6 W/mL). Recovery of ~110 mg/gVSS with thermal treatment + 0.5% Na_2CO_3. Recovery of ~25 mg/gVSS with cation exchange resin. Recovery of 100 mg/gVSS with formamide + NaOH.	High degree of recovery in the chemical or thermo-chemical treatments	High operating costs. High consumption of chemicals. Regeneration of the spent resin	[189]
Alkali treatment, ultrasonication and acid treatment	Activated secondary sludge from the paper mill	Protein	Recovery of 90.1% with H_2SO_4, 72.4% with HCl and 59.7% with ammonium sulphate.	High yield of protein recovery and lower heavy metal toxicity.	High operating costs.	[186]
Ultrasonication without or with additives (EDTA and Tween)	Activated sludge before gravity thickening	Protein	Domestic wastewater: 264 ± 10 mg BSA/g VSS were obtained. Synthetic wastewater: 191 ± 45 mg BSA/g VSS were obtained.	Maximal extracellular protein release with minimal contamination by intracellular proteins.	High operating cost due to energy consumption and/or the use of additives.	[187]

4.7. Hydrolytic Enzymes

Hydrolytic enzymes, such as amylase, phosphatase, lipase, protease glucosidase, and aminopeptidase, among others, have been found to be present in activated sludge and are responsible for the biodegradation of organic matter during the aerobic digestion [192]. Besides, enzymes are widely used in different fields, including agriculture, detergents, pulp and paper, cosmetics, dairy, etc. For example, lipases hydrolyse lipids in order to obtain glycerol and fatty acids, which can be used in the production of soap [193]. Protease can be employed as a target for therapeutic agents against important diseases, such as malaria or cancer, and amylase can be employed in starch saccharification, because it can hydrolyse starch into glucose [194]. Therefore, the valorisation of sludge to produce enzymes is a promising strategy to alleviate the environmental impact of its generation. There is no standardised methodology to extract valuable enzymes from activated sludge. In that sense, different chemical and physical methods have been proposed, depending on their nature (see Table 10).

Regarding lipase, Nabarlatz et al. [195] carried out its extraction from activated sludge using stirring or ultrasonication in combination with additives: non-ionic detergent (Triton X-100), cation exchange resins (CER), or buffer. It was found that the use of additives improved enzyme recovery when stirring was used. When ultrasonication was employed, extraction time and operating conditions were key factors, whereas the addition of a detergent was less influential.

Considering protease, Frolund et al. [196] were able to extract exopeptidase using CER and Triton X-100. CER was highly effective in the extraction of the enzyme in the floc matrix. Jung et al. [197] studied the use of a disrupting chamber (Dyno mill) for the extraction of two differently activated sludge: (i) cultivated in laboratory, and (ii) from WWTP, giving special attention to the effect of temperature on the extraction. The enzyme was recovered by means of precipitation with ammonium sulphate. Proteases activity varied from 75.8 ± 10.2 U/mg to 116.1 ± 4.2 U/mg at 75 °C as a function of the origin of the sludge. Gessesse et al. [198] analysed different methodologies: (i) Triton X-100, (ii) CER, (iii) ultrasonication in buffer, and (iv) ultrasonication with Triton X-100, obtaining the highest activity in this case. Therefore, it seemed that the combination of ultrasounds with additives is the ideal methodology for protease extraction.

In the case of amylases, it was recovered by ultrasonic CER and an additive. If CER is employed, around 2000 U/g VSS (volatile suspended solids) can be recovered, whereas, with Triton X-100, its concentration is a crucial factor, since 1600 U/g VSS were obtained with 0.1% Triton X-100 and 1300 with 2%Triton X-100 [199].

Table 10. Process employed for hydrolytic enzymes recovery from sewage sludge.

Process	Raw Material	Resource Recovered	Main Results	Advantages	Disadvantages	References
Stirring or ultrasonication with additives (CER, Triton X-100 or buffer)	Activated sludge after gravity thickening	Enzymes: lipase	15.5 lipase units/g VSS (stirring at pH 7.5 + 0.48 g/mL CER + 0.5% Triton X-100 (*v/v*)). 21 lipase units/g VSS (ultrasonication without additives) 25 lipase units/g VSS (ultrasonication with 2.0% Triton X-100 (*v/v*)).	High efficacy of ultrasonication in absence of additives.	High operating cost due to energy consumption and/or the use of additives.	[195]
Stirring or ultrasonication with additives (CER, Triton X-100 or buffer)	Activated sludge after gravity thickening	Enzymes: protease	335 lipase units/g VSS (ultrasonication using 0.1% Triton X-100 (*v/v*)).	Great quality of the final product.	High operating cost due to energy consumption and/or the use of additives.	[198]
Disrupting chamber (Dyno mill)	Activated sludge cultivated in the laboratory and from WWTP	Enzymes: protease	Protease activity: in cultivated sludge 116.1 ± 4.2 U/mg at 75 °C; in WWTP sludge 75.8 ± 10.2 U/mg at 50 °C Stored at −20 °C from 1 month decreased the protease activity around 32%.	No additives are needed.	High operating cost due to energy consumption.	[197]
Extraction with additives (CER and Triton X-100)	Activated sludge after gravity thickening	Enzymes: protease	High activity was observed for leucine aminopeptidase (0.015 µmol/mg VS/h)	No physical treatment was required for cell disruption.	High operating cost due to the use of additives.	[196]
Extraction with additives (CER and/or Triton X-100)	Activated sludge before gravity thickening	Enzymes: amylase	Triton X-100 concentration key factor: lower recoveries increasing detergent concentration	Great quality of the final product.	High operating cost due to the use of additives.	[199]

5. Towards Energy Self-Sufficiency

A wastewater treatment plant (WWTP) consumes from 0.45 to 1.25 kWh/m^3, depending on its size and the treatments taking place in it. This energy is usually supplied to the plant externally, from conventional energy supply stations [200,201]. However, according to Gude [202], municipal wastewater (WW) has a total amount of energy of up to 9.7 kWh/m^3. Assuming that the average energy consumption in WWTPs is 0.85 kWh/m^3, the water can contain up to 12 times more energy than what is needed for its treatment [202]. This energy contained in wastewater can be divided into chemical (from the organic load present in the water), thermal, and potential energy, and it can be recovered from the water through different technologies (Table 11).

In addition, energy can be recovered from the sewage sludge. The sludge generated during the treatment processes contains about 60% of the energy initially present in WW, so it is frequently used for energy recovery [203].

Currently, energy recovery technologies are used, primarily, in large-scale treatment plants, biogas recovery being the most widely used one [204].

Table 11. Technologies used for energy recovery from wastewater.

Process	Form of Recovery	Advantages	Disadvantages	Reference
Anaerobic digestion	Biogas	Great energy recovery Can produce both thermal and electrical energy	High investment needed	[140]
Transesterification	Biodiesel	Well-known technology Low-cost raw material High lipid content	Need for drying pre-treatment Risk of soap formation during transesterification Need for refining before its use as a fuel	[205]
Gasification pyrolysis	Hydrogen Syngas	High energy content Environmentally friendly	Need for pre-treatments/catalyst to increase H_2/CO ration	[206]
Microbial fuel cells	Electric power	Substrate directly transformed into electric energy No need for energy input Can operate in different ranges of T, pH and with different types of biomass	High investment cost Significant efficiency decrease at low temperatures	[207]
Heat pump	Thermal energy	Constant temperature of WW throughout the year	Surface fowling Final heat destination needs to be close to the WWTTP	[208]
Hydropower	Electric power	No associated greenhouse effect emissions Cost-effective option Energy generation can be adjusted to WWTP demand curve	Low flexibility to face significant flow drops	[209]

5.1. Biogas Recovery

Anaerobic digestion is one of the most widespread sludge stabilization processes in the current market. This process is capable of transforming the organic load present in the sewage sludge into biogas, which is a mixture of methane (50%–70%), carbon dioxide (30%–50%), and traces of other gases, such as nitrogen or hydrogen [210]. This process is usually carried out in an anaerobic environment, although aerobic digestion is also possible.

Methane, which is the significant component of biogas, is considered a greenhouse gas. Sending the sludge produced in a WWTP directly to a landfill would result in the release of this gas into the atmosphere by natural routes [140]. Moreover, thanks to its calorific value, biogas can be used for electricity generation, heat production, and as a fuel for vehicles among other uses. Due to that, it seems logical to implement anaerobic digesters that facilitate its collection which is highly beneficial [211].

The cost of electricity in a treatment plant is about 80% of the total operating cost. Even if the biogas can be sold, it is interesting to use the methane produced in the digesters in the treatment plant itself, to feed the gas engines and produce both electrical and thermal energy. In some cases, it can become one of the primary sources of energy in the WWTP [140].

The conversion of organic matter into biogas is a complex process that involves four different phases of biochemical reactions: hydrolysis, acidogenesis, acetogenesis, and methanogenesis. In case of the latter, the products of all the previous phases are transformed into the final products: methane and carbon dioxide [212].

Currently, anaerobic digestion is combined with other techniques to promote the generation of biogas and/or increasing the percentage of methane in it [203]. Within this framework, the co-digestion of food waste and wastewater sludge has been proposed. Some studies on this technique carried out through the concept of circular energy indicate that this technique is a right solution, generating significant energy savings in the overall operating cost of the plant [213].

However, the use of pre-treatments that increase methane production is most common. In fact, the objective of such pre-treatments could be described as: (1) to increase the digestibility of organic matter; (2) increase the rate of hydrolysis; (3) increase the percentage of methane in the mixture; (4) reduce sludge viscosity and reduce pumping costs [212].

A large variety of pre-treatments that can be applied, but all of them can be encompassed in one of the following categories:

- physical: electrical pulses, microwaves, ultrasound, etc. [214];
- chemical: ozonation, hydrogen peroxide, wet air oxidation, sulphite, etc. [17,215];
- biological [216];
- thermal [217];
- combined pre-treatments [218].

Among these, the most commonly applied are physical and thermal pre-treatments. [219] found out that methane production was increased by 21%–31% when using a heat pre-treatment. Moreover, by means of thermal pre-treatment, Liu et al. [220] achieved an increase in methane production of 51%. Moreover, by combining the use of microwaves (600 W) and hydrogen peroxide (0.2 g/g total solids), methane production increased by 20%, while reducing fluid viscosity [220].

5.2. Biodiesel Production

The growing scarcity of fossil fuel reserves, together with the environmental benefits of biodiesel (such as the decrease in SO_x and CO_2 emissions), is the reason why the interest in this biofuel has grown significantly in recent years [140].

Commonly, vegetable oils are used for biodiesel (BD) production. However, the cultivation of plant species with the only purpose of producing BD has raised awareness about the competition of these crops with the food industry [205]. In addition, the cost of BD production is high (mainly due to the cost of raw materials), which does not make it competitive with petroleum-derived diesel.

Therefore, finding a non-edible and low-cost raw material is necessary to increase biodiesel competitiveness. Among the available options, sewage sludge, which might seem as industrial waste, increasingly receives more attention. Its high lipid content, the large quantities generated in water treatment plants, and its low cost, make it an ideal raw material for this purpose [221].

This biofuel is produced by transesterification of refined triglycerides and methanol (MeOH). This reaction usually takes place in the presence of a homogeneous catalyst, either acidic or basic (e.g., H_2SO_4, KOH or NaOH) [222]. The higher the oil content of the sewage sludge, the more BD it can produce; hence, it is recommended to select microorganisms for the treatments in the WWTP based on their ability to produce oil [211].

Depending on the origin of the wastewater, the sludge will have a different lipidic composition. Therefore, to know the specific treatments that will work best in each case, it is necessary to analyse the lipid content every time [223]. Finally, a high proportion of free fatty acids (FFAs) (which can account for up to 70% of the lipid content) causes problems during transesterification due to soap formation when conventional basic catalysts, such as NaOH, are used. Thus, acid catalysts are normally preferred in these cases [224]. However, by using these catalysts, the reactions slow down considerably. Although some studies that focus on the synthesis of biodiesel through non-catalytic transesterification [222], the vast majority of researchers strive to find new catalysts that allow for high efficiency in the production of biodiesel in a short time [225].

5.3. Hydrogen and Syngas Production

Along with biogas, biohydrogen is one of the biofuels that can be recovered from wastewater, being a great alternative to conventional fuels, as its energy content is 2.75 times higher than that of hydrocarbon fuel [140].

Hydrogen can be considered the most environmentally friendly fuel since its combustion only generates water. Although it is the most abundant element in nature, it cannot be found in molecular form (H_2) due to its high reactivity, so it must be generated industrially [226].

Despite its great properties, very few processes can produce only the hydrogen. On the contrary, most recovery techniques produce syngas, which is a gaseous blend in which H_2 and CO are the major constituents [206]. There are currently various techniques for obtaining syngas, gasification and pyrolysis being the most widely used ones [227]. In general, due to the high water content in the sludge (80%), drying pre-treatments are required before these treatments, which considerably increases the cost of this process and involves an energy input that is counterproductive, given the objective of energy recovery [228]. To avoid this previous step, many investigations are currently focused on using the supercritical water gasification (SCWG) process. Water, when in extreme conditions beyond the critical point ($Tc \geq 374$ °C, $Pc \geq 22.1$ MPa), has unique properties, which makes it an excellent non-polar solvent that can dissolve various organic substances present in the sludge, thus, increasing the effectiveness of hydrogen production. Achieving optimal operating conditions may be quite costly, so the use of catalysts is a key factor in reducing the activation energy of the reactions. The catalysts could also increase the effectiveness of gasification by being H_2 selective [226].

Other processes that can be used to generate H_2 from wastewater, although not so widespread, are electrohydrolysis [229] and biological treatments, such as microbial fermentation under dark or photo fermentation [230].

As previously discussed, the most common way to obtain biohydrogen is through the production of syngas, which implies that, nowadays, most of the efforts in this field are concentrated in obtaining a gaseous mixture with the highest proportion of hydrogen possible. Sludge pre-treatment has been one of the most commonly used methods for this purpose. Wang et al. [231] employed ultraviolet radiation as pre-treatment, and an 80.6% increase in the amount of hydrogen as compared to the untreated sludge has been observed. Ultrasound and heat have also been tested for this purpose, obtaining excellent results. Elbeshbishy et al. [232] applied an ultrasonic pre-treatment, achieving a 120% increase in hydrogen production.

To enhance the efficiency of gasification, it has been proven that including an oxidizing agent, such as a vapour stream, increases hydrogen production [233]. In addition, using catalysts has also shown high efficiency in syngas production. Zhang et al. [234] tested the impacts of Fe/Ca-based conditioners on syngas generation and have found that the final H_2 yield was improved by 51.2% when iron was used, and 76.5% for CaO, catalysed process at 1273 K, compared to catalyst-free treatment.

Catalysts can also be used in combination with the SCWG process. Hantoko et al. [235] showed that activated carbon significantly improves syngas production. Additionally, Yan et al. (2019) studied the effect of the addition of Ni/Al_2O_3 catalysts. In this case, both the overall efficiency of the reactions and the proportion of H_2 in the final mixture improved substantially, as the later increased by 146% compared to the experiment without catalyst [236].

Other processes, such as the use of adsorbents to retain CO_2 and promote hydrogen generation, are also being investigated with considerable success. [233]. Although appreciable advances have been noticed in the recent years, it is still necessary to work hard to obtain a high-quality biohydrogen that can be used as fuel, which indeed would be a great alternative to currently used fossil fuels.

5.4. Microbial Fuel Cell

In addition to the production of biofuels and biogas, it is possible to recover wastewater energy from organic matter directly as electricity thanks to bioelectrochemical systems [202,237]. These processes, directly applicable both on water and sewage sludge, lead not only to electricity generation, but also to the removal of contaminants [238–242]. Among these processes, several authors have highlighted Microbial fuel Cell (MCF) as a potential solution for problems related to water and energy [243].

MFCs are composed of an anaerobic chamber (known as anode) and an aerobic chamber or cathode. Both chambers have an electrode and are separated by a membrane known as the cation exchange membrane (CEM). CEM has the function of regulating the passage of protons from one chamber to the other. The substrate, rich in organic components, is introduced into the anode and oxidized by microorganisms, leading to the generation of electrons and protons. The electrons generated in the process are transferred to the anode and, subsequently, travel to the cathode through an external circuit generating bioelectricity. Simultaneously the protons move from one chamber to the other through a polymer electrolyte membrane. Finally, electrons and protons combine with oxygen in the cationic chamber to produce a water molecule [244]. The voltage generated in an MFC is of the order of several hundred of millivolts. When the fraction of electron charge that contributes to electricity generation is 40% and the hydraulic retention times is 20 h, the potential of energy that can be recovered from wastewater by MFC can reach 0.65 kWh/m^3 [211].

The efficiency of this process depends on many different parameters, such as pH, temperature, the type of substrate that is degraded, and its quantity, salinity, type of microorganisms, electrode material, etc. [140]. Among these factors, the selection of microorganisms is a fundamental design factor in CFM. More than 35 different species have been reported as exoelectrogen in MFCs, including *Escherichia*, *Pseudomonas*, *Geobacter*, *Clostridum*, *Shewanela*, and *Lactococcus lactis* [245]. Nowadays, many efforts are focused on obtaining a biofilm with the optimal conditions of conductivity and penetrability through the genetic alteration of microorganisms.

Despite having many strengths, this technology has some important disadvantages that must be addressed to generalize its adoption in the real life. The main drawback of the MFCs is their investment cost, about 30 times higher than that of other sludge treatments, as well as the internal resistance that restricts power generation [207]. This high cost is mainly due to the high price of the materials used to manufacture the electrodes and membranes. Another drawback is that, although it can operate in a wide range of temperatures, very low temperatures imply a significant decrease in its efficiency since the metabolic rate of the microorganisms decreases [213,246]. Finally, the electricity generated by this system is not capable of supplying enough energy to guarantee the continuous operation of an electrical instrument. This issue, however, can be solved by linking MFCs to increase voltage or current generation [247].

To overcome the disadvantages, research in this field is focused on genetic modifications of microorganisms and development of electrodes and membranes, by aiming at better and cheaper materials. As an example, Zhang et al. [18] have demonstrated the efficiency of the cylindrical cell, instead of cubic ones, with a larger cathode area, increasing power density by 52% while reducing internal resistance by 60.9%.

5.5. Heat Pumps: Thermal Energy Recovery

The energy extracted by a heat pump can come from different sources, such as air, water, or earth. In recent decades, wastewater has been positioned as a strong candidate for the source of thermal energy in heat pumps, although the available heat is described as low-quality heat. In this case, the system is called wastewater source heat pump (WWSHP) [248]. Currently, more than 500 WWSHP are used around the world. The amount of energy that can be obtained in this way is much greater than that obtained from chemical energy [249].

A WWSHP is composed of a heat/cold distribution system, the main heat pump cycle and a heat exchanger. A basic scheme of the heat pump cycle, which consists of a compressor, a condenser, an expansion device, an evaporator, and auxiliary equipment. The heat is extracted from the wastewater through the evaporator and transferred to the medium, which is heated through the condenser [249].

The place where the recovered heat is used must not be more than 3–5 km away [248]. Thus, the best option is onsite consumption of the recovered energy. It can be used to heat digesters or even for sludge drying. Moreover, it can be used outside to heat nearby buildings or greenhouses [250]. Another factor that significantly affects the efficiency of the process is the variation in the flow and temperature of the effluent since, currently, the most widespread heat pumps have a single operating speed. To overcome this problem, Chae and Ren have created a fixed inverter hybrid heat pump that has managed to increase energy efficiency by more than 15% by adjusting the operating speed to match the flow rate [251].

Finally, the most important problem while applying this technology is the fouling of the surface of the exchanger due to the composition of the wastewater, which is primarily responsible for lowering the efficiency of the equipment considerably. Therefore, the outlet water is the most viable option for energy recovery, since it is relatively clean. Despite working only with treated water, fouling is still significant, so most of the efforts are focused on developing materials and configurations that prevent or delay the fouling of heat exchangers [252,253].

5.6. Hydropower

The kinetic energy contained in the wastewater can also be recovered in its path along with a WWTP. This is possible thanks to the installation of hydropower stations in the system. Their main elements are a turbine placed perpendicular to the water flow and an associated energy generator, to transform mechanical energy into electrical power. Due to the lower water flow available in a WWTP compared to rivers or waterfalls, the most appropriate systems would be mini and micro-hydropower (MHP) [254].

The generation of hydraulic energy has several advantages that must be considered. First, it is an environmentally friendly option, since it is a renewable energy source that has no associated greenhouse effect emissions of any kind. Additionally, compared to other renewable energies, such as solar or wind, it stands out because it can generate non-stop energy throughout the year, regardless of weather conditions. On the other hand, it is considered a cost-effective option since it allows to take advantage of the existing infrastructure for its installation, which reduces the dimensions of the necessary civil works [255]. Finally, the generation of hydraulic energy can be adjusted to the WWTP demand curve since consumption peaks correspond to the times when the highest water flow is collected and vice versa [209].

Despite the good prospects, the number of WWTP plants that have hydraulic power generators is quite small, due to the number of suspended solids contained in WW, which can damage the mechanism [209]. This problem can be minimized by locating the power generation system at the exit of the plant, where the water is cleaner [254].

However, the main issue faced by this technology is that most of the already installed MHP have very little flexibility to face significant flow drops, thus, hampering operation under the severe diurnal and seasonal flow variations typical of many WWTPs. To deal with this problem, new turbine models have been developed, as reported by [256], who have studied the behaviour of a variable flow turbine throughout a year. This system is capable of working in a flow range that goes from 53% to 123% of the design one, which allowed to take advantage of almost 96% of the effluent generated in 1 year, which yielded more than 68.1 MWh.

A study conducted on 100 plants in Ireland and the United Kingdom concluded that only larger and high-flow plants are viable for the installation of a hydraulic energy recovery system [257].

5.7. Real Examples of Self-Sufficient WWTP

All of the processes described so far can help to achieve energy self-sufficiency in a water treatment plant. Although most plants that apply these technologies do so only to provide a percentage of the total energy needed to carry out the treatments, it is possible to achieve energy self-sufficiency and even positive energy balance (especially in medium/large-sized plants). Table 12 shows some WWTP examples with these characteristics.

Nowadays, plants that have reached (or are close to achieving) energy self-sufficiency mainly rely on the production of biogas from anaerobic digestion [258].

Shen et al. [210], have performed an analysis of the current landscape by detecting 12 treatment plants in Europe and the United States that have achieved energy self-sufficiency. Among those, two plants located in Austria stand out as they exhibit a positive energy balance, which allows them to sell part of the energy produced. The WWTPs of Strass TP and Wolfgangsee-Ischl, respectively, generate 6.3% and 7% more energy than they use [259].

East Bay Municipal Utility District (EBMUD) wastewater facility (located in Oakland, CA) became the first energy-neutral WWTP in North America in 2012. Nowadays, EBMUD is producing electricity to meet 126% of the WWTP's electric power demand and the surplus electricity is supplied to the grid. Moreover, in the USA, the Sheboygan Wastewater Treatment Plant is nearly 100% energy self-sufficient thanks to its co-digestion program and cogeneration capacity [258].

Table 12. Energy self-sufficient WWTPs. Adapted from (Gu et al., 2017).

Name of WWTP	Country	Capacity $\cdot 10^3$ (m^3/d)	Energy Generation Technique	% Energy Self-Sufficiency	Reference
Grevesmuhlen	Germany	15	Anaerobic digester	100	[260]
Wolfgangsee-Ischl	Austria	19	Anaerobic digester	107	[259]
Strass im Zillertal	Austria	22	Anaerobic digester	106	[261,262]
Marselisborg	Denmark	33	Anaerobic digester	150	[204]
Gloversville-Johnstown Join	USA	41	Anaerobic digester	100	[263]
Sheboygan Regional	USA	41	Anaerobic digester	100	[264]
Gresham	USA	49	Anaerobic digester	100	[210]
Zürich Werdhölzli	Switzerland	253	Anaerobic digester	126	[265]
East Bay municipal Utility District	USA	264	Anaerobic digester	100	[210]
Point Loma	USA	662	Anaerobic digester	100	[266]

More surprising is the case of the Marselisborg WWTP in Denmark, since it generates more than 150% of its energy requirements, so it can also be considered as a power plant. This plant, with a treatment capacity of 12 million cubic meters per year, invested € 3M to optimize its treatment process and improve biogas production [204].

6. Outlook and Concluding Remarks

The main purpose of this review is to discuss the different opportunities and challenges that are prevalent to apply circular economy in the water sector. The consumption model of modern society is the first step towards avoiding the exertion of pressure over natural resources, and this change can be applied in all of the industrial activities. A circular economy has been applied to water, for some decades, when wastewater has been reclaimed and reused. However, this is always done by countries with hydric stress in order to increase the water resources. In this regard, the application of reclamation wastewater facilities, even in regions or countries without water scarcity, is recommended. This is mainly due to the positive environmental impacts of reclamation, which also minimizes the capture of water from natural sources. Despite the benefits of reclamation, possible associated risks might appear as a consequence of the presence of pathogen germs, emerging micropollutants, and antibiotic resistance genes. The control of these risks will ensure the increased use of reclaimed water. There are some other consolidated applications of a circular economy in the water sector, such as nutrients and energy recovery. Both strategies are widely implemented in new generation treatment plants, although research is mainly focused on the complete recovery of nutrients and the accomplishment of energy self-sufficiency. Despite the fact that the above-mentioned aspects are not only opportunities, but also realities in the water sector, future wastewater treatment plants should be designed as biorefineries, which do not only reclaim water as the product, but also facilitate valorisation and recovery of high-added value substances. Hence, society must change its view on wastewater, from being a waste effluent to a "stream rich of different valuable substances" with a tremendous economic impact and value. Moreover, sewage sludge is also potentially recoverable, by the extraction/recovery of substances or the generation of new products, such as adsorbents, bioplastics, or construction materials. In all of the cases, the recovery of substances or valorisation of the effluents help to decrease the pressure over natural resources. Nevertheless, efforts should be made to evolve technologies that would be able to bring down the high costs linked with these processes. However, a change of mindset, together with appropriate policies, should help achieve a fully circular economy in the water sector.

Finally, to conclude, Table 13 sums up the current situation of the wastewater sector regarding the implementation of a circular economy. With the help of an analysis of strengths, weaknesses, opportunities and threats (SWOT analysis), some challenges and opportunities have been identified. The strengths of each available technology, which have been discussed in the review, consist mainly in the possibility of obtaining benefits from something that has always been considered a waste, while decreasing pressure over natural resources. Even so, some weaknesses must be addressed, such as the considerable initial financial investment and operational costs, or the lack of adaptability of some treatments to different working conditions. However, the constant and quick development of new technologies driven by the global trend to promote greener processes is expected to help to overcome these deficiencies. Finally, it must not be forgotten that, for a circular economy to become a reality, a change of mindset is needed, together with the commitment of governments and private companies.

Table 13. SWOT analysis of the wastewater sector regarding the implementation of circular economy.

STRENGTHS	WEAKNESSES
• Obtain benefits from something that has always been considered a waste • Reduce/eliminate energy consumption • Large number of alternative technologies • Constant availability of wastewater • Decrease of pressure over natural resources	• Considerable initial financial outlay • Lack of adaptability • Little experience in large-scale implementation
OPPORTUNITIES	**THREATS**
• Global trend to promote greener processes • Due to water scarcity, wastewater reclamation and reuse plays a fundamental role • Quick development of new technologies	• Political changes • General distrust towards the use of reclaimed water • Recovery costs for some substances might end up being higher than their synthesis or extraction

Author Contributions: S.G.-R.: Writing—review & editing, Conceptualization. P.O.: Writing—review & editing, Conceptualization, Validation. E.R.: Conceptualization, Validation and Supervision. D.N.S.: Review & editing, Conceptualization. J.R.-C.: Writing—review & editing, Supervision, Project administration, Conceptualization, Formal analysis. All authors have read and agreed to the published version of the manuscript.

Funding: Universidad Politécnica de Madrid through the seed research project with Asian Institutions (ASIA-18-Z77QCN-2-CWL7CN).

Acknowledgments: Authors acknowledge the financial support from the Universidad Politécnica de Madrid through the seed research project with Asian Institutions (ASIA-18-Z77QCN-2-CWL7CN). Jorge Rodríguez-Chueca acknowledges Universidad Politécnica de Madrid through the research project APOYO-JOVENES-X5PKL6-88-KZ46KU. Likewise, S. Guerra-Rodríguez also acknowledges UPM for support through the pre-doctoral contract of the "Programa Propio".

Conflicts of Interest: The authors declare no conflict of interest.

References

1. ONU. Resolución aprobada por la Asamblea General el 28 de julio de 2010. 64/292. *El Derecho Hum. Agua Saneam.* **2010**, *660*, 9–11.
2. Hailu, D.; Rendtorff-Smith, S.; Gankhuyag, U.; Ochieng, C. *Toolkit and Guidance for Preventing and Managing Land and Natural Resources Conflict*; The United Nations Interagency Framework Team for Preventive Action: Nairobi, Kenya, 2012. [CrossRef]
3. SDG. *Sustainable Development Goal 6 Synthesis Report 2018 on Water and Sanitation*; United Nations: New York, NY, USA, 2018. [CrossRef]
4. Gassert, F.; Reig, P.; Luo, T.; Maddocks, A. *Aqueduct Country and River Basin Rankings: A Weighted Aggregation of Spatially Distinct Hydrological Indicators*; World Resources Institute: Washington, DC, USA, December 2013. Available online: wri.org/publication/aqueduct-country-river-basin-rankings (accessed on 15 December 2019).
5. COM. *Closing the Loop—An EU Action Plan for the Circular Economy*; COM: Brussels, Belgium, 2015. [CrossRef]
6. Salgot, M.; Folch, M. Tratamento de águas residuais e reutilização de agua. *Curr. Opin. Environ. Sci. Health* **2018**, *2*, 64–74. [CrossRef]
7. Ricart, S.; Rico, A.M. Assessing technical and social driving factors of water reuse in agriculture: A review on risks, regulation and the yuck factor. *Agric. Water Manag.* **2019**, *217*, 426–439. [CrossRef]
8. Levine, A.D.; Asano, T. Recovering sustainable water from wastewater. *Environ. Sci. Technol.* **2004**, *38*, 201A–208A. [CrossRef]
9. Lyu, S.; Chen, W.; Zhang, W.; Fan, Y.; Jiao, W. Wastewater reclamation and reuse in China: Opportunities and challenges. *J. Environ. Sci.* **2016**, *39*, 86–96. [CrossRef]
10. Maryam, B.; Büyükgüngör, H. Wastewater reclamation and reuse trends in Turkey: Opportunities and challenges. *J. Water Process Eng.* **2019**, *30*, 100501. [CrossRef]

11. Amann, A.; Zoboli, O.; Krampe, J.; Rechberger, H.; Zessner, M.; Egle, L. Environmental impacts of phosphorus recovery from municipal wastewater. *Resour. Conserv. Recycl.* **2018**, *130*, 127–139. [CrossRef]
12. Bouzas, A.; Martí, N.; Grau, S.; Barat, R.; Mangin, D.; Pastor, L. Implementation of a global P-recovery system in urban wastewater treatment plants. *J. Clean. Prod.* **2019**, *227*, 130–140. [CrossRef]
13. Malila, R.; Lehtoranta, S.; Viskari, E.L. The role of source separation in nutrient recovery—Comparison of alternative wastewater treatment systems. *J. Clean. Prod.* **2019**, *219*, 350–358. [CrossRef]
14. Wu, Y.; Luo, J.; Zhang, Q.; Aleem, M.; Fang, F.; Xue, Z.; Cao, J. Potentials and challenges of phosphorus recovery as vivianite from wastewater: A review. *Chemosphere* **2019**, *226*, 246–258. [CrossRef]
15. Gikas, P. Towards energy positive wastewater treatment plants. *J. Environ. Manag.* **2017**, *203*, 621–629. [CrossRef] [PubMed]
16. Grasham, O.; Dupont, V.; Camargo-Valero, M.A.; García-Gutiérrez, P.; Cockerill, T. Combined ammonia recovery and solid oxide fuel cell use at wastewater treatment plants for energy and greenhouse gas emission improvements. *Appl. Energy* **2019**, *240*, 698–708. [CrossRef]
17. Zan, F.; Zeng, Q.; Hao, T.; Ekama, G.A.; Hao, X.; Chen, G. Achieving methane production enhancement from waste activated sludge with sulfite pretreatment: Feasibility, kinetics and mechanism study. *Water Res.* **2019**, *158*, 438–448. [CrossRef] [PubMed]
18. Zhang, M.; Ma, Z.; Zhao, N.; Zhang, K.; Song, H. Increased power generation from cylindrical microbial fuel cell inoculated with P. aeruginosa. *Biosens. Bioelectron.* **2019**, *141*, 111394. [CrossRef] [PubMed]
19. Hemalatha, M.; Sravan, J.S.; Min, B.; Venkata Mohan, S. Microalgae-biorefinery with cascading resource recovery design associated to dairy wastewater treatment. *Bioresour. Technol.* **2019**, *284*, 424–429. [CrossRef] [PubMed]
20. Kaushik, A.; Basu, S.; Raturi, S.; Batra, V.S.; Balakrishnan, M. Recovery of antioxidants from sugarcane molasses distillery wastewater and its effect on biomethanation. *J. Water Process Eng.* **2018**, *25*, 205–211. [CrossRef]
21. Ochando-Pulido, J.M.; Corpas-Martínez, J.R.; Martinez-Ferez, A. About two-phase olive oil washing wastewater simultaneous phenols recovery and treatment by nanofiltration. *Process Saf. Environ. Prot.* **2018**, *114*, 159–168. [CrossRef]
22. Oliveira, G.A.; Carissimi, E.; Monje-Ramírez, I.; Velasquez-Orta, S.B.; Rodrigues, R.T.; Ledesma, M.T.O. Comparison between coagulation-flocculation and ozone-flotation for Scenedesmus microalgal biomolecule recovery and nutrient removal from wastewater in a high-rate algal pond. *Bioresour. Technol.* **2018**, *259*, 334–342. [CrossRef]
23. Peiter, F.S.; Hankins, N.P.; Pires, E.C. Evaluation of concentration technologies in the design of biorefineries for the recovery of resources from vinasse. *Water Res.* **2019**, *157*, 483–497. [CrossRef]
24. Kim, E.J.; Baek, K. Selective recovery of ferrous oxalate and removal of arsenic and other metals from soil-washing wastewater using a reduction reaction. *J. Clean. Prod.* **2019**, *221*, 635–643. [CrossRef]
25. Min, X.; Luo, X.; Deng, F.; Shao, P.; Wu, X.; Dionysiou, D.D. Combination of multi-oxidation process and electrolysis for pretreatment of PCB industry wastewater and recovery of highly-purified copper. *Chem. Eng. J.* **2018**, *354*, 228–236. [CrossRef]
26. Swain, N.; Mishra, S. A review on the recovery and separation of rare earths and transition metals from secondary resources. *J. Clean. Prod.* **2019**, *220*, 884–898. [CrossRef]
27. Xu, X.; Yang, Y.; Zhao, X.; Zhao, H.; Lu, Y.; Jiang, C.; Shi, J. Recovery of gold from electronic wastewater by Phomopsis sp. XP-8 and its potential application in the degradation of toxic dyes. *Bioresour. Technol.* **2019**, *288*, 121610. [CrossRef] [PubMed]
28. Wang, Y.; Mei, X.; Ma, T.; Xue, C.; Wu, M.; Ji, M.; Li, Y. Green recovery of hazardous acetonitrile from high-salt chemical wastewater by pervaporation. *J. Clean. Prod.* **2018**, *197*, 742–749. [CrossRef]
29. Zeng, G.; Ling, B.; Li, Z.; Luo, S.; Sui, X.; Guan, Q. Fluorine removal and calcium fluoride recovery from rare-earth smelting wastewater using fluidized bed crystallization process. *J. Hazard. Mater.* **2019**, *373*, 313–320. [CrossRef]
30. Raheem, A.; Sikarwar, V.S.; He, J.; Dastyar, W.; Dionysiou, D.D.; Wang, W.; Zhao, M. Opportunities and challenges in sustainable treatment and resource reuse of sewage sludge: A review. *Chem. Eng. J.* **2018**, *337*, 616–641. [CrossRef]

31. Laura, F.; Tamara, A.; Müller, A.; Hiroshan, H.; Christina, D.; Serena, C. Selecting sustainable sewage sludge reuse options through a systematic assessment framework: Methodology and case study in Latin America. *J. Clean. Prod.* **2020**, *242*, 118389. [CrossRef]
32. Babiker, E.; Al-Ghouti, M.A.; Zouari, N.; McKay, G. Removal of boron from water using adsorbents derived from waste tire rubber. *J. Environ. Chem. Eng.* **2019**, *7*, 102948. [CrossRef]
33. Dai, Y.; Sun, Q.; Wang, W.; Lu, L.; Liu, M.; Li, J.; Zhang, Y. Utilizations of agricultural waste as adsorbent for the removal of contaminants: A review. *Chemosphere* **2018**, *211*, 235–253. [CrossRef]
34. Li, J.; Yu, G.; Xie, S.; Pan, L.; Li, C.; You, F.; Wang, Y. Immobilization of heavy metals in ceramsite produced from sewage sludge biochar. *Sci. Total Environ.* **2018**, *628–629*, 131–140. [CrossRef]
35. Mo, J.; Yang, Q.; Zhang, N.; Zhang, W.; Zheng, Y.; Zhang, Z. A review on agro-industrial waste (AIW) derived adsorbents for water and wastewater treatment. *J. Environ. Manag.* **2018**, *227*, 395–405. [CrossRef] [PubMed]
36. Chen, Y.-D.; Bai, S.; Li, R.; Su, G.; Duan, X.; Wang, S.; Ren, N.-Q.; Ho, S.-H. Magnetic biochar catalysts from anaerobic digested sludge: Production, application and environment impact. *Environ. Int.* **2019**, *126*, 302–308. [CrossRef] [PubMed]
37. Huang, Z.; Wang, T.; Shen, M.; Huang, Z.; Chong, Y.; Cui, L. Coagulation treatment of swine wastewater by the method of in-situ forming layered double hydroxides and sludge recycling for preparation of biochar composite catalyst. *Chem. Eng. J.* **2019**, *369*, 784–792. [CrossRef]
38. Tao, S.; Yang, J.; Hou, H.; Liang, S.; Xiao, K.; Qiu, J.; Hu, J.; Liu, B.; Yu, W.; Deng, H. Enhanced sludge dewatering via homogeneous and heterogeneous Fenton reactions initiated by Fe-rich biochar derived from sludge. *Chem. Eng. J.* **2019**, *372*, 966–977. [CrossRef]
39. Zhu, S.; Wang, W.; Xu, Y.; Zhu, Z.; Liu, Z.; Cui, F. Iron sludge-derived magnetic Fe 0 /Fe 3 C catalyst for oxidation of ciprofloxacin via peroxymonosulfate activation. *Chem. Eng. J.* **2019**, *365*, 99–110. [CrossRef]
40. Sun, B.; Yuan, Y.; Li, H.; Li, X.; Zhang, C.; Guo, F.; Liu, X.; Wang, K.; Zhao, X.S. Waste-cellulose-derived porous carbon adsorbents for methyl orange removal. *Chem. Eng. J.* **2019**, *371*, 55–63. [CrossRef]
41. Wen, H.; Gu, L.; Yu, H.; Qiao, X.; Zhang, D.; Ye, J. Radical assisted iron impregnation on preparing sewage sludge derived Fe/carbon as highly stable catalyst for heterogeneous Fenton reaction. *Chem. Eng. J.* **2018**, *352*, 837–846. [CrossRef]
42. Kazner, C. Water Reclamation Technologies for Safe Managed Aquifer Recharge. *Water Intell. Online* **2012**, *11*. [CrossRef]
43. Confederación Hidrográfica del Segura. Available online: https://www.chsegura.es/chs/cuenca/resumendedatosbasicos/recursoshidricos/reutilizacion.html (accessed on 15 November 2019).
44. World Water Assessment Programme (WWAP). *The UN World Water Development Report 2017*; Wastewater; The Untapped Resource: Paris, France, 2017.
45. Jiménez, B.; Asano, T. Water reclamation and reuse around the world. In *Water Reuse—An International Survey of Current Practice, Issues and Needs*; IWA Publishing: London, UK, 2008. [CrossRef]
46. Jodar-Abellan, A.; López-Ortiz, M.I.; Melgarejo-Moreno, J. Wastewater Treatment and Water Reuse in Spain. Current Situation and Perspectives. *Water* **2019**, *11*, 1551. [CrossRef]
47. Ormad, M.P. *Reutilización de Aguas Residuales Urbanas. Cátedra Mariano López Navarro*; Universidad de Zaragoza: Zaragoza, Spain, 2011.
48. Lefebvre, O. Beyond NEWater: An insight into Singapore's water reuse prospects. *Curr. Opin. Environ. Sci. Health* **2018**, *2*, 26–31. [CrossRef]
49. Woltersdorf, L.; Scheidegger, R.; Liehr, S.; Döll, P. Municipal water reuse for urban agriculture in Namibia: Modeling nutrient and salt flows as impacted by sanitation user behavior. *J. Environ. Manag.* **2016**, *169*, 272–284. [CrossRef] [PubMed]
50. Técnica y Proyectos Sociedad Anónima (TYPSA). Updated Report on Wastewater Reuse in the European Union. April 2013. Available online: https://ec.europa.eu/environment/water/blueprint/pdf/Final%20Report_Water%20Reuse_April%202013.pdf (accessed on 20 December 2019).
51. Behera, S.K.; Kim, H.W.; Oh, J.E.; Park, H.S. Occurrence and removal of antibiotics, hormones and several other pharmaceuticals in wastewater treatment plants of the largest industrial city of Korea. *Sci. Total Environ.* **2011**, *409*, 4351–4360. [CrossRef] [PubMed]
52. Kosma, C.I.; Lambropoulou, D.A.; Albanis, T.A. Investigation of PPCPs in wastewater treatment plants in Greece: Occurrence, removal and environmental risk assessment. *Sci. Total Environ.* **2014**, *466*, 421–438. [CrossRef] [PubMed]

53. Kulkarni, P.; Olson, N.D.; Paulson, J.N.; Pop, M.; Maddox, C.; Claye, E.; Goldstein, R.E.R.; Sharma, M.; Gibbs, S.; Mongodin, E.F.; et al. Conventional wastewater treatment and reuse site practices modify bacterial community structure but do not eliminate some opportunistic pathogens in reclaimed water. *Sci. Total Environ.* **2018**, *639*, 1126–1137. [CrossRef]
54. Čelić, M.; Gros, M.; Farré, M.; Barceló, D.; Petrović, M. Pharmaceuticals as chemical markers of wastewater contamination in the vulnerable area of the Ebro Delta (Spain). *Sci. Total Environ.* **2019**, *652*, 952–963. [CrossRef]
55. Cerqueira, F.; Matamoros, V.; Bayona, J.; Elsinga, G.; Hornstra, L.M.; Piña, B. Distribution of antibiotic resistance genes in soils and crops. A field study in legume plants (*Vicia faba* L.) grown under different watering regimes. *Environ. Res.* **2019**, *170*, 16–25. [CrossRef]
56. Jasim, S.Y.; Saththasivam, J.; Loganathan, K.; Ogunbiyi, O.O.; Sarp, S. Reuse of Treated Sewage Effluent (TSE) in Qatar. *J. Water Process Eng.* **2016**, *11*, 174–182. [CrossRef]
57. López, A.; Rodríguez-Chueca, J.; Mosteo, R.; Gómez, J.; Rubio, E.; Goñi, P.; Ormad, M.P. How does urban wastewater treatment affect the microbial quality of treated wastewater? *Process Saf. Environ. Prot.* **2019**, *130*, 22–30. [CrossRef]
58. Mosteo, R.; Ormad, M.P.; Goñi, P.; Rodríguez-Chueca, J.; García, A.; Clavel, A. Identification of pathogen bacteria and protozoa in treated urban wastewaters discharged in the Ebro River (Spain): Water reuse possibilities. *Water Sci. Technol.* **2013**, *68*, 575–583. [CrossRef]
59. Ajonina, C.; Buzie, C.; Ajonina, I.U.; Basner, A.; Reinhardt, H.; Gulyas, H.; Liebau, E.; Otterpohl, R. Occurrence of cryptosporidium in a wastewater treatment plant in north germany. *J. Toxicol. Environ. Health Part A Curr. Issues* **2012**, *75*, 1351–1358. [CrossRef]
60. Li, Z.; Liu, X.; Huang, Z.; Hu, S.; Wang, J.; Qian, Z.; Feng, J.; Xian, Q.; Gong, T. Occurrence and ecological risk assessment of disinfection byproducts from chlorination of wastewater effluents in East China. *Water Res.* **2019**, *157*, 247–257. [CrossRef] [PubMed]
61. Rodríguez-Chueca, J.; Ormad, M.P.; Mosteo, R.; Sarasa, J.; Ovelleiro, J.L. Conventional and Advanced Oxidation Processes Used in Disinfection of Treated Urban Wastewater. *Water Environ. Res.* **2015**, *87*, 281–288. [CrossRef] [PubMed]
62. Aguas, Y.; Hincapie, M.; Martínez-Piernas, A.B.; Agüera, A.; Fernández-Ibáñez, P.; Nahim-Granados, S.; Polo-López, M.I. Reclamation of Real Urban Wastewater Using Solar Advanced Oxidation Processes: An Assessment of Microbial Pathogens and 74 Organic Microcontaminants Uptake in Lettuce and Radish. *Environ. Sci. Technol.* **2019**, *53*, 9705–9714. [CrossRef] [PubMed]
63. Nahim-Granados, S.; Pérez, J.S.; Polo-Lopez, M.I. Effective solar processes in fresh-cut wastewater disinfection: Inactivation of pathogenic E. coli O157:H7 and Salmonella enteritidis. *Catal. Today* **2018**, *313*, 79–85. [CrossRef]
64. Nahim-Granados, S.; Oller, I.; Malato, S.; Pérez, J.S.; Polo-Lopez, M.I. Commercial fertilizer as effective iron chelate (Fe3+-EDDHA) for wastewater disinfection under natural sunlight for reusing in irrigation. *Appl. Catal. B: Environ.* **2019**, *253*, 286–292. [CrossRef]
65. Rodríguez-Chueca, J.; Ormad, M.P.; Mosteo, R.; Canalis, S.; Ovelleiro, J.L. Escherichia coli Inactivation in Fresh Water Through Photocatalysis with TiO2-Effect of H2O2 on Disinfection Kinetics. *Clean Soil Air Water* **2016**, *44*, 515–524. [CrossRef]
66. Valero, P.; Verbel, M.; Silva-Agredo, J.; Mosteo, R.; Ormad, M.P.; Torres-Palma, R.A. Electrochemical advanced oxidation processes for Staphylococcus aureus disinfection in municipal WWTP effluents. *J. Environ. Manag.* **2017**, *198*, 256–265. [CrossRef]
67. Rodríguez-Chueca, J.; García-Cañibano, C.; Lepistö, R.J.; Encinas, Á.; Pellinen, J.; Marugán, J. Intensification of UV-C tertiary treatment: Disinfection and removal of micropollutants by sulfate radical based Advanced Oxidation Processes. *J. Hazard. Mater.* **2019**, *372*, 94–102. [CrossRef]
68. Rodríguez-Chueca, J.; Guerra-Rodríguez, S.; Raez, J.M.; López-Muñoz, M.J.M.-J.; Rodríguez, E. Assessment of different iron species as activators of $S_2O_8{}^{2-}$ and HSO_{5-} for inactivation of wild bacteria strains. *Appl. Catal. B Environ.* **2019**, *248*, 54–61. [CrossRef]
69. Ferreira, L.C.; Castro-Alférez, M.; Nahim-Granados, S.; Polo-López, M.I.; Lucas, M.S.; Li Puma, G.; Fernández-Ibáñez, P. Inactivation of water pathogens with solar photo-activated persulfate oxidation. *Chem. Eng. J.* **2020**, *12*, 9542–9561. [CrossRef]

70. Ponce-Robles, L.; Oller, I.; Polo-López, M.I.; Rivas-Ibáñez, G.; Malato, S. Microbiological evaluation of combined advanced chemical-biological oxidation technologies for the treatment of cork boiling wastewater. *Sci. Total Environ.* **2019**, *687*, 567–576. [CrossRef] [PubMed]
71. Rodríguez-Chueca, J.; Mesones, S.; Marugán, J. Hybrid UV-C/microfiltration process in membrane photoreactor for wastewater disinfection. *Environ. Sci. Pollut. Res.* **2018**, *26*, 36080–36087. [CrossRef] [PubMed]
72. Symonds, E.M.; Verbyla, M.E.; Lukasik, J.O.; Kafle, R.C.; Breitbart, M.; Mihelcic, J.R. A case study of enteric virus removal and insights into the associated risk of water reuse for two wastewater treatment pond systems in Bolivia. *Water Res.* **2014**, *65*, 257–270. [CrossRef] [PubMed]
73. De Sanctis, M.; Del Moro, G.; Chimienti, S.; Ritelli, P.; Levantesi, C.; Di Iaconi, C. Removal of pollutants and pathogens by a simplified treatment scheme for municipal wastewater reuse in agriculture. *Sci. Total Environ.* **2017**, *580*, 17–25. [CrossRef]
74. Shingare, R.P.; Nanekar, S.V.; Thawale, P.R.; Karthik, R.; Juwarkar, A.A. Comparative study on removal of enteric pathogens from domestic wastewater using Typha latifolia and Cyperus rotundus along with different substrates. *Int. J. Phytoremediation* **2017**, *19*, 899–908. [CrossRef]
75. Voumard, M.; Giannakis, S.; Carratalà, A.; Pulgarin, C. *E. coli*—MS2 bacteriophage interactions during solar disinfection of wastewater and the subsequent post-irradiation period. *Chem. Eng. J.* **2019**, *359*, 1224–1233. [CrossRef]
76. Kuzmanović, M.; Ginebreda, A.; Petrović, M.; Barceló, D. Risk assessment based prioritization of 200 organic micropollutants in 4 Iberian rivers. *Sci. Total Environ.* **2015**, *503*, 289–299. [CrossRef]
77. Barceló, D.; Petrovic, M. Challenges and achievements of LC-MS in environmental analysis: 25 years on. *TrAC Trends Anal. Chem.* **2007**, *26*, 2–11. [CrossRef]
78. Virkutyte, J.; Varma, R.S.; Jegatheesan, V. *Treatment of Micropollutants in Water and Wastewater*; IWA Publishing: London, UK, 2010.
79. Besha, A.T.; Gebreyohannes, A.Y.; Tufa, R.A.; Bekele, D.N.; Curcio, E.; Giorno, L. Removal of emerging micropollutants by activated sludge process and membrane bioreactors and the effects of micropollutants on membrane fouling: A review. *J. Environ. Chem. Eng.* **2017**, *5*, 2395–2414. [CrossRef]
80. Gros, M.; Blum, K.M.; Jernstedt, H.; Renman, G.; Rodríguez-Mozaz, S.; Haglund, P.; Anderssion, P.L.; Wiberg, K.; Ahrens, L. Screening and prioritization of micropollutants in wastewaters from on-site sewage treatment facilities. *J. Hazard. Mater.* **2017**, *328*, 37–45. [CrossRef]
81. Ibáñez, M.; Borova, V.; Boix, C.; Aalizadeh, R.; Bade, R.; Thomaidis, N.S.; Hernández, F. UHPLC-QTOF MS screening of pharmaceuticals and their metabolites in treated wastewater samples from Athens. *J. Hazard. Mater.* **2017**, *323*, 26–35. [CrossRef] [PubMed]
82. Backhaus, T.; Karlsson, M. Screening level mixture risk assessment ofpharmaceuticals in STP effluents. *Water Res.* **2014**, *49*, 157–165. [CrossRef] [PubMed]
83. Esplugas, S.; Bila, D.M.; Krause, L.G.T.; Dezotti, M. Ozonation and advanced oxidation technologies to remove endocrine disrupting chemicals (EDCs) and pharmaceuticals and personal care products (PPCPs) in water effluents. *J. Hazard. Mater.* **2007**, *149*, 631–642. [CrossRef] [PubMed]
84. WHO. *Guidelines for Drinking-Water Quality*; WHO: Geneva, Switzerland, 2011.
85. WHO. *Pharmaceuticals in Drinking Water*; WHO Press: Geneva, Switzerland, 2012. [CrossRef]
86. Gaffney, J.; Almeida, V.; Rodrigues, C.; Ferreira, A.; Benoliel, E.; Cardoso, M.J.; Vale, V. Occurrence of pharmaceuticals in a water supply system and related human health risk assessment. *Water Res.* **2015**, *72*, 199–208. [CrossRef] [PubMed]
87. Schwab, B.W.; Hayes, E.P.; Fiori, J.M.; Mastrocco, F.J.; Roden, N.M.; Cragin, D.; Meyerhoff, R.D.; D'Aco, V.J.; Anderson, P.D. Human pharmaceuticals in US surface waters: A human health risk assessment. *Regul. Toxicol. Pharmacol.* **2005**, *42*, 296–312. [CrossRef] [PubMed]
88. Rodríguez-Chueca, J.; Garcia-Cañibano, C.; Sarro, M.; Encinas, A.; Medana, C.; Fabbri, D.; Calza, P.; Marugán, J. Evaluation of transformation products from chemical oxidation of micropollutants in wastewater by photoassisted generation of sulfate radicals. *Chemosphere* **2019**, *226*, 509–519. [CrossRef] [PubMed]
89. Rodríguez-Chueca, J.; Laski, E.; García-Cañibano, C.; Martín de Vidales, M.J.; Encinas, Á.; Kuch, B.; Marugán, J. Micropollutants removal by full-scale UV-C/sulfate radical based Advanced Oxidation Processes. *Sci. Total Environ.* **2018**, *630*, 1216–1225. [CrossRef]

90. Rodríguez-Chueca, J.; Mediano, A.; Pueyo, N.; García-Suescun, I.; Mosteo, R.; Ormad, M.P. Degradation of chloroform by Fenton-like treatment induced by electromagnetic fields: A case of study. *Chem. Eng. Sci.* **2016**, *159*, 89–96. [CrossRef]
91. Jaén-Gil, A.; Buttiglieri, G.; Benito, A.; Gonzalez-Olmos, R.; Barceló, D.; Rodríguez-Mozaz, S. Metoprolol and metoprolol acid degradation in UV/H2O2 treated wastewaters: An integrated screening approach for the identification of hazardous transformation products. *J. Hazard. Mater.* **2019**, *380*, 120851. [CrossRef]
92. te Brinke, E.; Reurink, D.M.; Achterhuis, I.; de Grooth, J.; de Vos, W.M. Asymmetric polyelectrolyte multilayer membranes with ultrathin separation layers for highly efficient micropollutant removal. *Appl. Mater. Today* **2020**, *18*, 100471. [CrossRef]
93. Fenyvesi, É.; Barkács, K.; Gruiz, K.; Varga, E.; Kenyeres, I.; Záray, G.; Szente, L. Removal of hazardous micropollutants from treated wastewater using cyclodextrin bead polymer—A pilot demonstration case. *J. Hazard. Mater.* **2020**, *383*, 121181. [CrossRef] [PubMed]
94. Ruppelt, J.P.; Pinnekamp, J.; Tondera, K. Elimination of micropollutants in four test-scale constructed wetlands treating combined sewer overflow: Influence of filtration layer height and feeding regime. *Water Res.* **2020**, *169*, 115214. [CrossRef] [PubMed]
95. Asif, M.B.; Fida, Z.; Tufail, A.; van de Merwe, J.P.; Leusch, F.D.L.; Pramanik, B.K.; Price, W.E.; Hai, F.I. Persulfate oxidation-assisted membrane distillation process for micropollutant degradation and membrane fouling control. *Sep. Purif. Technol.* **2019**, *222*, 321–331. [CrossRef]
96. Malvestiti, J.A.; Cruz-Alcalde, A.; López-Vinent, N.; Dantas, R.F.; Sans, C. Catalytic ozonation by metal ions for municipal wastewater disinfection and simulataneous micropollutants removal. *Appl. Catal. B Environ.* **2019**, *259*, 118104. [CrossRef]
97. Lim, M.; Ahmad, R.; Guo, J.; Tibi, F.; Kim, M.; Kim, J. Removals of micropollutants in staged anaerobic fluidized bed membrane bioreactor for low-strength wastewater treatment. *Process Saf. Environ. Prot.* **2019**, *127*, 162–170. [CrossRef]
98. Harrabi, M.; Varela Della Giustina, S.; Aloulou, F.; Rodriguez-Mozaz, S.; Barceló, D.; Elleuch, B. Analysis of multiclass antibiotic residues in urban wastewater in Tunisia. *Environ. Nanotechnol. Monit. Manag.* **2018**, *10*, 163–170. [CrossRef]
99. Lucas, D.; Castellet-Rovira, F.; Villagrasa, M.; Badia-Fabregat, M.; Barceló, D.; Vicent, T.; Caminal, G.; Sarrà, M.; Rodríguez-Mozaz, S. The role of sorption processes in the removal of pharmaceuticals by fungal treatment of wastewater. *Sci. Total Environ.* **2018**, *610*, 1147–1153. [CrossRef]
100. Ferre-Aracil, J.; Valcárcel, Y.; Negreira, N.; de Alda, M.L.; Barceló, D.; Cardona, S.C.; Navarro-Laboulais, J. Ozonation of hospital raw wastewaters for cytostatic compounds removal. Kinetic modelling and economic assessment of the process. *Sci. Total Environ.* **2016**, *556*, 70–79. [CrossRef]
101. Boonnorat, J.; Kanyatrakul, A.; Prakhongsak, A.; Honda, R.; Panichnumsin, P.; Boonapatcharoen, N. Effect of hydraulic retention time on micropollutant biodegradation in activated sludge system augmented with acclimatized sludge treating low-micropollutants wastewater. *Chemosphere* **2019**, *230*, 606–615. [CrossRef]
102. Kora, E.; Theodorelou, D.; Gatidou, G.; Fountoulakis, M.S.; Stasinakis, A.S. Removal of polar micropollutants from domestic wastewater using a methanogenic – aerobic moving bed biofilm reactor system. *Chem. Eng. J.* **2020**, *382*, 122983. [CrossRef]
103. Vatankhah, H.; Riley, S.M.; Murray, C.; Quiñones, O.; Steirer, K.X.; Dickenson, E.R.V.; Bellona, C. Simultaneous ozone and granular activated carbon for advanced treatment of micropollutants in municipal wastewater effluent. *Chemosphere* **2019**, *234*, 845–854. [CrossRef] [PubMed]
104. Guillossou, R.; Le Roux, J.; Mailler, R.; Vulliet, E.; Morlay, C.; Nauleau, F.; Gasperi, J.; Rocher, V. Organic micropollutants in a large wastewater treatment plant: What are the benefits of an advanced treatment by activated carbon adsorption in comparison to conventional treatment? *Chemosphere* **2019**, *218*, 1050–1060. [CrossRef] [PubMed]
105. Bertagna Silva, D.; Cruz-Alcalde, A.; Sans, C.; Giménez, J.; Esplugas, S. Performance and kinetic modelling of photolytic and photocatalytic ozonation for enhanced micropollutants removal in municipal wastewaters. *Appl. Catal. B Environ.* **2019**, *249*, 211–217. [CrossRef]
106. Cruz-Alcalde, A.; Esplugas, S.; Sans, C. Abatement of ozone-recalcitrant micropollutants during municipal wastewater ozonation: Kinetic modelling and surrogate-based control strategies. *Chem. Eng. J.* **2019**, *360*, 1092–1100. [CrossRef]

107. Ministerio para la Transición Ecológica y el Reto Demográfico de España (MITECO). Guía Para la Aplicación del R.D. 1620/2007 por el que se Establece el Régimen Jurídico de la Reutilización de las Aguas Depuradas (Ministerio). 2010. Available online: https://www.miteco.gob.es/es/agua/publicaciones/GUIA%20RD%201620_2007__tcm30-213764.pdf (accessed on 15 December 2019).
108. McConville, J.R.; Kvarnström, E.; Jönsson, H.; Kärrman, E.; Johansson, M. Source separation: Challenges & opportunities for transition in the swedish wastewater sector. *Resour. Conserv. Recycl.* **2017**, *120*, 144–156. [CrossRef]
109. Rao, K.; Otoo, M.; Drechsel, P.; Hanjra, M.A. Resource recovery and reuse as an incentive for a more viable sanitation service chain. *Water Altern.* **2017**, *10*, 493–512.
110. Gottardo Morandi, C.; Wasielewski, S.; Mouarkech, K.; Minke, R.; Steinmetz, H. Impact of new sanitation technologies upon conventional wastewater infrastructures. *Urban Water J.* **2018**, *15*, 526–533. [CrossRef]
111. Cieślik, B.; Konieczka, P. A review of phosphorus recovery methods at various steps of wastewater treatment and sewage sludge management. The concept of "no solid waste generation" and analytical methods. *J. Clean. Prod.* **2017**, *142*, 1728–1740. [CrossRef]
112. Chrispim, M.; Scholz, M.; Nolasco, M.A. Phosphorus recovery from municipal wastewater treatment: Critical review of challenges and opportunities for developing countries. *J. Environ. Manag.* **2019**, *248*, 109268. [CrossRef]
113. Tchobanoglous, G.; Stensel, H.D.; Tsuchihashi, R.; Burton, F.; Abu-Orf, M.; Bowden, G.; Pfrang, W. *Metcalf & Eddy/AECOM: Tratamento de Efluentes e Recuperação de Recursos*, 5th ed.; AMGH Editora Ltda (Portugal): Porto Alegre, Brazil, 2015.
114. Khiewwijit, R.; Temmink, H.; Rijnaarts, H.; Keesman, K.J. Energy and nutrient recovery for municipal wastewater treatment: How to design a feasible plant layout? *Environ. Model. Softw.* **2015**, *68*, 156–165. [CrossRef]
115. Egle, L.; Rechberger, H.; Krampe, J.; Zessner, M. Phosphorus recovery from municipal wastewater: An integrated comparative technological, environmental and economic assessment of P recovery technologies. *Sci. Total Environ.* **2016**, *571*, 522–542. [CrossRef] [PubMed]
116. Mehta, C.M.; Khunjar, W.O.; Nguyen, V.; Tait, S.; Batstone, D.J. Technologies to Recover Nutrients from Waste Streams: A Critical Review. *Crit. Rev. Environ. Sci. Technol.* **2015**, *45*, 385–427. [CrossRef]
117. Johir, M.A.H.; George, J.; Vigneswaran, S.; Kandasamy, J.; Grasmick, A. Removal and recovery of nutrients by ion exchange from high rate membrane bio-reactor (MBR) effluent. *Desalination* **2011**, *275*, 197–202. [CrossRef]
118. Kocatürk-Schumacher, N.P.; Bruun, S.; Zwart, K.; Jensen, L.S. Nutrient Recovery From the Liquid Fraction of Digestate by Clinoptilolite. *Clean Soil Air Water* **2017**, *45*, 1500153. [CrossRef]
119. Kabbe, C. Inventories Phosphorus Recycling Strategies and Technology (Web Document). P-Rex, 10.1.19. 2017. Available online: https://phosphorusplatform.eu/activities/p-recovery-technology-inventory (accessed on 2 January 2020).
120. Desmidt, E.; Ghyselbrecht, K.; Zhang, Y.; Pinoy, L.; Van der Bruggen, B.; Verstraete, W.; Rabaey, K.; Meesschaert, B. Global Phosphorus Scarcity and Full-Scale P-Recovery Techniques: A Review. *Crit. Rev. Environ. Sci. Technol.* **2015**, *45*, 336–384. [CrossRef]
121. Sartorius, C.; von Horn, J.; Tettenborn, F. Phosphorus Recovery from Wastewater-Expert Survey on Present Use and Future Potential. *Water Environ. Res. A Res. Publ. Water Environ. Fed.* **2012**, *84*, 313–322. [CrossRef]
122. McNamara, C.J.; Anastasiou, C.C.; O'Flaherty, V.; Mitchell, R. Bioremediation of olive mill wastewater. *Int. Biodeterior. Biodegrad.* **2008**, *61*, 127–134. [CrossRef]
123. Dermeche, S.; Nadour, M.; Larroche, C.; Moulti-Mati, F.; Michaud, P. Olive mill wastes: Biochemical characterizations and valorization strategies. *Process Biochem.* **2013**, *48*, 1532–1552. [CrossRef]
124. Yangui, A.; Abderrabba, M. Towards a high yield recovery of polyphenols from olive mill wastewater on activated carbon coated with milk proteins: Experimental design and antioxidant activity. *Food Chem.* **2018**, *262*, 102–109. [CrossRef]
125. Kiai, H.; Raiti, J.; El-Abbassi, A.; Hafidi, A. Recovery of phenolic compounds from table olive processing wastewaters using cloud point extraction method. *J. Environ. Chem. Eng.* **2018**, *6*, 1569–1575. [CrossRef]
126. Kalogerakis, N.; Politi, M.; Foteinis, S.; Chatzisymeon, E.; Mantzavinos, D. Recovery of antioxidants from olive mill wastewaters: A viable solution that promotes their overall sustainable management. *J. Environ. Manag.* **2013**, *128*, 749–758. [CrossRef] [PubMed]

127. Gopinatha Kurup, G.; Adhikari, B.; Zisu, B. Recovery of proteins and lipids from dairy wastewater using food grade sodium lignosulphonate. *Water Resour. Ind.* **2019**, *22*, 100114. [CrossRef]
128. Armstrong, W.; Kennedy, A.; Wan, X.; Atiba, J.; McLaren, C.; Meyskens, F. Single dose administration of Bowman–Birk Inhibitor Concentrate (BBIC) in patients with oral leukoplakia. *Cancer Epidemiol. Biomark. Prev. A Publ. Am. Assoc. Cancer Res. Cosponsored Am. Soc. Prev. Oncol.* **2000**, *9*, 43–44.
129. Li, X.; Long, J.; Hua, Y.; Chen, Y.; Kong, X.; Zhang, C. Protein recovery and anti-nutritional factor removal from soybean wastewater by complexing with a high concentration of polysaccharides in a novel quick-shearing system. *J. Food Eng.* **2019**, *241*, 1–9. [CrossRef]
130. Yu, Z.; Chen, Y.; Feng, D.; Qian, Y. Process Development, Simulation, and Industrial Implementation of a New Coal-Gasification Wastewater Treatment Installation for Phenol and Ammonia Removal. *Ind. Eng. Chem. Res.* **2010**, *49*, 2874–2881. [CrossRef]
131. Guo, C.; Tan, Y.; Yang, S.; Qian, Y. Development of phenols recovery process with novel solvent methyl propyl ketone for extracting dihydric phenols from coal gasification wastewater. *J. Clean. Prod.* **2018**, *198*, 1632–1640. [CrossRef]
132. Gui, X.; Xu, W.; Cao, H.; Ning, P.; Zhang, Y.; Li, Y.; Sheng, Y. A novel phenol and ammonia recovery process for coal gasification wastewater altering the bacterial community and increasing pollutants removal in anaerobic/anoxic/aerobic system. *Sci. Total Environ.* **2019**, *661*, 203–211. [CrossRef]
133. Gadipelly, C.; Pérez-González, A.; Yadav, G.D.; Ortiz, I.; Ibáñez, R.; Rathod, V.K.; Marathe, K.V. Pharmaceutical Industry Wastewater: Review of the Technologies for Water Treatment and Reuse. *Ind. Eng. Chem. Res.* **2014**, *53*, 11571–11592. [CrossRef]
134. Zhang, W.; He, G.; Gao, P.; Chen, G. Development and characterization of composite nanofiltration membranes and their application in concentration of antibiotics. *Sep. Purif. Technol.* **2003**, *30*, 27–35. [CrossRef]
135. Shahtalebi, A.; Sarrafzadeh, M.; Rahmati, M. Application of nanofiltration membrane in the separation of amoxicillin from pharmaceutical waste water. *Iran. J. Environ. Health Sci. Eng.* **2001**, *8*, 109–116.
136. Tripathi, P.; Neti, N.; Chavhan, C.; Pophali, G.; Kashyap, S.; Lokhande, S.; Gan, L. Treatment of refractory nano-filtration reject from a tannery using Pd-catalyzed wet air oxidation. *J. Hazard. Mater.* **2013**, *261C*, 63–71. [CrossRef] [PubMed]
137. Commission, E. *Environmental, Economic and Social Impacts of the Use of Sewage Sludge on Land; Final Report; Part III: Project Interim Reports*; Milieu Ltd.: Bruxelles, Belgium, 2008.
138. Cieślik, B.M.; Namieśnik, J.; Konieczka, P. Review of sewage sludge management: Standards, regulations and analytical methods. *J. Clean. Prod.* **2015**, *90*, 1–15. [CrossRef]
139. Zhang, Q.; Hu, J.; Lee, D.-J.; Chang, Y.; Lee, Y.-J. Sludge treatment: Current research trends. *Bioresour. Technol.* **2017**, *243*, 1159–1172. [CrossRef] [PubMed]
140. Tyagi, V.K.; Lo, S.L. Sludge: A waste or renewable source for energy and resources recovery? *Renew. Sustain. Energy Rev.* **2013**, *25*, 708–728. [CrossRef]
141. Sharma, S.; Meenu, P.S.; Asha Latha, R.; Shashank, B.S.; Singh, D.N. Characterization of Sediments from the Sewage Disposal Lagoons for Sustainable Development. Advances in Civil Engineering Materials. *ASTM Int.* **2016**, *5*, 1–23. [CrossRef]
142. Sharma, S.; Singh, D.N. Characterization of Sediments for Sustainable Development: State of the Art. *Mar. Georesour. Geotechnol.* **2015**, *33*, 447–465. [CrossRef]
143. Remy, C.; Boulestreau, M.; Warneke, J.; Jossa, P.; Kabbe, C.; Lesjean, B. Evaluating new processes and concepts for energy and resource recovery from municipal wastewater with life cycle assessment. *Water Sci. Technol.* **2015**, *73*, 1074–1080. [CrossRef]
144. Günther, S.; Grunert, M.; Müller, S. Overview of recent advances in phosphorus recovery for fertilizer production. *Eng. Life Sci.* **2018**, *18*, 434–439. [CrossRef]
145. Kaikake, K.; Sekito, T.; Dote, Y. Phosphate recovery from phosphorus-rich solution obtained from chicken manure incineration ash. *Waste Manag.* **2009**, *29*, 1084–1088. [CrossRef]
146. Mulchandani, A.; Westerhoff, P. Recovery opportunities for metals and energy from sewage sludges. *Bioresour. Technol.* **2016**, *215*, 215–226. [CrossRef]
147. Zhang, L.; Zhu, Z.; Zhang, R.; Zheng, C.; Zhang, H.; Qiu, Y.; Zhao, J. Extraction of copper from sewage sludge using biodegradable chelant EDDS. *J. Environ. Sci.* **2008**, *20*, 970–974. [CrossRef]
148. Veeken, A.H.M.; Hamelers, H.V.M. Removal of heavy metals from sewage sludge by extraction with organic acids. *Water Sci. Technol.* **1999**, *40*, 129–136. [CrossRef]

149. Yesil, H.; Tugtas, A.E. Removal of heavy metals from leaching effluents of sewage sludge via supported liquid membranes. *Sci. Total Environ.* **2019**, *693*, 133608. [CrossRef] [PubMed]
150. Li, R.; Zhai, Z.; Li, Y.; Yang, T.; Chen, Y. Kinetic study of heavy metals Cu and Zn removal during sewage sludge ash calcination in air and N2 atmospheres. *J. Hazard. Mater.* **2018**, *347*, 227–232. [CrossRef]
151. Tang, J.; He, J.; Liu, T.; Xin, X.; Hu, H. Removal of heavy metal from sludge by the combined application of a biodegradable biosurfactant and complexing agent in enhanced electrokinetic treatment. *Chemosphere* **2017**, *189*, 599–608. [CrossRef]
152. Koshy, N.; Singh, D.N. Fly ash Zeolites for Water Treatment Applications. *J. Environ. Chem. Eng.* **2016**, *4*, 1460–1472. [CrossRef]
153. Dai, Y.; Zhang, N.; Xing, C.; Cui, Q.; Sun, Q. The adsorption, regeneration and engineering applications of biochar for removal organic pollutants: A review. *Chemosphere* **2019**, *223*, 12–27. [CrossRef]
154. Smith, K.M.; Fowler, G.D.; Pullket, S.; Graham, N.J.D. Sewage sludge-based adsorbents: A review of their production, properties and use in water treatment applications. *Water Res.* **2009**, *43*, 2569–2594. [CrossRef]
155. Zhai, Y.; Wei, X.-X.; Zeng, G.-M. Effect of pyrolysis temperature and hold time on the characteristic parameters of adsorbent derived from sewage sludge. *J. Environ. Sci.* **2004**, *16*, 683–686.
156. Bagreev, A.; Bandosz, T.J.; Locke, D.C. Pore structure and surface chemistry of adsorbents obtained by pyrolysis of sewage sludge-derived fertilizer. *Carbon* **2001**, *39*, 1971–1979. [CrossRef]
157. Inguanzo, M.; Menéndez, J.A.; Fuente, E.; Pis, J.J. Reactivity of pyrolyzed sewage sludge in air and CO_2. *J. Anal. Appl. Pyrolysis* **2001**, *58–59*, 943–954. [CrossRef]
158. Rio, S.; Faur-Brasquet, C.; Le Coq, L.; Le Cloirec, P. Structure Characterization and Adsorption Properties of Pyrolyzed Sewage Sludge. *Environ. Sci. Technol.* **2005**, *39*, 4249–4257. [CrossRef] [PubMed]
159. Kong, L.; Tian, S.; Luo, R.; Liu, W.; Tu, Y.; Xiong, Y. Demineralization of sludge-based adsorbent by post-washing for development of porosity and removal of dyes. *J. Chem. Technol. Biotechnol.* **2013**, *88*, 1473–1480. [CrossRef]
160. Rio, S.; Le Coq, L.; Faur, C.; Le Cloirec, P. Production of porous carbonaceous adsorbent from physical activation of sewage sludge: Application to wastewater treatment. *Water Sci. Technol.* **2006**, *53*, 237–244. [CrossRef] [PubMed]
161. Ros, A.; Lillo-Ródenas, M.A.; Fuente, E.; Montes-Morán, M.A.; Martín, M.J.; Linares-Solano, A. High surface area materials prepared from sewage sludge-based precursors. *Chemosphere* **2006**, *65*, 132–140. [CrossRef]
162. Shen, W.; Guo, Q.; Yang, X.; Liu, Y.; Song, Y.; Cheng, J. Adsorption of Methylene Blue in Acoustic and Magnetic Fields by Porous Carbon Derived from Sewage Sludge. *Adsorpt. Sci.* **2006**, *24*, 433. [CrossRef]
163. Ros, A.; Lillo-Ródenas, M.A.; Canals-Batlle, C.; Fuente, E.; Montes-Morán, M.A.; Martin, M.J.; Linares-Solano, A. A New Generation of Sludge-Based Adsorbents for H2S Abatement at Room Temperature. *Environ. Sci. Technol.* **2007**, *41*, 4375–4381. [CrossRef]
164. Lillo-Ródenas, M.A.; Ros, A.; Fuente, E.; Montes-Morán, M.A.; Martin, M.J.; Linares-Solano, A. Further insights into the activation process of sewage sludge-based precursors by alkaline hydroxides. *Chem. Eng. J.* **2008**, *142*, 168–174. [CrossRef]
165. Law, K.; Leung, Y.C.; Chua, H.; Wai-Hung, L.; Yu, P. Production of Polyhydroxybutyrate by Bacillus Species Isolated from Municipal Activated Sludge. *Appl. Biochem. Biotechnol.* **2001**, *91–93*, 515–524. [CrossRef]
166. Takabatake, H.; Satoh, H.; Mino, T.; Matsuo, T. Recovery of biodegradable plastics from activated sludge process. *Water Sci. Technol.* **2000**, *42*, 351–356. [CrossRef]
167. Krishna, C.; Van Loosdrecht, M.C.M. Effect of temperature on storage polymers and settleability of activated sludge. *Water Res.* **1999**, *33*, 2374–2382. [CrossRef]
168. Tyagi, R.D.; Surampalli, R.Y.; Yan, S.; Zhang, T.; Kao, C.M.; Lohani, B.N. *Sustainable Sludge Management: Production of Value Added Products*; American Society of Civil Engineers: Reston, VA, USA, 2009. [CrossRef]
169. Morgan-Sagastume, F.; Valentino, F.; Hjort, M.; Cirne, D.; Karabegovic, L.; Gerardin, F.; Johansson, P.; Karlsson, A.; Magnusson, P.; Alexandersson, T.; et al. Polyhydroxyalkanoate (PHA) production from sludge and municipal wastewater treatment. *Water Sci. Technol. A J. Int. Assoc. Water Pollut. Res.* **2014**, *69*, 177–184. [CrossRef] [PubMed]
170. Reddy, C.S.K.; Ghai, R.; Kalia, V.C. Polyhydroxyalkanoates: An overview. *Bioresour. Technol.* **2003**, *87*, 137–146. [CrossRef]
171. Chang, Z.; Long, G.; Zhou, J.L.; Ma, C. Valorization of sewage sludge in the fabrication of construction and building materials: A review. *Resour. Conserv. Recycl.* **2020**, *154*, 104606. [CrossRef]

172. Xu, W.; Xu, J.; Liu, J.; Li, H.; Cao, B.; Huang, X.; Li, G. The utilization of lime-dried sludge as resource for producing cement. *J. Clean. Prod.* **2014**, *83*, 286–293. [CrossRef]
173. Lin, K.-L.; Lin, D.F.; Luo, H.L. Influence of phosphate of the waste sludge on the hydration characteristics of eco-cement. *J. Hazard. Mater.* **2009**, *168*, 1105–1110. [CrossRef]
174. Tay, J.H.; Show, K.Y. Properties of Cement Made from Sludge. *J. Environ. Eng.* **1991**, *117*, 236–246. [CrossRef]
175. Rezaee, F.; Danesh, S.; Tavakkolizadeh, M.; Mohammadi-Khatami, M. Investigating chemical, physical and mechanical properties of eco-cement produced using dry sewage sludge and traditional raw materials. *J. Clean. Prod.* **2019**, *214*, 749–757. [CrossRef]
176. Valls, S.; Vàzquez, E. Stabilisation and solidification of sewage sludges with Portland cement. *Cem. Concr. Res.* **2000**, *30*, 1671–1678. [CrossRef]
177. Hamood, A.; Khatib, J.M.; Williams, C. The effectiveness of using Raw Sewage Sludge (RSS) as a water replacement in cement mortar mixes containing Unprocessed Fly Ash (u-FA). *Constr. Build. Mater.* **2017**, *147*, 27–34. [CrossRef]
178. Suchorab, Z.; Barnat-Hunek, D.; Franus, M.; Łagód, G. Mechanical and Physical Properties of Hydrophobized Lightweight Aggregate Concrete with Sewage Sludge. *Materials* **2016**, *9*, 317. [CrossRef] [PubMed]
179. Tuan, B.L.A.; Hwang, C.-L.; Lin, K.-L.; Chen, Y.-Y.; Young, M.-P. Development of lightweight aggregate from sewage sludge and waste glass powder for concrete. *Constr. Build. Mater.* **2013**, *47*, 334–339. [CrossRef]
180. Hwang, J.; Zhang, L.; Seo, S.; Lee, Y.-W.; Jahng, D. Protein recovery from excess sludge for its use as animal feed. *Bioresour. Technol.* **2008**, *99*, 8949–8954. [CrossRef] [PubMed]
181. More, T.T.; Yadav, J.S.S.; Yan, S.; Tyagi, R.D.; Surampalli, R.Y. Extracellular polymeric substances of bacteria and their potential environmental applications. *J. Environ. Manag.* **2014**, *144*, 1–25. [CrossRef]
182. Chen, X.; Li, C.; Ji, X.; Zhong, Z.; Li, P. Recovery of protein from discharged wastewater during the production of chitin. *Bioresour. Technol.* **2008**, *99*, 570–574. [CrossRef]
183. Pilli, S.; Bhunia, P.; Yan, S.; LeBlanc, R.J.; Tyagi, R.D.; Surampalli, R.Y. Ultrasonic pretreatment of sludge: A review. *Ultrason. Sonochem.* **2011**, *18*, 1–18. [CrossRef]
184. Gao, J.; Weng, W.; Yan, Y.; Wang, Y.; Wang, Q. Comparison of protein extraction methods from excess activated sludge. *Chemosphere* **2020**, *249*, 126107. [CrossRef]
185. Navia, R.; Soto, M.; Vidal, G.; Bornhardt, C.; Diez, C. Alkaline Pretreatment of Kraft Mill Sludge to Improve Its Anaerobic Digestion. *Bull. Environ. Contam. Toxicol.* **2003**, *69*, 869–876. [CrossRef]
186. Pervaiz, M.; Sain, M. High-yield Protein Recovery from Secondary Sludge of Paper Mill Effluent and Its Characterization. *Bioresources* **2012**, *7*. [CrossRef]
187. Ras, M.; Girbal-Neuhauser, E.; Etienne, P.; Dominique, L. A high yield multi-method extraction protocol for protein quantification in activated sludge. *Bioresour. Technol.* **2008**, *99*, 7464–7471. [CrossRef]
188. Suárez-Iglesias, O.; Urrea, J.L.; Oulego, P.; Collado, S.; Díaz, M. Valuable compounds from sewage sludge by thermal hydrolysis and wet oxidation. A review. *Sci. Total Environ.* **2017**, *584–585*, 921–931. [CrossRef]
189. Feng, C.; Lotti, T.; Lin, Y.; Malpei, F. Extracellular polymeric substances extraction and recovery from anammox granules: Evaluation of methods and protocol development. *Chem. Eng. J.* **2019**, *374*, 112–122. [CrossRef]
190. Wu, B.; Chai, X.; Zhao, Y. Enhanced dewatering of waste-activated sludge by composite hydrolysis enzymes. *Bioprocess Biosyst. Eng.* **2016**, *39*, 627–639. [CrossRef] [PubMed]
191. Smith, S.H.; Rothman, H. Recycling sewage sludge as a food for farm animals: Some ecological and strategic implications for Great Britain. *Agric. Wastes* **1981**, *3*, 87–108. [CrossRef]
192. Karn, S.; Kumar, A. Protease, lipase and amylase extraction and optimization from activated sludge of pulp and paper industry. *Indian J. Exp. Biol.* **2019**, *57*, 201–205.
193. Hoq, M.M.; Yamane, T.; Shimizu, S.; Funada, T.; Ishida, S. Continuous hydrolysis of olive oil by lipase in microporous hydrophobic membrane bioreactor. *J. Am. Oil Chem. Soc.* **1985**, *62*, 1016–1021. [CrossRef]
194. Gupta, R.; Gigras, P.; Mohapatra, H.; Goswami, V.K.; Chauhan, B. Microbial α-amylases: A biotechnological perspective. *Process Biochem.* **2003**, *38*, 1599–1616. [CrossRef]
195. Nabarlatz, D.; Vondrysova, J.; Jenicek, P.; Stüber, F.; Font, J.; Fortuny, A.; Fabregat, A.; Bengoa, C. Hydrolytic enzymes in activated sludge: Extraction of protease and lipase by stirring and ultrasonication. *Ultrason. Sonochem.* **2010**, *17*, 923–931. [CrossRef]
196. Frolund, B.; Griebe, T.; Nielsen, P.H. Enzymatic activity in the activated-sludge floc matrix. *Appl. Microbiol. Biotechnol.* **1995**, *43*, 755–761. [CrossRef]

197. Jung, J.; Xing, X.; Matsumoto, K. Recoverability of protease released from disrupted excess sludge and its potential application to enhanced hydrolysis of proteins in wastewater. *Biochem. Eng. J.* **2002**, *10*, 67–72. [CrossRef]
198. Gessesse, A.; Dueholm, T.; Petersen, S.B.; Nielsen, P.H. Lipase and protease extraction from activated sludge. *Water Res.* **2003**, *37*, 3652–3657. [CrossRef]
199. Karn, S.K.; Kumar, A. Sludge: Next paradigm for enzyme extraction and energy generation. *Prep. Biochem. Biotechnol.* **2019**, *49*, 105–116. [CrossRef] [PubMed]
200. Liu, Y.J.; Gu, J.; Liu, Y. Energy self-sufficient biological municipal wastewater reclamation: Present status, challenges and solutions forward. *Bioresour. Technol.* **2018**, *269*, 513–519. [CrossRef]
201. Smith, K.; Liu, S.; Hu, H.Y.; Dong, X.; Wen, X. Water and energy recovery: The future of wastewater in China. *Sci. Total Environ.* **2018**, *637–638*, 1466–1470. [CrossRef] [PubMed]
202. Gude, V.G. Wastewater treatment in microbial fuel cells - An overview. *J. Clean. Prod.* **2016**, *122*, 287–307. [CrossRef]
203. Cano, R.; Pérez-Elvira, S.I.; Fdz-Polanco, F. Energy feasibility study of sludge pretreatments: A review. *Appl. Energy* **2015**, *149*, 176–185. [CrossRef]
204. Diaz-Elsayed, N.; Rezaei, N.; Guo, T.; Mohebbi, S.; Zhang, Q. Wastewater-based resource recovery technologies across scale: A review. *Resour. Conserv. Recycl.* **2019**, *145*, 94–112. [CrossRef]
205. Olkiewicz, M.; Plechkova, N.V.; Earle, M.J.; Fabregat, A.; Stüber, F.; Fortuny, A.; Font, J.; Bengoa, C.; Capafons, J.F. Biodiesel production from sewage sludge lipids catalysed by Brønsted acidic ionic liquids. *Appl. Catal. B Environ.* **2016**, *181*, 738–746. [CrossRef]
206. Lv, P.; Yuan, Z.; Wu, C.; Ma, L.; Chen, Y.; Tsubaki, N. Bio-syngas production from biomass catalytic gasification. *Energy Convers. Manag.* **2007**, *48*, 1132–1139. [CrossRef]
207. He, L.; Du, P.; Chen, Y.; Lu, H.; Cheng, X.; Chang, B.; Wang, Z. Advances in microbial fuel cells for wastewater treatment. *Renew. Sustain. Energy Rev.* **2017**, *71*, 388–403. [CrossRef]
208. Shen, C.; Lei, Z.; Wang, Y.; Zhang, C.; Yao, Y. A review on the current research and application of wastewater source heat pumps in China. *Therm. Sci. Eng. Prog.* **2018**, *6*, 140–156. [CrossRef]
209. Bousquet, C.; Samora, I.; Manso, P.; Rossi, L.; Heller, P.; Schleiss, A.J. Assessment of hydropower potential in wastewater systems and application to Switzerland. *Renew. Energy* **2017**, *113*, 64–73. [CrossRef]
210. Shen, Y.; Linville, J.L.; Urgun-Demirtas, M.; Mintz, M.M.; Snyder, S.W. An overview of biogas production and utilization at full-scale wastewater treatment plants (WWTPs) in the United States: Challenges and opportunities towards energy-neutral WWTPs. *Renew. Sustain. Energy Rev.* **2015**, *50*, 346–362. [CrossRef]
211. Gherghel, A.; Teodosiu, C.; De Gisi, S. A review on wastewater sludge valorisation and its challenges in the context of circular economy. *J. Clean. Prod.* **2019**, *228*, 244–263. [CrossRef]
212. Elalami, D.; Carrere, H.; Monlau, F.; Abdelouahdi, K.; Oukarroum, A.; Barakat, A. Pretreatment and co-digestion of wastewater sludge for biogas production: Recent research advances and trends. *Renew. Sustain. Energy Rev.* **2019**, *114*, 109287. [CrossRef]
213. Do, M.H.; Ngo, H.H.; Guo, W.S.; Liu, Y.; Chang, S.W.; Nguyen, D.D.; Nghiem, L.; Ni, B.J. Challenges in the application of microbial fuel cells to wastewater treatment and energy production: A mini review. *Sci. Total Environ.* **2018**, *639*, 910–920. [CrossRef] [PubMed]
214. Gil, A.; Siles, J.A.; Martín, M.A.; Chica, A.F.; Estévez-Pastor, F.S.; Toro-Baptista, E. Effect of microwave pretreatment on semi-continuous anaerobic digestion of sewage sludge. *Renew. Energy* **2018**, *115*, 917–925. [CrossRef]
215. Bhoite, G.M.; Vaidya, P.D. Iron-catalyzed wet air oxidation of biomethanated distillery wastewater for enhanced biogas recovery. *J. Environ. Manag.* **2018**, *226*, 241–248. [CrossRef]
216. Ding, H.H.; Chang, S.; Liu, Y. Biological hydrolysis pretreatment on secondary sludge: Enhancement of anaerobic digestion and mechanism study. *Bioresour. Technol.* **2017**, *244*, 989–995. [CrossRef]
217. Carlsson, M.; Lagerkvist, A.; Morgan-Sagastume, F. Energy balance performance of municipal wastewater treatment systems considering sludge anaerobic biodegradability and biogas utilisation routes. *J. Environ. Chem. Eng.* **2016**, *4*, 4680–4689. [CrossRef]
218. Zhang, B.; Ji, M.; Wang, F.; Li, R.; Zhang, K.; Yin, X.; Li, Q. Damage of EPS and cell structures and improvement of high-solid anaerobic digestion of sewage sludge by combined ($Ca(OH)_2$ + multiple-transducer ultrasonic) pretreatment. *RSC Adv.* **2017**, *7*, 22706–22714. [CrossRef]

219. Ruffino, B.; Campo, G.; Genon, G.; Lorenzi, E.; Novarino, D.; Scibilia, G.; Zanetti, M. Improvement of anaerobic digestion of sewage sludge in a wastewater treatment plant by means of mechanical and thermal pre-treatments: Performance, energy and economical assessment. *Bioresour. Technol.* **2015**, *175*, 298–308. [CrossRef] [PubMed]
220. Liu, X.; Xu, Q.; Wang, D.; Zhao, J.; Wu, Y.; Liu, Y.; Ni, B.-J.; Wang, Q.; Zeng, G.; Li, X.; et al. Improved methane production from waste activated sludge by combining free ammonia with heat pretreatment: Performance, mechanisms and applications. *Bioresour. Technol.* **2018**, *268*, 230–236. [CrossRef] [PubMed]
221. Di Maria, F.; Micale, C.; Contini, S. Energetic and environmental sustainability of the co-digestion of sludge with bio-waste in a life cycle perspective. *Appl. Energy* **2016**, *171*, 67–76. [CrossRef]
222. Jung, J.-M.; Oh, J.-I.; Kim, J.-G.; Kwon, H.-H.; Park, Y.-K.; Kwon, E.E. Valorization of sewage sludge via non-catalytic transesterification. *Environ. Int.* **2019**, *131*, 105035. [CrossRef]
223. Choi, O.K.; Park, J.Y.; Kim, J.K.; Lee, J.W. Bench-scale production of sewage sludge derived-biodiesel (SSD-BD)and upgrade of its quality. *Renew. Energy* **2019**, *141*, 914–921. [CrossRef]
224. Wu, X.; Zhu, F.; Qi, J.; Zhao, L. Biodiesel Production from Sewage Sludge by Using Alkali Catalyst Catalyze. *Procedia Environ. Sci.* **2016**, *31*, 26–30. [CrossRef]
225. Zhang, R.; Zhu, F.; Dong, Y.; Wu, X.; Sun, Y.; Zhang, D.; Zhang, T.; Han, M. Function promotion of SO42−/Al2O3–SnO2 catalyst for biodiesel production from sewage sludge. *Renew. Energy* **2020**, *147*, 275–283. [CrossRef]
226. Ibrahim, A.B.A.; Akilli, H. Supercritical water gasification of wastewater sludge for hydrogen production. *Int. J. Hydrog. Energy* **2019**, *44*, 10328–10349. [CrossRef]
227. Chen, S.; Sun, Z.; Zhang, Q.; Hu, J.; Xiang, W. Steam gasification of sewage sludge with CaO as CO2 sorbent for hydrogen-rich syngas production. *Biomass Bioenergy* **2017**, *107*, 52–62. [CrossRef]
228. Zhu, J.; Yang, Y.; Yang, L.; Zhu, Y. High quality syngas produced from the co-pyrolysis of wet sewage sludge with sawdust. *Int. J. Hydrog. Energy* **2018**, *43*, 5463–5472. [CrossRef]
229. Yarımtepe, C.C.; Türen, B.; Oz, N.A. Hydrogen production from municipal wastewaters via electrohydrolysis process. *Chemosphere* **2019**, *231*, 168–172. [CrossRef] [PubMed]
230. Usman, T.M.; Banu, J.R.; Gunasekaran, M.; Kumar, G. Biohydrogen production from industrial wastewater: An overview. *Bioresour. Technol. Rep.* **2019**, *7*, 100287. [CrossRef]
231. Wang, H.; Fang, M.; Fang, Z.; Bu, H. Effects of sludge pretreatments and organic acids on hydrogen production by anaerobic fermentation. *Bioresour. Technol.* **2010**, *101*, 8731–8735. [CrossRef] [PubMed]
232. Elbeshbishy, E.; Hafez, H.; Nakhla, G. Enhancement of biohydrogen producing using ultrasonication. *Renew. Energy* **2010**, *35*, 6184–6193. [CrossRef]
233. Yang, X.; Tian, S.; Kan, T.; Zhu, Y.; Xu, H.; Strezov, V.; Jiang, Y. Sorption-enhanced thermochemical conversion of sewage sludge to syngas with intensified carbon utilization. *Appl. Energy* **2019**, *254*, 113663. [CrossRef]
234. Zhang, Q.; Liu, H.; Zhang, X.; Lu, G.; Wang, J.; Hu, H.; Yao, H. Effect of Fe/Ca-based composite conditioners on syngas production during different sludge gasification stages: Devolatilization, volatiles homogeneous reforming and heterogeneous catalyzing. *Int. J. Hydrog. Energy* **2017**, *42*, 29150–29158. [CrossRef]
235. Hantoko, D.; Kanchanatip, E.; Yan, M.; Weng, Z.; Gao, Z.; Zhong, Y. Assessment of sewage sludge gasification in supercritical water for H2-rich syngas production. *Process Saf. Environ. Prot.* **2019**, *131*, 63–72. [CrossRef]
236. Yan, M.; Liu, J.; Hantoko, D.; Kanchanatip, E.; Grisdanurak, N.; Cai, Y.; Gao, Z. Hydrogen-rich syngas production by catalytic cracking of tar in wastewater under supercritical condition. *Int. J. Hydrogen Energy* **2019**, *44*, 19908–19919. [CrossRef]
237. Tang, J.; Zhang, C.; Shi, X.; Sun, J.; Cunningham, J.A. Municipal wastewater treatment plants coupled with electrochemical, biological and bio-electrochemical technologies: Opportunities and challenge toward energy self-sufficiency. *J. Environ. Manag.* **2019**, *234*, 396–403. [CrossRef]
238. Chen, Z.; Zhang, S.; Zhong, L. Simultaneous sulfide removal, nitrogen removal and electricity generation in a coupled microbial fuel cell system. *Bioresour. Technol.* **2019**, *291*, 121888. [CrossRef] [PubMed]
239. Goenka, R.; Mukherji, S.; Ghosh, P.C. Characterization of electrochemical behaviour of Escherichia coli MTCC 1610 in a microbial fuel cell. *Bioresour. Technol. Rep.* **2018**, *3*, 67–74. [CrossRef]
240. Kipf, E.; Erben, J.; Zengerle, R.; Gescher, J.; Kerzenmacher, S. Systematic investigation of anode materials for microbial fuel cells with the model organism G. sulfurreducens. *Bioresour. Technol. Rep.* **2018**, *2*, 29–37. [CrossRef]

241. Nikhil, G.N.; Krishna Chaitanya, D.N.S.; Srikanth, S.; Swamy, Y.V.; Venkata Mohan, S. Applied resistance for power generation and energy distribution in microbial fuel cells with rationale for maximum power point. *Chem. Eng. J.* **2018**, *335*, 267–274. [CrossRef]
242. Zhao, W.; Chen, S. Critical parameters selection in polarization behavior analysis of microbial fuel cells. *Bioresour. Technol. Rep.* **2018**, *3*, 185–190. [CrossRef]
243. Meena, R.; Kannah, R.Y.; Sindhu, J.; Ragavi Kumar, G.; Gunasekaran, M.; Banu, J.R. Trends and resource recovery in biological wastewater treatment system. *Bioresour. Technol. Rep.* **2019**, *7*, 100235. [CrossRef]
244. Kumar, S.S.; Kumar, V.; Malyan, S.K.; Sharma, J.; Mathimani, T.; Maskarenj, M.S.; Ghosh, P.C.; Pugazhendhi, A. Microbial fuel cells (MFCs) for bioelectrochemical treatment of different wastewater streams. *Fuel* **2019**, *254*, 115526. [CrossRef]
245. Kracke, F.; Vassilev, I.; Krömer, J.O. Microbial electron transport and energy conservation - The foundation for optimizing bioelectrochemical systems. *Front. Microbiol.* **2015**, *6*, 575. [CrossRef]
246. Mei, X.; Xing, D.; Yang, Y.; Liu, Q.; Zhou, H.; Guo, C.; Ren, N. Adaptation of microbial community of the anode biofilm in microbial fuel cells to temperature. *Bioelectrochemistry* **2017**, *117*, 29–33. [CrossRef]
247. Palanisamy, G.; Jung, H.Y.; Sadhasivam, T.; Kurkuri, M.D.; Kim, S.C.; Roh, S.H. A comprehensive review on microbial fuel cell technologies: Processes, utilization, and advanced developments in electrodes and membranes. *J. Clean. Prod.* **2019**, *221*, 598–621. [CrossRef]
248. Hao, X.; Li, J.; van Loosdrecht, M.C.M.; Jiang, H.; Liu, R. Energy recovery from wastewater: Heat over organics. *Water Res.* **2019**, *161*, 74–77. [CrossRef] [PubMed]
249. Hepbasli, A.; Biyik, E.; Ekren, O.; Gunerhan, H.; Araz, M. A key review of wastewater source heat pump (WWSHP) systems. *Energy Convers. Manag.* **2014**, *88*, 700–722. [CrossRef]
250. Đurđević, D.; Balić, D.; Franković, B. Wastewater heat utilization through heat pumps: The case study of City of Rijeka. *J. Clean. Prod.* **2019**, *231*, 207–213. [CrossRef]
251. Chae, K.J.; Ren, X. Flexible and stable heat energy recovery from municipal wastewater treatment plants using a fixed-inverter hybrid heat pump system. *Appl. Energy* **2016**, *179*, 565–574. [CrossRef]
252. Shen, C.; Lei, Z.; Lv, G.; Ni, L.; Deng, S. An experimental investigation on a novel WWSHP system with the heat recovery through the evaporation of wastewater using circulating air as a medium. *Energy Build.* **2019**, *191*, 117–126. [CrossRef]
253. Shen, C.; Yang, L.; Wang, X.; Jiang, Y.; Yao, Y. An experimental and numerical study of a de-fouling evaporator used in a wastewater source heat pump. *Appl. Therm. Eng.* **2014**, *70*, 501–509. [CrossRef]
254. Sari, M.A.; Badruzzaman, M.; Cherchi, C.; Swindle, M.; Ajami, N.; Jacangelo, J.G. Recent innovations and trends in in-conduit hydropower technologies and their applications in water distribution systems. *J. Environ. Manag.* **2018**, *228*, 416–428. [CrossRef]
255. Ak, M.; Kentel, E.; Kucukali, S. A fuzzy logic tool to evaluate low-head hydropower technologies at the outlet of wastewater treatment plants. *Renew. Sustain. Energy Rev.* **2017**, *68*, 727–737. [CrossRef]
256. Chae, K.J.; Kim, I.S.; Ren, X.; Cheon, K.H. Reliable energy recovery in an existing municipal wastewater treatment plant with a flow-variable micro-hydropower system. *Energy Convers. Manag.* **2015**, *101*, 681–688. [CrossRef]
257. Power, C.; McNabola, A.; Coughlan, P. Development of an evaluation method for hydropower energy recovery in wastewater treatment plants: Case studies in Ireland and the UK. *Sustain. Energy Technol. Assess.* **2014**, *7*, 166–177. [CrossRef]
258. Gu, Y.; Li, Y.; Li, X.; Luo, P.; Wang, H.; Robinson, Z.P.; Li, F. The feasibility and challenges of energy self-sufficient wastewater treatment plants. *Appl. Energy* **2017**, *204*, 1463–1475. [CrossRef]
259. Nowak, O.; Keil, S.; Fimml, C. Examples of energy self-sufficient municipal nutrient removal plants. *Water Sci. Technol.* **2011**, *64*, 1–6. [CrossRef] [PubMed]
260. Schwarzenbeck, N.; Pfeiffer, W.; Bomball, E. Can a wastewater treatment plant be a powerplant? A case study. *Water Sci. Technol.* **2008**, *57*, 1555–1561. [CrossRef] [PubMed]
261. Wett, B.; Buchauer, K.; Fimml, C. Energy self-sufficiency as a feasible concept for wastewater treatment systems. In Proceedings of the IWA Leading Edge Technology Conference, Asian Water, Singapore, 21–24 September 2007.
262. Crawford, G.V. *Best Practices for Sustainable Wastewater Treatment: Initial Case Study Incorporating European Experience and Evaluation Tool Concept*; IWA Publishing: London, UK, 2010; p. 9.

263. Ostapczuk, R.E.; Bassette, P.C.; Dassanayake, C.; Smith, J.E.; Bevington, G. Achieving zero net energy utilization at municipal WWTPs: The gloversville-johnstown joint WWTP experience. *Proc. Water Environ. Fed.* **2011**, *2011*, 1191–1200. [CrossRef]
264. Willis, J.; Stone, L.; Durden, K.; Beecher, N.; Hemenway, C.; Greenwood, R. Barriers to biogas use for renewable energy. *Water Environ. Res. Found* **2012**, *11*. [CrossRef]
265. Shi, C.Y. *Mass Flow and Energy Efficiency of Municipal Wastewater Treatment Plants*; IWA Publishing: London, UK, 2011.
266. Wiser, J.; Schettler, J.; Willis, J. Evaluation of Combined Heat and Power Technologies for Wastewater Facilities. 2010. Available online: http://www.cwwga.org/documentlibrary/121_EvaluationCHPTechnologiespreliminary%5B1%5D.pdf (accessed on 2 January 2020).

Review

Use of Ultrasound as an Advanced Oxidation Process for the Degradation of Emerging Pollutants in Water

Ana L. Camargo-Perea [1], Ainhoa Rubio-Clemente [2,3,*] and Gustavo A. Peñuela [2]

[1] Grupo de Investigación en Remediación Ambiental y Biocatálisis, Instituto de Química, Universidad de Antioquia UdeA, Calle 70, No. 52-21 Medellín, Colombia; ana.camargo@udea.edu.co

[2] Grupo GDCON, Facultad de Ingeniería, Sede de Investigaciones Universitarias (SIU), Universidad de Antioquia UdeA, Calle 70, No. 52-21 Medellín, Colombia; gustavo.penuela@udea.edu.co

[3] Facultad de Ingeniería, Tecnológico de Antioquia–Institución Universitaria TdeA, Calle 78b, No. 72A-220 Medellín, Colombia

* Correspondence: ainhoa.rubio@tdea.edu.co or ainhoarubioclem@gmail.com; Tel.: +57-421-965-70

Received: 4 February 2020; Accepted: 11 March 2020; Published: 9 April 2020

Abstract: Emerging pollutants are compounds of increased environmental importance and, as such there is interest among researchers in the evaluation of their presence, continuity and elimination in different environmental matrices. The present work reviews the available scientific data on the degradation of emerging pollutants, mainly pharmaceuticals, through ultrasound, as an advanced oxidation process (AOP). This study analyzes the influence of several parameters, such as the nature of the pollutant, the ultrasonic frequency, the electrical power, the pH, the constituents of the matrix and the temperature of the solution on the efficiency of this AOP through researches previously reported in the literature. Additionally, it informs on the application of the referred process alone and/or in combination with other AOPs focusing on the treatment of domestic and industrial wastewaters containing emerging pollutants, mainly pharmaceuticals, as well as on the economic costs associated with and the future perspectives that make ultrasound a possible candidate to solve the problem of water pollution by these emerging pollutants.

Keywords: emerging pollutants; advanced oxidation process; water pollution; ultrasound

---

1. Introduction

Emerging contaminants (ECs) are chemical products, both natural and synthetic ones, that comprise a wide range of chemical compounds, including medical and recreational drugs, personal care products, steroids, hormones, surfactants, perfluorinated compounds, flame retardants, dyes, plasticizers and industrial additives [1–3]. The presence of ECs in the environment was not measured or controlled in the past because they did not cause concern and, in general terms, there were no studies demonstrating a health risk to humankind and living beings. Additionally, the use of ECs was not as high as it is currently; and they were not detected in water, since advances in instrumental analytical chemistry have only recently permitted their quantification at ultra-trace and trace concentrations [4,5], i.e., at concentrations from ng L^{-1} to μg L^{-1} [3,6]. Indeed, in the last years, ECs have been identified and quantified in effluents from wastewater treatment plant effluents, surface water, groundwater and even drinking water [3,5,7,8].

It is important to note that ECs can have harmful effects both on the environment where they are located and on human health. Nowadays, the toxicity ascribed to the presence of these pollutants on the environment has not been fully evaluated [8]; nevertheless, more and more eco-toxicological studies are being conducted [9]. In fact, the presence of ECs has been reported to represent a serious risk to both the environment and human health due to direct and/or indirect exposure [3,10], since they can negatively influence algae, invertebrates and fish, as well as ecosystem dynamics and community structure [11,12]. It has been found that ECs can act as endocrine disruptors and alter the reproduction cycles, water transport and osmoregulation processes of biota [13,14]. Other emerging pollutants have antimicrobial activity, leading to bacteria resistance to commonly used antibiotics [5] and, subsequently, resulting in worldwide spread of diseases. Additionally, ECs can be bioaccumulated [8], changing cellular reactions in vital organs, such as liver, kidney and gills [15]. Other studies have reported gene expression changes in organisms exposed to ECs [16].

It has been proven that some ECs are persistent pollutants that are hardly degraded by conventional processes [8,17,18]. For this reason, the implementation of new technologies to guarantee their removal is proposed [3,7,17,19].

Advanced Oxidation Processes (AOPs) have been evaluated as an option for the degradation of a variety of organic pollutants in waters [1,20]. These processes are characterized by a wide number of radical reactions, most of which involve chemical agents along with a source of ultraviolet (UV) radiation [21]. These radicals attack a large number of recalcitrant organic compounds such as ECs and, since they are not very selective, they become an excellent precursor to the conversion of a wide range of pollutants.

Several works have been carried out assisted by AOPs in order to evaluate their efficiency in degrading CEs. AOPs consist of the formation of the free hydroxyl radicals (HO•), which are capable of oxidizing toxic and/or recalcitrant organic compounds into more biodegradable and less dangerous products, such as oxidized species and short chain hydrocarbons of low molecular weight like formaldehyde and aliphatic acids [22], among other innocuous products; thus, they provide an improvement to the treatability of AOP effluents [17]. In fact, photocatalytic degradation has been conducted in the presence of UV radiation and photosensitizers including TiO_2, H_2O_2 and persulfate, among other chemical agents, obtaining very positive results [23–25]. Likewise, photo-Fenton and ozonation at basic pH have been proven to be highly efficient in the degradation of this type of pollutants [26,27]. These advanced systems, therefore, offer a solution to the problem of EC environmental accumulation and resistance to biological degradation, in contrast to other processes, such as conventional physical or chemical processes [17,25].

Nevertheless, it should be noted that, among the different AOPs used in the treatment of ECs present in water, the use of ultrasound (US) has been reported to be a highly efficient process, not only in the removal of this kind of contaminants, but also in their degradation [28,29] and the conversion of other recalcitrant pollutants [26] and microbial load [30] in water. Likewise, the use of US, as an advanced oxidation process, is environmentally "clean" since it does not require the addition of chemicals to the aqueous medium in order to achieve its EC degradation target, and does not generate waste [31] like Fenton and photo-Fenton. Consequently, the use of US waves is an alternative option for the conversion of recalcitrant ECs.

Considering the above, this article reviews and discusses the contributions of researches on the degradation of ECs, especially pharmaceuticals, due to their potential risks to human and other living beings, in aqueous media through US, as an advanced oxidation technology, considering the presence or absence of catalysts or dissolved gases, among other parameters, influencing the efficiency of the aforementioned process. Additionally, the application of this process is described focusing on domestic and industrial wastewater containing ECs, as well as the economic cost estimation associated with the future perspectives related to its implementation alone or in combination with other AOPs.

2. Ultrasound Process

The US process has been reported as a very efficient AOP for the degradation of ECs present in water [32–36]. Additionally, it can overcome the limitations ascribed to the use of other AOPs commonly used for water treatment. It is noteworthy to mention that, by using the US process, mass transfer within the reaction medium is improved, as well as the EC degradation reaction rates. Additionally, the consumption of chemicals, such as oxidizing and catalyzing agents, is reduced and no sludge is generated [22,37].

As part of this review, the state-of-the-art of the implementation of US, as an advanced oxidation process, is analyzed based on several works reported in the literature. First of all, the fundamentals of the process are described to continue with the factors mainly influencing the efficiency of the process. Afterwards, a number of examples are provided in order to have a general idea of the versatility of the advanced oxidation technology alone and/or in combination with other AOPs to efficiently degrade persistent compounds such as ECs. Finally, the capital and operation and maintenance costs are mentioned, and the future perspectives related to the application of the process are highlighted.

2.1. Operation Fundamentals

Aqueous medium sonolysis involves the production of waves through sound at a specific frequency, with compression and expansion cycles, leading to the formation of cavitation bubbles. These bubbles grow by the diffusion of vapor or gas from the liquid medium, reaching an unstable size that provokes their violent implosion, which in turn generates very high temperatures and pressures, approximately 4200 degrees K and 975 bar, producing the so-called "hot spots" that allow the decomposition of the water molecule to generate HO• [9], which is capable of oxidizing recalcitrant pollutants such as ECs with its high oxidation potential (2.8 V) [38], leading to the degradation of the toxic compounds and producing innocuous products, such as H_2O, carbon dioxide (CO_2) and inorganic ions.

Equations (1)–(4) show the decomposition of water and other molecules commonly dissolved in water by sonochemical waves [9,39], being the HO•, as well as the hydroperoxyl radicals (HO_2•), the main species that oxidizes the organic compounds present in the aqueous medium.

$$H_2O \overset{)))}{\rightarrow} H^{\bullet} + HO^{\bullet} \tag{1}$$

$$O_2 \overset{)))}{\rightarrow} 2O^{\bullet} \tag{2}$$

$$N_2 \overset{)))}{\rightarrow} 2N^{\bullet} \tag{3}$$

$$H^{\bullet} + O_2 \overset{)))}{\rightarrow} HO_2^{\bullet} \tag{4}$$

The cavitation bubbles are produced in two ways, symmetrically and asymmetrically. The difference between these is the support provided by a rigid surface (for instance, the surface of the reactor) for the bubbles to be formed. This difference has a direct influence on the way in which the bubbles implode, and thus on the release of pressure and temperature into the medium, resulting in the rupture of the water molecule and the formation of HO• [9]. The symmetrical bubbles release energy in all directions around their surface, while the asymmetrical ones generate an eruption of the liquid, mainly on the parts of the bubbles that are far away from the surfaces, forming long-range "micro-jets" that go to the solid surfaces [28].

There are three reaction zones in the solution during the ultrasonic treatment process: (a) inside the cavitation bubble, (b) the bubble/water interface and (c) within the bulk solution [28,32,40]. In each of these zones, different reactions occur that favor the decomposition of pollutants. Hydrophobic, non-polar and/or volatile compounds react inside the cavitation bubbles and at the bubble/water interface, while hydrophilic and/or non-volatile pollutants react within the bulk solution [28,41–43].

Inside the cavitation bubbles, the reaction of the pollutant can occur in two ways: pyrolysis of the highly volatile compounds, or chemical reaction with the free HO• formed. At the bubble/water interface, the reaction occurs by pyrolysis and, fundamentally, by a reaction with the HO• that are formed from implosion and tend to diffuse throughout the solution medium, reacting with the compounds that are present at the interface. Within the solution, decomposition occurs only by reaction with HO•, which are released into the aqueous medium through implosion of the cavitation bubbles [9].

When free radicals reach the aqueous solution, they can recombine, as expressed in Equations (5)–(7), or react with hydroxyl ions (HO^-) (Equation (8)), resulting in a decrease of the system oxidation potential.

$$HO_2^\bullet + HO_2^\bullet \rightarrow H_2O_2 + O_2 + O_2(a^1\,\Delta g) \quad k = 8.3 \times 10^5 \tag{5}$$

$$HO^\bullet + HO^\bullet \rightarrow H_2O + 1/2(O_2 + O_2(a^1\,g)) \quad k = 5.5 \times 10^9 \tag{6}$$

$$HO^\bullet + HO_2^\bullet \rightarrow H_2O + O_2 + O_2(a^1\,\Delta g) \quad k = 7.1 \times 10^9 \tag{7}$$

$$HO_2^\bullet + HO^- \rightarrow O_2^\bullet + H_2O \quad k = 10^{10} \tag{8}$$

However, from Equation (8), superoxide radicals ($O_2\bullet^-$) are formed, as well as from the decomposition of $HO_2\bullet$, as described by Equation (9), which also contribute to the degradation of emerging organic compounds, although in a smaller proportion than by HO• [38]. Additionally, in acidic medium, $O_2\bullet^-$ can react with protons (H^+) to form $HO_2\bullet$ (Equation (10)). Both of the free radicals can recombine, as represented in Equation (11), resulting in the production of HO_2^-, which in turn can be involved in HO• quenching (Equation (12)).

$$HO_2^\bullet \rightarrow H^+ + O_2^\bullet \quad k = 7.5 \times 10^6 \tag{9}$$

$$H^+ + O_2^\bullet \rightarrow HO_2^\bullet \quad k = 5.1 \times 10^{10} \tag{10}$$

$$HO_2^\bullet + O_2^\bullet \rightarrow HO_2^- + O_2 \quad k = 9.7 \times 10^7 \tag{11}$$

$$HO^\bullet + HO_2^- \rightarrow HO_2^\bullet + HO^- \quad k = 7.5 \times 10^9 \tag{12}$$

Hydrogen peroxide (H_2O_2) can also be formed in the US process, as described in Equation (5). In spite of the fact that H_2O_2 can scavenge HO• or be decomposed (Equations (13)–(15), respectively), it can be involved in the oxidation of ECs, as well as on the production of a higher amount of HO•, when US process is combined with UV radiation.

$$HO^\bullet + H_2O_2 \rightarrow H_2O + HO_2^\bullet \quad k = 3 \times 10^7 \tag{13}$$

$$H_2O_2 \rightarrow HO_2^- + H^+ \quad k = 2 \times 10^{-2} \tag{14}$$

$$HO_2^- + H^+ \rightarrow H_2O_2 \quad k = 10^{10} \tag{15}$$

The reaction rate constants for the reactions expressed in Equations (5)–(15) were taken from Pavlovna et al. [44], demonstrating that, in general terms and according to the values of the reaction rate constants, the free radicals are easily formed through the US waves. As mentioned previously, these free radicals can react with the target pollutant; however, they can also recombine or be quenched by other compounds found in water such as the natural constituents of the matrix, making the reaction of the hydrophilic compounds within the solution less efficient and slower [45]. In this regard, in order to avoid side reactions of the US oxidation system, the optimization of the operating parameters or factors influencing the most the oxidation potential of the system must be conducted. This would subsequently allow the reduction of the economic costs associated with the studied advanced oxidation process for a more efficient degradation of the ECs of interest.

2.2. *Efficiency of the Ultrasound Process*

The US process must consider the control and variation of the different operating parameters, including the ultrasonic frequency, the electrical power and the pH and temperature of the solution [46,47], in order to be optimized with the subsequent reduction in the costs associated with the process performance. The nature of the contaminant of interest and the constituents of the water matrix must also be considered during the US-assisted AOP optimization procedure since they are involved in the efficiency of the process. In addition to these factors, the type and the geometry of the sonochemical reactor must be considered.

2.2.1. Reaction Zones—The Nature of the Emerging Pollutant

In the ultrasonic radiation process, as indicated above, three reaction zones are recognized for the degradation of compounds: the cavitation bubble, the bubble–water interface and the bulk solution [28,43,48]. The process by which degradation occurs differs from zone to zone. Hydrophilic substances are located within the solution, non-volatile hydrophobic compounds are mainly housed in the bubble–water interface, and volatile substances are commonly located within the cavitation bubble [22].

Inside the cavitation bubble, the degradation reaction of the contaminant occurs by pyrolysis; on the other hand, in the bubble–water interface, the main reaction mechanism is by the attack of free radicals, such as HO•, which are immediately formed by the implosion of the cavitation bubbles; finally, in the bulk solution the reaction occurs directly with the free radicals that reach this zone [28].

According to different investigations, in the US process, the degradation of volatile compounds occur in two zones: in the bubble–liquid interface, through the reaction with the HO• released from the implosion, and/or inside the bubble, directly by pyrolysis [22,41]. The rates of destruction of volatile contaminants depend on the physical and chemical conditions within the bubble, specifically the hydrophobic and volatile nature of these compounds [31,41]. On the other hand, it has been shown that the reaction rate constant of US degradation of volatile compounds decreases with increasing initial concentration, indicating that the relationship between the concentration of a volatile compound in the cavitation bubble and its concentration in the solution will influence the rate of ultrasonic reaction, considering that the collapse temperature depends on the specific heat ratio of the gas mixture [41,48–51].

Hydrophobic compounds such as carbamazepine (CBZ), which has a Henry's constant of approximately 1.08×10^{-10} atm. m^3 mol^{-1} and a moderate solubility in water [9], can be mainly housed in the bubble–water interface, but it is also found within the solution, allowing the protagonist of its degradation to be the HO•, which are immediately formed from the implosion both of the cavitation bubbles and the bubbles that travel within the solution [9].

To evaluate the zone and the way in which a compound is degraded, Nie et al. [28] have implemented the so-called "scavengers" of the HO•. In an experiment where the US process was used to degrade the pharmaceutical diclofenac (DCF), isopropyl alcohol and terephthalic acid were used to inhibit the reaction of the target compound with HO•, functioning as quenchers. The acid was considered to react with free radicals in the bulk solution, while the alcohol reacted both at the bubble–water interface and in the bulk solution. In this regard, the authors verified that when only the acid was added, the degradation of the compound was inhibited. However, when the alcohol was used exclusively as an inhibitor, degradation of the target EC was considerably reduced. It was, therefore, concluded that oxidation of DCF occurred mainly by HO• in the supercritical interface, especially when water was saturated with air and oxygen (O_2). Nonetheless, under argon (Ar)- and nitrogen (N_2)- saturated conditions, DCF degradation occurred within the cavitation bubbles and/or the bulk solution.

In a study carried out by Kidak and Dogan [52], where the degradation of alachlor through the US process was evaluated, it was concluded that due to the physical properties of the compound, such as the water solubility limit (140 mg L^{-1} at 20 °C), vapor pressure (negligible), Henry's constant (3.2×10^{-8} to 1.2×10^{-10} atm-m^3 mol^{-1}), octanol–water partition coefficient (Log K_{ow} = 2.63–3.53) and its positive ionization, the compound was housed in the bubble–liquid interface, indicating that the degradation was due to the HO• recently formed from the implosion of the cavitation bubbles. The degradation obtained of the target compound was near 100% with a frequency of 575 kHz and an electrical power of 90 W.

Adityosulindro et al. [53] evaluated the degradation of ibuprofen (IBU) in order to ascertain the reaction zone in which the degradation of IBU was established, and whether it was due exclusively to HO•. For this purpose, they tested the sequestration of these radicals through two compounds, n-butanol, which is a short chain alcohol with partial solubility in water that is expected to react with the radicals housed in the bubble–liquid interface; and acetic acid, which should react with the free radicals in the bulk solution due to it is a completely miscible compound. The results obtained indicated that, indisputably, IBU reacted with the HO• recently formed during the implosion of the cavitation bubbles, which means that it is a compound housed in the interfacial zone [53]. The same conclusion was reached by Méndez-Arriaga et al. [42], who attributed the degradation of IBU to the HO• recently produced, since IBU is considered to be housed at the bubble–water interface due to its Henry's constant (1.5×10^{-7} atm m^3 mol^{-1}), low solubility in water (21 mg L^{-1}) and octanol–water partition coefficient (3.9).

In the case of acetaminophen (ACP), a polar compound with high solubility (12.5 mg mL^{-1}), Villaroel et al. [54] reported that this contaminant was degraded in a greater proportion within the bulk, estimating that its behavior would be that of a hydrophilic substrate. Nonetheless, in this investigation, it was concluded that ACP can be housed both in the bulk solution and in the bubble–water interface, attributing its degradation to the HO• formed during the implosion of the cavitation bubbles. Based on the aforementioned authors' estimations, the hydrophilic or hydrophobic behavior of the target compound was more related to the initial pH value of the solution at which the study was carried out.

2.2.2. Ultrasonic Frequency

The frequency with which ultrasonic waves are produced can range from 20 to 10,000 kHz, and the US process is divided into three regions: low, high and very high frequency [22]. In Table 1, the frequency ranges used in the ultrasonic oxidation process are listed.

Table 1. Frequency ranges used in the ultrasonic process. Taken from [22].

Name	Ultrasound Range (kHz)
Very high	5000–10,000
High	200–1000
Low	20–100

Ultrasonic frequency is a fundamental parameter in the performance of US process, since the size and duration of the cavitation bubble, the violence of the implosion and, therefore, the production of HO• depend considerably on it [9,55].

The number of cavitation bubbles and bubble collapses increases with rising frequency. However, it is important to note that the bubbles generated at high frequencies are small, and release less energy than low frequency bubbles generated by a single pulse [6,56,57]. In addition, the escape of more HO• is inferred, before recombining, when faster collapses occur [9,58]. In this sense, the optimal frequency is determined by the integral efficiency of the energy discharge, which depends on the quantity, size and lifetime of the bubbles. It is noteworthy to mention that the optimal frequency varies according to the different compound to be treated [52,59].

Rao et al. [9] chose two frequency values (200 and 400 kHz) to determine the optimal one for the degradation of CBZ. The first of these values was more effective for the degradation of the target compound. This result was ascribed to the differences in calorimetric powers obtained for both frequencies under the same electrical power (100 W), resulting in a higher calorimetric power for the 200 kHz frequency. This can be attributed to what was previously explained, i.e., each EC requires an optimal frequency at which its degradation will be favored, which depends on its physicochemical properties. This optimal frequency will also be influenced by the geometry of the reactor since, as mentioned above, it will depend on the formation of symmetrical or asymmetrical cavitation bubbles.

On the other hand, in the research carried out by Güyer and Ince [23], different levels of ultrasonic frequency were evaluated in the US process of the DCF. The results obtained allowed the conclusion that the maximal rates of DCF degradation were reached at a frequency of 861 kHz and the minimal ones at 1145 kHz (carrying out tests with values of 577, 861 and 1145 kHz). The improvement between the 577 and 861 kHz was due to the fact that the latter reduced the size of the bubbles, leading to a greater number of bubbles and active oscillations, which contributed to the generation of HO• improvement. However, the highest frequency evaluated this efficiency was reduced due to the fact that the "optimal" frequency related to the reactor configuration was surpassed [53,56].

2.2.3. Electrical Power

The electrical power supplied to the ultrasonic transducer is a critical parameter that can largely determine the performance of the US process [9].

For Jiang et al. [41], the increase in ultrasonic power in the degradation of volatile compounds such as chlorobenzene, 1, 4-dichlorobenzene and 1-chloronaphthalene caused an increase in the cavitation energy, decreasing the cavitation limit and increasing the amount of bubbles produced. This resulted in a rise in the rate of degradation of this type of compounds, considering that the bubbles formed had enough energy to pyrolyze the tested pollutants. This is justified by the fact that volatile compounds are pyrolyzed within the cavitation bubbles, so the more bubbles formed, the more spaces for these compounds to react.

In a study carried out by Tran et al. [18], sonochemical efficiency was evaluated by means of calorimetric tests to determine the optimal power and to propose an experimental design in order to degrade the drug CBZ. It was determined that powers between 20 and 40 W favored sonochemical efficiencies, unlike what happened with powers of 10 W. This finding was attributed to the fact that as the power increased, so did the ultrasonic energy of the reactor, which caused the pulsation and collapse of the bubbles to be generated at a faster rate, resulting in a greater number of cavitation bubbles.

It is important to note that the effect of ultrasonic power and oxidizing species can be influenced by bubble dynamics [18]. The results reported by Gogate et al. [60] indicated that the size, number, lifetime and pressure of the bubbles were a complex function of the power dissipation rate. The research conducted by these authors explains the results obtained by Tran et al. [18], since they found that by increasing the power, the number of cavitation bubbles rose and, consequently, the production of HO• increased. In this way, the degradation of the target compound, CBZ in this case, was directly increased. Similar results were observed in the work carried out by Madhavan et al. [61] for DCF, who studied the degradation of this compound under a frequency of 213 kHz, a temperature of 25 °C, a variation of power density between 16–55 mW mL^{-1} and a concentration of the pharmaceutical compound of interest of 0.07 mM. The same conclusion was also reached by Rao et al. [9], who studied the degradation of CBZ at pH 6, a frequency of 200 kHz and a power variation between 20 and 100 W, obtaining a higher degradation of CBZ at 100 W.

In the work carried out by Naddeo et al. [6], the degradation of DCF during the US process was evaluated. It was determined that, by increasing the power density from 100 to 400 W L^{-1}, the concentration of the contaminant decreased, making US the most efficient process. This result supports the theory developed in the work described above, i.e., the greater the potency, the greater the degradation percentage of the compound under study.

On the other hand, the combination of different levels of the parameters that influence the degradation of a compound in the US process must be considered. In this context, when the response surface methodology was used to determine the optimal operating levels of CBZ ultrasonic treatment, [18] it was observed that the treatment time had a more statistically significant impact on the efficiency of CBZ removal in comparison with the electrical power, as efficient degradation of the contaminant at lesser powers (10–40 W) required more treatment time. This fact is ultimately reflected in the use of electricity and, therefore, in higher operating costs associated with the application of the oxidation process.

compete with IBU for HO• and that the latter was capable of reacting first with the oxidizing agent. It is important to highlight that the experimentation was carried out at acid pH, which could favor the location of IBU in the interface zone, making it more competitive when reacting with the HO• formed from the implosion of the cavitation bubbles.

Rao et al. [9] compared the degradation of CBZ contained in synthetic water with that of an effluent from a municipal wastewater treatment plant, evaluating the efficiency of two processes: ultrasonic irradiation alone, and in combination with photolysis using UV radiation emitting at a wavelength of 254 nm. The results showed that, for the US process, the constituents of the real wastewater matrix had no influence on the degradation of CBZ when compared with the results for distilled water. On the other hand, in the combined process, the wastewater matrix increased the efficiency of the studied drug degradation. This can be attributed to the photolysis of certain compounds contained in the wastewater that provide the oxidizing agent and favor the degradation of CBZ. As a matter of fact, the referred authors gave the example of nitrate ions (NO_3^-).

In the research conducted by Villaroel et al. [54], the influence of ionic constituents of water on the degradation of ACP (82.69 μmol L^{-1} and 1.65 μmol L^{-1}), at a power of 60 W and ultrasonic frequency of 600 kHz, was evaluated. The results obtained in distilled water and in synthetic water containing calcium ions (Ca^{2+}), magnesium ions (Mg^{2+}), sulphates ions (SO_4^{2-}), bicarbonates ions (HCO_3^-), Cl^-, potassium ions (K^+) and fluorides ions (F^-) were compared. The results indicated that, for the lowest concentration of ACP, a more pronounced acceleration of degradation was observed when this occurred in water with similar ion content than in distilled water. The authors attributed this to the high content of HCO_3^-, which was likely to be the protagonist in the formation of the carbonate radical (HCO_3•) when reacting with HO• radicals, being HCO_3•, a contributor to the degradation of the target EC.

With regard to the use of dissolved gases and their influence on the degradation of organic ECs, in the work conducted by Nie et al. [28], whose objective was to degrade DCF through the US process, it was observed that under saturated air, O_2 and Ar, a complete mineralization of nitrogen and a partial mineralization of carbon was achieved. When oxygenation was added to the reaction solution, HO_2• was formed. Although these radicals do not have an oxidation potential as high as HO•, as mentioned previously, HO_2• can contribute to the degradation of the compounds of interest [6].

On the other hand, it must be highlighted that when chlorine atoms are part of the target EC structure, they are transformed to Cl^- through the reaction of the pollutant of interest with HO• or by pyrolysis in the US process [41,49]. Therefore, the release of Cl^- occurs during the sonochemical degradation of chlorinated compounds, which was attributed to the rapid excision of the carbon–chlorine bonds by high temperature combustion occurring within the cavitation bubbles or at the bubble–liquid interface. Cl^-, as indicated above, can reduce the oxidation potential of the process.

Under this scenario, studies aiming at examining the efficiency of the US process are required to be conducted by using real matrices due to the natural constituents of the water matrices can positively or negatively influence the degradation percentages and reaction rates of the ECs of interest.

2.2.6. Temperature of the Solution

According to some authors, temperature variation in the US process directly influences cavitation intensity due to the changes in the physicochemical properties of the compound and the type of cavities formed, which can affect the kinetic velocity constant of the degradation reaction [65].

Al-Hamadani et al. [31] indicated that certain parameters were affected by increasing the temperature in the US process. First, it was found that cavitation energy decreased, as well as the threshold limit of the energy required to produce cavitation. In addition, it was found that the amount of dissolved gas was reduced, leading to the transfer of organic molecules from the bulk solution to the bubble–water interfacial region. Finally, the vapor pressure increased, causing the cavitation bubbles to contain more water vapor. Furthermore, the aforementioned authors, who evaluated the degradation of SFX and IBU through US, evidenced the temperature influence on the oxidation process. Temperatures between 15 and 55 °C were tested and it was concluded that, when this parameter was increased, the degradation of the studied compounds rose, as a rise in the temperature of the bulk caused the cavitation threshold to lower, which contributed to the formation of a greater number of cavitation bubbles and, therefore, to a greater amount of HO•. However, these authors pointed out that other works have shown an adverse effect of temperature on the degradation of the contaminant. These findings can be attributed to the fact that the surface tension and viscosity of the solution increase, generating cavitation bubbles with less intensity due to a rise in the vapor pressure of the liquid.

2.3. Application of Ultrasound Process to Water Treatment

Table 2 compiles several relevant research works related to the treatment of ECs through ultrasound as an AOP alone or in combination with other physical-chemical and advanced oxidation technologies.

Table 2. Summary of works related to the removal of emerging pollutants through the ultrasound process and its combination with other physical-chemical and advanced oxidation processes.

Process	Ref.	Pollutant/Type of Water	Operating Conditions	Found Results
US	[28]	DCF/Synthetic water	Co DCF: 0.05 mM. Frequency: 585 kHz. Power intensity 160 W L^{-1} pH: 7 Situations: air saturation, argon, oxygen and nitrogen. Temperature: 4 °C Glass cylindrical reactor of 750 mL connected to transducer Working volume: 500 mL. Treatment time: 60 min. HO• scavenger agents: Isopropyl alcohol and terephthalic acid. Co H_2O_2: 0.5 and 5 mM.	The elimination of DCF (without scavenger) and the formation of chloride ions were established as first-order reactions. Dichlorination rates, under all gas saturation conditions, were 1 to 2 times higher than DCF degradation rates. Dichlorination was a major reaction pathway during ultrasonic degradation of DCF; it developed within the solution by HO• attacks. There was only a partial mineralization in the 4 gas saturation conditions. The lowest peroxide concentration allowed a higher rate of degradation of the DCF.
US	[52]	Alachlor/ Synthetic water	Co Alachlor: 100 µg L^{-1} Frequency: 575, 861 y 1141 kHz. Electric power: 45, 60 and 90 W. Reactor: Glass cylindrical reactor of 500 mL Temperature: 25 °C. Treatment time: 90 min. pH: 7	Alachlor degradation was a pseudo-first order kinetics. A 100% degradation of alachlor and a mineralization of 25% was achieved, in 60 minutes of treatment, with a frequency of 575 kHz and a power of 90 W. The intermediate products from degradation of each tested power were analyzed, identifying their abundance in the samples.
US	[66]	Rosaniline (PRA) and ethyl violet (EV)	[PRA] and [EV]: 10 ppm Frequency: 350 kHz Electrical Power: 60 W. Treatment time: 30 min. Presence of ions: Cl^-, NO_3^-, SO_4^{2-}, CO_3^{2-}.	A complete degradation of EV and PRA was observed with a first order pseudo velocity constant. A good COD removal of 97% and 92%, respectively, was observed for EV and PRA after 3 h. The rate constants were higher with the addition of chloride ions in the case of EV and were not altered in the case of PRA. The improved degradation of EV in the presence of chloride is probably due to the salting effect and the reaction of the secondary radicals. EV degradation decreased from 100% to 80% with an increase in carbonate ion concentration from 0 to 100 ppm. In the case of PRA, a significant improvement in degradation was observed with the addition of CO_3^{2-}.

Table 2. *Cont.*

Process	Ref.	Pollutant/Type of Water	Operating Conditions	Found Results
US	[33]	Benzophenone-3 (BP-3)/ Synthetic water	Treatment time: 10 min Frequency: 574, 856 and 1134 kHz. Electrical Power: 100–200 W L^{-1}. [BP-3]: 1 ppm. Temperature: 25 ± 2 °C. Relationship of pulse time and silence time: PT/ST.	574 kHz or a lower frequency value is optimal for degradation of BP-3. The optimum power density level was 200 W L^{-1}. A maximum degradation level of 79.2% was obtained for EP = 200 W L^{-1}, a PT/ST ratio of 10 and frequency 574 kHz. The degradation was almost the same for all PT/ST ratios from 3 to 12.
US	[34]	Triclosan (TCS)/ Synthetic water	Treatment time: 60 min. Frequency: 215, 373, 574, 856 and 1134 kHz. Electrical Power: 40, 76, 140 and 200 W L^{-1} [TCS]: 1 mg L^{-1}. Temperature: 25 ± 2 °C. Treatment volume: 300 mL.	The 574 kHz frequency had the highest degradation rates. With 574 kHz, at 40 W L^{-1}, 88% of TCS degraded in 60 min, while at 140 W L^{-1}, TCS degraded completely in less than 25 min. The highest TCS degradation rate was obtained at the highest power density level of the equipment, 200 W L^{-1}. It was shown that the only variable that had statistical significance and an effect on degradation after 10 min was the power density.
US	[35]	Bisphenol-A/ Synthetic water	Frequency: 300 kHz. Electrical Power: 80 W. Treatment volume: 300 mL. [BPA]: 0.12 and 300 μM. pH: 8.3 [HCO_3^-]: 12–500 mg L^{-1} Temperature: 21 °C. Addition: Cl^-, SO_4^{2-} and HPO_4^{2-} [6 mM].	The addition of HCO_3^-, in the range of 12–500 mg L^{-1} did not have a significant effect on the BPA degradation rate. The bicarbonate concentration had a significant effect for the 0.12 BPA concentration: a higher bicarbonate concentration produced higher initial decomposition rates. Solutions containing ions other than bicarbonate showed significantly lower degradation rates. The bicarbonate/carbonate solution produced a significantly improved degradation rate of BPA.
US	[54]	Acetaminophen (ACP)/ Synthetic water and mineral water	Frequency: 600 kHz. Electrical Power: 20–60 W. Treatment volume: 300 mL. [ACP]: 82.69 μM. pH: 3–12. Temperature: 20 ± 1 °C. Addition: glucose, oxalic acid, propan-2-ol and hexan-1-ol.	The ultrasonic degradation in acidic medium (pH 3.0–5.6) is greater than that obtained in basic aqueous solutions (pH 9.5–12.0). The degradation of ACP would increase if its hydrophobicity is favored. The degradation rate increases with increasing acoustic power. The substrate degradation rate increases with increasing initial substrate concentration to a plateau. The presence of organic compounds negatively affects the sonochemical degradation efficiency of ACP, except glucose. A positive effect of mineral water was observed when the ACP concentration decreased 50 times (1.65 μM).
US	[36]	1-H- Benzotriazole (1HB)	[1HB]: 41.97–167.88 μM. Presence of oxygen, nitrogen, ozone and radical scavengers	With the increase in concentration, the degradation rate of 1HB also increased by 40%. A high applied ultrasonic power improved the degree of elimination of 1HB. The initial degradation rate accelerated in the presence of ozone and oxygen, but was inhibited by nitrogen. The most favorable pH for degradation was an acid medium. The removal of more than 90% of the contaminant was achieved

Table 2. *Cont.*

Process	Ref.	Pollutant/Type of Water	Operating Conditions	Found Results
US/Electro-oxidation (EO)	[27]	IBU/ Synthetic water and sewage	Co IBU Synthetic: 10 mg L^{-1} Increase in conductivity Na_2SO_4 0.01 mol L^{-1}. Co IBU Municipal: 20, 100 μg L^{-1} and 10 mg L^{-1}. pH residual municipal: 6.6. Frequency: 520 kHz. Electric power: 10–40 W. Current densities: 3.6–35.7 mA cm^{-2}. Cylindrical reactor with a cathode and an anode immersed in the solution. Temperature: 5–40 °C. Working volume: 3 L. Treatment time: 30–180 min.	The best constant for speed and efficiency of degradation was obtained with the US/EO, process, followed by EO alone and then US alone. 84.74% elimination of the IBU was achieved with US/EO. In the EO process, HO• can be generated on the surface of the electrode, then the US increases the mass transfer between these and the contaminants. Between 10–40 °C there were no significant differences in the degradation of IBU. Intensity of the current and treatment time are the most influential factors. Optimum conditions are: 110 min treatment, 4.09 A and 20 W. In municipal sewage, 90% of IBU was removed.
US O_3 O_3/US US/UV O_3/UV US/O_3/UV	[62]	Azo dyes (AD), Endocrine Disrupting Compounds (EDC) and pharmaceuticals (PHAC)/ Synthetic water	Reactor 1: horn-type sonicator. Capacity of 100 mL. Frequency 20 kHz. Power: 0.46 W mL^{-1}. Reactor 2: plate-type sonicator. Frequency: 577, 866, 1100 kHz. Power intensity: 0.23 w mL^{-1}. Use US + O_3. Reactor 3: Ultrasonic bath. Frequency: 200 kHz. Power: 0.07 W mL^{-1}. Reactor 4: tailor-made hexagonal glass reactor coupled with 3 UV lamps (254 nm). Frequency: 520 kHz. Power: 0.19 W mL^{-1}.	AD degradation is faster by O_3/US. The UV/US process was very effective in degrading AD. With the addition of H_2O_2 a better discoloration was obtained. The rate of AD decomposition is faster in the presence of solid particles. EDCs had better degradation at alkaline pH and low frequency. At acidic pH, degradation was improved by adding Fenton or O_3 processes. For PHAC, ultrasonic processes were more efficient at high frequencies and acid pH.
US/Zn^0	[46]	DCF/Synthetic water	Co DCF: 10 mg L^{-1}. Reactor: Beakers, ultrasound probe. Working volume: 100 mL. pH: 2–7. Frequency: 20 kHz Power: 30–300 W. Treatment time: 30 min. Addition of Zn^0	At acid pH, the US process accompanied with Zn^0 was more efficient, while adding Zn^0 alone and experimenting with the US alone did not result in further degradation of DCF. At pH higher than 2 the DCF was not eliminated. At pH 2, degradation of 80.92% was achieved in 15 min. Process of US/Zn^0. There were no significant differences in degradation at different Zn^0 concentrations and different power densities. Dichlorination was the degradation pathway. The main aspect of this reaction, together with the Zn^0 reduction, was the $O_2{\bullet}^-$.
US Fenton/US	[53]	IBU/Synthetic water and municipal sewage	Co IBU: 20 mg L^{-1}. pH: 2–8. Power density: 25–100 W L^{-1}. Frequency: 12–862 kHz. Addition of H_2O_2. Addition of Iron (Fe). HO scavenger agents: n-butanol and acetic acid. Reactor: 1 L glass. Ultrasound probe, cup horn type. Temperature: 25 °C.	At alkaline pH the degradation rate decreased significantly. The addition of H_2O_2 did not contribute to thedegradation of IBU by the US process. The sono-Fenton process was more efficient in eliminating the IBU than both processes separately. In the sono-Fenton process no significant influence on the degradation of the IBU was achieved by varying the power density in the studied range. In the municipal sewage the degradation was more effective with the combined processes, with results similar to those obtained with synthetic water. However, the efficiency of the individual US process decreased.

Table 2. *Cont.*

Process	Ref.	Pollutant/Type of Water	Operating Conditions	Found Results
US US/UV	[9]	CBZ/Synthetic water	Co CBZ: 0.00625–0.1 mM. Sonolytic Reactor: 500 mL Cylindrical glass beaker Frequency: 200 and 400 kHz. Power: 20–100 W. Temperature: 20 °C. pH: 2–11. Photolytic reactor: Camera with two low-pressure Hg lamps, 253.7 nm. Combined reactor: Assembly of the sonolytic reactor inside the photolytic reactor.	CBZ degradation follows a pseudo-first order kinetics. Faster degradation rate and greater removal with a frequency of 200 kHz. When methanol was applied as HO• sequestering agent, there was no significant drug removal. The HO• was the protagonist of the degradation. As electrical power increased, CBZ degradation increased. SO_4^{2-} and NO_3^- hindered the transfer of electrons during oxidation. The degradation of CBZ with UV radiation alone was negligible. The UV/US process achieved the highest CBZ removal. Twenty-one reaction intermediates were detected.
US/Single-walled carbon nanotubes	[31]	SFX and IBU/ Synthetic water	Co SFX and IBU: 10 µM. Single-walled carbon nanotubes (SCN). Stainless steel reactor. Frequency: 1000 kHz Power: 180 W pH: 3.5–7–9.5. Temperature: 15 to 55 °C. Reaction time: 60 min. Working volume: 1 L.	As the temperature increased, the cavitation threshold decreased, bubble formation increased together with the amount of HO•. At pH values below the pKa of the compounds, complete degradation was obtained within 50–60 minutes. At higher pH values, complete degradation was not achieved. In the presence of the SCN the degradation and the speed constant of the same was favored. The adsorption capacity of the SCN favored the removal of the compounds.
US/EO	[29]	CBZ /Synthetic water	Working volume: Reactor 1: 1 L and Reactor 2: 100 L. Cathode and anode in the form of expanded metal plates. Anode: Ti/PbO_2 Cathode: Ti Electric current: 1–15 A. Type of water: Potable (from the tap). Co CBZ: 10 mg L^{-1}. Na_2SO_4: 0.01 mol L^{-1} Temperature: 20 °C. Ceramic transducer: diameter 4 cm. Frequency: 520 kHz. Power: between 10 and 40 W. Reaction time: between 90 and 180 min.	The combined US/EO process offered the best kinetic velocity constant. The degree of synergy, in the combination of the processes, rose with the increase in US power. As the current intensity increased, the depurative capacity rose. CBZ degradation was greater when the two processes (US and EO) were implemented simultaneously than separately. There was a 99.5% degradation of CBZ with the combined process.
US/O_2 /Fe	[67]	Metazachlor (MTZ)/Synthetic water	Generator US: 20 kHz. Titanium alloy probe. Co: 10 µM MTZ. pH: 3.0. Temperature: 22 °C. Presence or absence of dissolved oxygen. Presence or absence of nitrogen. Treatment time: 120 min. Addition of powdered ferric oxyhydroxide 50 mg L^{-1}.	MTZ degradation followed a pseudo-first order kinetics. The saturation of water with oxygen favored the degradation of MTZ. Excess oxygen can capture H• and avoid recombination with HO•. With the addition of ferric oxide and the recombination of HO• to produce H_2O_2, the Fenton process is generated in the middle of sonolysis. The application of US made the iron leaching process three times faster than conventional mechanical agitation, allowing better contact between the liquid and solid phases. 97% of MTZ was degraded with the addition of ferric oxide. The velocity constant was twice than that of US process alone.

Table 2. *Cont.*

Process	Ref.	Pollutant/Type of Water	Operating Conditions	Found Results
US/Additives	[68]	Oxacillin (OXA)/ Synthetic water	Working volume: 250 mL Electrical power: 60 W. Frequency: 275 kHz. Temperature: 20 °C. Mannitol and calcium carbonate were used as additives	In the presence of additives, OXA was efficiently removed. The sonochemical process was able to completely degrade the antibiotic, generating solutions without Antimicrobial Activity. The contaminant did not mineralize even after 360 min.
US/O_3	[32]	Benzophenone-3 (Bp3)/ Synthetic water	Frequency: 20 kHz. Electrical power: 55.9 W. Temperature: 25 °C. Working volume: 200 mL [Bp3]: 3.9 mg L^{-1}. pH: 2, 6.5 and 10. O_3: 0.5 mL min^{-1}. N_2 y O_2: 800 mL min^{-1}. Presence of nitrate, chloride and bicarbonate ions [5 mmol L^{-1}].	Increasing the electrical power also increases the degradation of Bp3. At a lower pH (2) a more effective degradation of Bp3 was observed. PKa Bp3: 8.06. The presence of O_2, O_3 and the combined process of US/O_3 improved the degradation of Bp3. Being faster US/O_3. Bicarbonate ions accelerated the degradation of Bp3.

Due to the demonstrated efficiency ascribed to the use of US-assisted AOPs in the degradation of ECs in water, it has been widely applied for tackling the problem of water pollution with these pollutants of growing concern [32,36,64]. As stated previously, it is highlighted that the water matrix is a topic of utmost importance when it comes to the evaluation of the pollutant removal capability through AOPs. In fact, in the literature, different works have been reported based on the elimination through US waves of various ECs commonly present in water matrices of different nature, from drinking water effluents to natural surface water, with domestic and industrial wastewaters being highly studied [40,69,70] due to the vast variety of compounds that can be found in these kinds of aqueous matrices.

For instance, Cetinkaya et al. [69] investigated the decolorization of textile waters using the sono-Fenton process, obtaining better results at pH 3, achieving 96% of color removal. The influence of ferrous ions (Fe^{2+}) concentration was analyzed, testing its variation between 0.05 g L^{-1} and 0.2 g L^{-1}. A color removal of 90% and 99% was observed with the lowest and the highest Fe^{2+} concentration, respectively. These results indicated that the sono-Fenton process required small amounts of Fe^{2+} to achieve high removals of the dyes. Additionally, H_2O_2 consumption was reduced by about 30% with the sono-Fenton process compared to the classic Fenton process. Furthermore, authors optimized operating parameters involved in the investigated AOP, achieving the highest removal of color at a frequency of 35 kHz, pH 3, 0.05 g L^{-1} of Fe^{2+}, 1.65 g L^{-1} of H_2O_2 and a treatment time of 60 min.

The removal of tetracycline (TC) has also been evaluated by Nasseri et al. [40] in a wastewater effluent by applying the US process. Some of the natural characteristics of the studied wastewater were: pH 7.9, chemical organic demand (COD) of 25 mg L^{-1}, HCO_3^- content of 164 mg L^{-1}, Cl^- of 92 mg L^{-1}, NO_3^- of 24 mg L^{-1} and Na^+ of 50 mg L^{-1}. A lower removal rate of TC, but in the same order of magnitude, in wastewater (1.25×10^{-2} min^{-1}) compared to that one obtained in ultrapure water (1.75×10^{-2} min^{-1}) was observed. These results may be ascribed to the negative influence of the water constituents, as explained previously; in this case, due to the high levels of organic matter, in terms of COD, which can prevent the formation of OH• and, subsequently, reduce the rate of TC degradation.

In turn, Serna-Galvis et al. [71] experimented with wastewater from El Salitre Treatment Plant, located in Bogotá (Colombia), with the objective of applying the sono-photo-Fenton/Oxalic Acid AOP for the removal of the following pharmaceuticals: DCF, CBZ, venlafaxine, ciprofloxacin, norfloxacin, valsartan, losartan, irbesartan, SFX, clarithromycin, azithromycin, erythromycin, metronidazole, trimethoprimine and clinimetropimine, as well as cocaine and its main metabolite benzoylecgonine. The operating conditions were: 300 mL of working volume, 88 W L^{-1} of power density, 375 kHz of frequency, 20 °C of temperature, a UVA lamp of 4 W, a Fe^{2+} content of 5 mg L^{-1} and an oxalic acid concentration of 2 mg L^{-1}. It was observed that the application of the sonochemical process alone led to the release of contaminants from suspended solids. The addition of Fe^{2+}, UVA light and oxalic acid to the US process significantly increased the elimination of the studied ECs in the effluent, thanks to the production of additional HO• through reactions between iron and the sonogenerated H_2O_2. It is important to note that the presence of oxalic acid makes iron more available for the formation of additional free radicals within the solution, causing the improvement of EC degradation.

With the aim of comparing the findings of degradation reported by US in wastewater, the work conducted by Vilardi et al. [70], where the efficiency of conventional and heterogeneous Fenton for the degradation of contaminants present in the wastewater of a tannery in terms of COD, total phenolic compounds (TP) and Cr(VI), is presented. The authors carried out the experimentation at large laboratory scale using a reactor with a volume of 7.4 L. It was concluded that the heterogeneous Fenton process was significantly more efficient with respect to the conventional one for the elimination of COD and TP, once the optimal values of the operating parameters were found. The percentages of COD and TP removal for the heterogeneous Fenton were 75.5 ± 2.1% and 85.1 ± 0.7%, respectively. Likewise, it was observed that a smaller amount of iron sludge was produced due to the heterogeneous Fenton process (17.5%) compared to that one achieved through the conventional Fenton process (21.6%), which is a key aspect for the feasible implementation of the process at industrial scale.

Although the heterogeneous Fenton process implemented above was demonstrated to produce relatively low amounts of sludge, a more environmentally safe process must be required to overcome the pollution of aqueous resources with recalcitrant contaminants. In this regard, the use of US as an AOP alone or in combination with other advanced oxidation technologies seem to be an attractive treatment option.

3. Future Perspectives

Although the application of US alone as an advanced oxidation technology to overcome the critical situation ascribed to ECs in aqueous environments has been demonstrated to be efficient, the coupling of US with other AOPs could improve the mineralization of emerging organic compounds [72,73] within a further reduced time of treatment. For this reason, the use of US hybrid techniques has been recently studied to improve EC mineralization results [6,23,45,73]. A clear example of this is the combination of sonolysis with the Fenton process. This combination, which is so-called sono-Fenton, could stimulate a faster conversion and/or mineralization of ECs. This is achieved through: firstly, higher generation of HO• [74]; secondly, an improved mixture and contact between HO• and the pollutants of interest [72,73], and thirdly, improved generation of Fe^{2+} [75].

Different strategies in addition to the combination of the Fenton process with sonolysis have been tested in the last years. An example of this is the work developed by Tran et al. [20], where the electro-oxidation (EO) process was combined with US. This combination was based on the fact that, initially, the formation of HO• is achieved on the wall of an electrode made up of a non-active material through the EO process, and the chemical exchange of these HO• with contaminants could then be improved due to the formation of the US waves and cavitation bubbles resulting from the US process. In this study, a higher kinetic velocity constant and a greater efficiency in the removal of IBU was obtained with the combined process of EO/US in comparison with the results obtained in each process independently. As a result, 90% of the IBU contained in samples of municipal sewage was removed using optimal parameter levels, such as the treatment time, the current intensity and the US power, which were determined through the response surface methodology. The beneficial results of the exposure of electrochemical cells to the effects of US power are related to the improved mass transport, increased current efficiencies, and continuous electrode surface activation [20,76,77]. These effects can be attributed to the rapid generation and collapse of the micro-bubbles within the electrolyte medium or near the electrode surface [20,77,78].

In turn, Ince's 2018 study [62] evaluated the degradation of toxic ECs through US in combination with other AOPs. In this study PCT, DCF and IBU were analyzed, finding that the degradation of the selected ECs was more efficient at high frequencies and acid pH. Degradation was further improved with the presence of solid catalysts, which provided surfaces that enhanced the formation of cavitation bubbles and, therefore, the performance of the oxidation processes. In the referred research, the use of iron nano- and micro-particles resulted in a higher rate of DCF elimination by using nano-particles [62], which was attributed to the synergy of US with these particles through the enrichment of massive surfaces with excessive sorption sites and cavitation nuclei. In addition, reactions at the bubble–liquid interface were intensified by the distortion of asymmetric shapes, the degree of which increases as particle size decreases [62]. On the other hand, the coupling of an ozonation system with UV radiation and sonication, with the optional addition of $FeSO_4$, completely degraded DCF [62]. Finally, this work compared the efficiency of the following AOPs: US, O_3/US, UV/US and O_3/US/UV. High removals of the drugs of interest were found in all the tested processes, reaching about 100% elimination accompanied by a mineralization between 40 and 60% of all the ECs with the combination of US, O_3 and UV radiation.

In the work developed by Rao et al. [9], sonolytic and photolytic AOPs were combined for the degradation of CBZ. The result was a significant improvement in the drug degradation compared to the results obtained when the processes were individually implemented. The reason for this fact was related to the formation of H_2O_2 resulting from the recombination of HO• from sonolysis. This oxidizing agent can be photolized by UV light and more HO• can be produced, which are the main contributors to CBZ degradation.

CBZ removal was also studied by Mohapatra et al. [79], through the US process, Fenton and ferro-sonication (a combination of $FeSO_4$ with the US process). It was found that the most efficient AOP was the Fenton process, with elimination percentages between 84–100%; this was followed by ferro-sonication, with values between 62–93%, while sonolysis only achieved CBZ elimination percentages between 22%–51%. The authors concluded that the higher the radiation intensity (5.8, 12.4 and 16 W cm^{-2}), the greater the elimination of the target drug. Moreover, according to their research, the resulting ranges of efficiency between one process and another were because $FeSO_4$ contributed to the formation of a greater amount of HO•.

Although Fenton process has been proven to be an efficient technology for the degradation of some ECs [70,79], residual sludge is produced, especially when the homogeneous Fenton process is applied [70]. In this regard, further studies are needed to give an alternative use to such as sludge, contributing to the so-called principles of the circular economy. In this regard, Vilardi et al. [80] treated a tannery wastewater with mixed-iron coated olive stone bio-sorbent particles in combination with H_2O_2. They found a COD removal efficiency of 58.4% and a TP removal of 59.2%, at H_2O_2/COD (w/w) equal to 0.875. The coated olive stones were regenerated with sodium hydroxide (NaOH) and oxalic acid ($C_2H_2O_4$) solutions after five cycles in order to enable their reuse.

In addition, considerate the circular economy principles, economic costs analysis must be carried out in order to discern whether an AOP tested at laboratory or pilot plant can be scale up for industrial application in real water effluents.

4. Cost Consideration

As reviewed, the efficiency of ultrasound has been demonstrated to degrade any kind of recalcitrant pollutants. However, there are limitations related to the economic costs associated with the use of this advance oxidation technology for the treatment of water containing toxic pollutants [39]. One such limitation is the cost, which can be divided into two groups: the capital or inversion costs, which consists of those costs associated with the manufacture of the sonochemical reactors and can be amortized over a span of years at a considered amortization rate [39], and the operation and maintenance costs. The economic cost estimation linked to the operation and maintenance labor include the part replacements, which mainly consists of the transducer element replacement and the tip or electronic circuit replacements. In fact, according to Mahamuni and Adewuyi [39], the part replacement costs are assumed to be 0.5% of the capital costs. Labor and analytical costs must also be considered when operation and maintenance costs are estimated. Labor costs include inspection, repair and replacement based on hours of service life of control panels, leakages and pressure gauge, among others. In turn, analytical costs consist of the costs related to the analysis of samples and, subsequently, the costs associated with the reactants and chemicals used for the sample analysis. Additionally, electrical costs, which can be based on the power consumption of the referred AOP, are of utmost importance since they are usually very high, especially in those countries where the cost of each kWatt is high. Hence, the use of renewable resources for generating electrical energy is an attractive option that is emerging for the advanced oxidation system to be implemented. As a matter of fact, Rubio-Clemente et al. [81] assessed the efficiency of the UV/H_2O_2 system powered by a photovoltaic (PV) system in a photochemical reactor at laboratory scale. According to the results reported by the authors, similar efficiency was observed between the oxidation system powered with energy from the electrical grid and that one generated using the PV cells implemented.

Another alternative for reducing the economic costs related to the use of ultrasound for treating polluted water is utilizing hybrid oxidation techniques by combining US with other AOP, including the use of oxidizing or catalyzing agents, such as ozone (O_3), H_2O_2, iron, titanium dioxide (TiO_2), wolfram trioxide (WO_3), zinc oxide (ZnO), etc., and electrochemistry to name just a few. In this regard, Expósito et al. [25] evaluated the efficiencies of mineralization in terms of total organic carbon (TOC) and CBZ removal by using the $US/UV/H_2O_2/Fe$ oxidation process at laboratory scale in a thin film UV reactor coupled to a 24 kHz 200 W direct immersion horn-type sonicator, obtaining efficiencies around 90%, which are higher than the efficiencies reached by the processes alone. In fact, a synergistic effect higher than 55% was found between the US process and UV irradiation.

However, although application of US hybrid techniques in some occasions can be more attractive for water treatment, Mahamuni and Adewuyi [39] reported that the costs associated with these treatment techniques are one to two orders of magnitude higher than when US is implemented alone. This can be ascribed to the costs linked to the additional chemicals used for the hybrid process to occur, i.e., the use of oxidizing agents such as O_3 and H_2O_2, or the catalyzing agents as iron salts, TiO_2, ZnO or WO_3, among others, as well as the adjustment of the pH of the solution if needed. Moreover, when US is used along with UV radiation, the costs associated with the replacements of the lamps and the electrical consumption of the lamps must be considered, as well as those ones related to the O_3 generator repair when O_3 is combined with US.

With this in mind, it can be concluded that the cost estimation studies based on pilot plants would be of high importance for to discern both the capital and the operation and maintenance costs related to the implementation of the US process. Furthermore, although high efficiencies can be obtained in a short period of time by using hybrid techniques with US, the economic costs associated with it are higher; therefore, further studies are needed to discern if the combination of US with another AOP is worth to be implemented under any circumstances. On the other hand, the type of pollutant plays a crucial role on the cost estimation procedures, since treating water containing hydrophobic pollutants has lower costs ascribed in comparison with those ones for treating compounds of hydrophilic nature [39].

5. Conclusions

After a critical review of the results found in the literature concerning the US process for the elimination of ECs, it is important to highlight the following conclusions:

- The US process is environmentally clean, as it does not produce chemical residues or sludge in comparison with other AOPs, such as Fenton and photo-Fenton processes, and other advanced oxidation technologies using catalysts, including TiO_2, ZnO and WO_3, among others.
- The nature of the pollutant is an issue of utmost concern when evaluating the efficiency of the ultrasound process, since hydrophobic, non-polar and/or volatile compounds react inside the cavitation bubbles and at the bubble/water interface, while hydrophilic and/or non-volatile pollutants react within the bulk solution.

- On the other hand, the operating parameters, such as the pH and the temperature of the solution, ultrasonic frequency, electrical power, dissolved gases and the nature and concentration of the pollutant, must be evaluated under a wide range, since the efficiency of the process depends on them. In this regard, the considered operating factors should be optimized in order to maximize the degradation of the pollutant of interest and minimize the operation and maintenance costs.
- The degradation efficiency of aqueous pollutants also depends, to a large extent, on the type of sonoreactor and the geometry of the system. Therefore, the optimization of the sonoreactor, in terms of geometry and type, is recommended to be carried out especially when scaling the US-assisted AOP up.
- Further researches are needed for evaluating the efficiency of the referred process in real water matrices since, as reviewed, aqueous matrix background can highly influence the efficiency of the oxidation system and, subsequently, the degradation of the pollutant to be studied.
- The combination of ultrasound with other advanced oxidation or conventional processes used for water treatment can offer a high percentage of removal and mineralization of the compound under study. However, the associated economic costs are commonly higher than when US is applied alone. Therefore, further studies based on the efficiency about the cost estimation of the US oxidation process alone and in combination with other AOPs are required, especially in pilot plants, to obtain a closer point of view for the advanced oxidation technology scale-up.

Author Contributions: Conceptualization, A.L.C.-P., A.R.-C., G.A.P.; investigation, A.L.C.-P., A.R.-C., G.A.P.; writing—original draft preparation, A.L.C.-P.; writing—review and editing, A.R.-C., G.A.P.; supervision, A.R.-C., G.A.P. All authors have read and agreed to the published version of the manuscript.

Funding: This research received no external funding.

Acknowledgments: To the Universidad de Antioquia for its commitment to education in the country and its stimulus "Student Instructor" that contributed to the development of this article. To the Research Group "Diagnóstico y Control de la Contaminación-GDCON" for its support in the development of this review.

Conflicts of Interest: The authors declare no conflict of interest.

References

1. Gil, M.J.; Soto, A.M.; Usma, J.I.; Gutiérrez, O.D. Contaminantes emergentes en aguas, efectos y posibles tratamientos. *Producción + Limpia* **2012**, *7*. Available online: https://www.hit2lead.com/ (accessed on 9 April 2020).
2. Zhang, Y.; Geißen, S.U.; Gal, C. Carbamazepine and diclofenac: Removal in wastewater treatment plants and occurrence in water bodies. *Chemosphere* **2008**, *73*, 1151–1161. [CrossRef] [PubMed]
3. Gogoi, A.; Mazumder, P.; Tyagi, V.K.; Tushara Chaminda, G.G.; An, A.K.; Kumar, M. Occurrence and fate of emerging contaminants in water environment: A review. *Groundw. Sustain. Dev.* **2018**, *6*, 169–180. [CrossRef]
4. Aristizabal-Ciro, C.; Botero-Coy, B.; López, F.; Peñuela, G.A. Monitoring pharmaceuticals and personal care products in reservoir water used for drinking water supply. *Environ. Sci. Pollut. Res.* **2017**, *24*, 7335–7347. [CrossRef] [PubMed]
5. Rozman, D.; Hrkal, Z.; Váňa, M.; Vymazal, J.; Boukalová, Z. Occurrence of pharmaceuticals in wastewater and their interaction with shallow aquifers: A case study of Horní Beřkovice, Czech Republic. *Water* **2017**, *9*, 218. [CrossRef]
6. Naddeo, V.; Belgiorno, V.; Ricco, D.; Kassinos, D. Degradation of diclofenac during sonolysis, ozonation and their simultaneous application. *Ultrason. Sonochem.* **2009**, *16*, 790–794. [CrossRef] [PubMed]
7. Hai, F.I.; Yang, S.; Asif, M.B.; Sencadas, V.; Shawkat, S.; Sanderson-Smith, M.; Gorman, J.; Xu, Z.Q.; Yamamoto, K. Carbamazepine as a possible anthropogenic marker in water: Occurrences, toxicological effects, regulations and removal by wastewater treatment technologies. *Water* **2018**, *10*, 107. [CrossRef]
8. Emmanouil, C.; Bekyrou, M.; Psomopoulos, C.; Kungolos, A. An Insight into Ingredients of Toxicological Interest in Personal Care Products and A Small-Scale Sampling Survey of the Greek Market: Delineating a Potential Contamination Source for Water Resources. *Water* **2019**, *11*, 2501. [CrossRef]

9. Rao, Y.; Yang, H.; Xue, D.; Guo, Y.; Qi, F.; Ma, J. Sonolytic and sonophotolytic degradation of Carbamazepine: Kinetic and mechanisms. *Ultrason. Sonochem.* **2016**, *32*, 371–379. [CrossRef]
10. Cleuvers, M. Mixture toxicity of the anti-inflammatory drugs diclofenac, ibuprofen, naproxen, and acetylsalicylic acid. *Ecotoxicol. Environ. Saf.* **2004**, *59*, 309–315. [CrossRef]
11. Jarvis, A.L.; Bernot, M.J.; Bernot, R.J. The effects of the psychiatric drug carbamazepine on freshwater invertebrate communities and ecosystem dynamics. *Sci. Total Environ.* **2014**, *496*, 461–470. [CrossRef] [PubMed]
12. Almeida, Â.; Calisto, V.; Esteves, V.I.; Schneider, R.J.; Soares, A.M.V.M.; Figueira, E.; Freitas, R. Presence of the pharmaceutical drug carbamazepine in coastal systems: Effects on bivalves. *Aquat. Toxicol.* **2014**, *156*, 74–87. [CrossRef] [PubMed]
13. Han, S.; Choi, K.; Kim, J.; Ji, K.; Kim, S.; Ahn, B.; Choi, K.; Khim, J.S.; Zhang, X.; Giesy, J.P. Endocrine disruption and consequences of chronic exposure to ibuprofen in Japanese medaka (Oryzias latipes) and freshwater cladocerans Daphnia magna and Moina macrocopa. *Aquat. Toxicol.* **2010**, *98*, 256–264. [CrossRef] [PubMed]
14. Gonzalez-Rey, M.; Bebianno, M.J. Does non-steroidal anti-inflammatory (NSAID) ibuprofen induce antioxidant stress and endocrine disruption in mussel Mytilus galloprovincialis? *Environ. Toxicol. Pharmacol.* **2012**, *33*, 361–371. [CrossRef]
15. Schmidt, W.; O'Rourke, K.; Hernan, R.; Quinn, B. Effects of the pharmaceuticals gemfibrozil and diclofenac on the marine mussel (Mytilus Spp.) and their comparison with standardized toxicity tests. *Mar. Pollut. Bull.* **2011**, *62*, 1389–1395. [CrossRef] [PubMed]
16. Guiloski, I.C.; Ribas, J.L.C.; da Silva Pereira, L.; Neves, A.P.P.; Silva de Assis, H.C. Effects of trophic exposure to dexamethasone and diclofenac in freshwater fish. *Ecotoxicol. Environ. Saf.* **2015**, *114*, 204–211. [CrossRef]
17. Rubio-Clemente, A.; Torres-Palma, R.A.; Peñuela, G.A. Removal of polycyclic aromatic hydrocarbons in aqueous environment by chemical treatments: A review. *Sci. Total Environ.* **2014**, *478*, 201–225. [CrossRef]
18. Tran, N.; Drogui, P.; Zaviska, F.; Brar, S.K. Sonochemical degradation of the persistent pharmaceutical carbamazepine. *J. Environ. Manag.* **2013**, *131*, 25–32. [CrossRef]
19. González, K.; Quesada, I.; Julcour, C.; Delmas, H.; Cruz, G.; Jáuregui, U.J. El empleo del ultrasonido en el tratamiento de aguas residuales. *Rev. CENIC Cienc. Químicas* **2010**, *41*, 1–11.
20. Tran, N.; Drogui, P.; Brar, S.K. Sonoelectrochemical oxidation of carbamazepine in waters: Optimization using response surface methodology. *J. Chem. Technol. Biotechnol.* **2015**, *90*, 921–929. [CrossRef]
21. Ikehata, K.; Naghashkar, N.J.; El-Din, M.G. Degradation of Aqueous Pharmaceuticals by Ozonation and Advanced Oxidation Processes: A Review. *Ozone Sci. Eng.* **2006**, *28*, 353–414. [CrossRef]
22. Torres-Palma, R.A.; Serna-Galvis, E.A. Chapter 7 Sonolysis. In *Advanced Oxidation Processes for Waste Water Treatment*; Ameta, S.C., Ameta, R., Eds.; Academic Press: Cambridge, MA, USA, 2018; pp. 177–213. [CrossRef]
23. Güyer, G.T.; Ince, N.H. Degradation of diclofenac in water by homogeneous and heterogeneous sonolysis. *Ultrason. Sonochem.* **2011**, *18*, 114–119. [CrossRef] [PubMed]
24. Lin, L.; Wang, H.; Xu, P. Immobilized TiO_2-reduced graphene oxide nanocomposites on optical fibers as high performance photocatalysts for degradation of pharmaceuticals. *Chem. Eng. J.* **2017**, *310*, 389–398. [CrossRef]
25. Expósito, A.J.; Patterson, D.A.; Monteagudo, J.M.; Durán, A. Sono-photo-degradation of carbamazepine in a thin falling film reactor: Operation costs in pilot plant. *Ultrason. Sonochem.* **2017**, *34*, 496–503. [CrossRef]
26. Kakavandi, B.; Ahmadi, M. Efficient treatment of saline recalcitrant petrochemical wastewater using heterogeneous UV-assisted sono-Fenton process. *Ultrason. Sonochem.* **2019**, *56*, 25–36. [CrossRef]
27. Tran, N.; Drogui, P.; Nguyen, L.; Brar, S.K. Optimization of sono-electrochemical oxidation of ibuprofen in wastewater. *J. Environ. Chem. Eng.* **2015**, *3*, 2637–2646. [CrossRef]
28. Nie, E.; Yang, M.; Wang, D.; Yang, X.; Luo, X.; Zheng, Z. Degradation of diclofenac by ultrasonic irradiation: Kinetic studies and degradation pathways. *Chemosphere* **2014**, *113*, 165–170. [CrossRef]
29. Tran, N.; Drogui, P.; Brar, S.K.; De Coninck, A. Synergistic effects of ultrasounds in the sonoelectrochemical oxidation of pharmaceutical carbamazepine pollutant. *Ultrason. Sonochem.* **2017**, *34*, 380–388. [CrossRef]
30. Rubio-Clemente, A.; Chica, E.; Peñuela, G. Total coliform inactivation in natural water by UV/H_2O_2, UV/US, and $UV/US/H_2O_2$ systems. *Environ. Sci. Pollut. Res.* **2019**, *26*, 4462–4473. [CrossRef]
31. Al-Hamadani, Y.A.J.; Chu, K.H.; Flora, J.R.V.; Kim, D.H.; Jang, M.; Sohn, J.; Yoon, Y. Sonocatalytical degradation enhancement for ibuprofen and sulfamethoxazole in the presence of glass beads and single-walled carbon nanotubes. *Ultrason. Sonochem.* **2016**, *32*, 440–448. [CrossRef]

32. Zúñiga-Benítez, H.; Soltan, J.; Peñuela, G.A. Application of ultrasound for degradation of benzophenone-3 in aqueous solutions. *Int. J. Environ. Sci. Technol.* **2016**, *13*, 77–86. [CrossRef]
33. Vega, L.P.; Gomez-Miranda, I.N.; Peñuela, G.A. Benzophenone-3 ultrasound degradation in a multifrequency reactor: Response surface methodology approach. *Ultrason. Sonochem.* **2018**, *43*, 201–207. [CrossRef]
34. Vega, L.P.; Soltan, J.; Peñuela, G.A. Sonochemical degradation of triclosan in water in a multifrequency reactor. *Environ. Sci. Pollut. Res. Int.* **2019**, *26*, 4450–4461. [CrossRef] [PubMed]
35. Pétrier, C.; Torres-Palma, R.; Combet, E.; Sarantakos, G.; Baup, S.; Pulgarin, C. Enhanced sonochemical degradation of bisphenol-A by bicarbonate ions. *Ultrason. Sonochem.* **2010**, *17*, 111–115. [CrossRef] [PubMed]
36. Zuñiga, H.; Soltan, J.; Peñuela, G.A. Ultrasonic degradation of 1-H-Benzotriazole in water. *Water Sci. Technol.* **2014**, *70*, 152–159. [CrossRef]
37. Ince, N.H.; Tezcanli, G.; Belen, R.K.; Apikyan, İ.G. Ultrasound as a catalyzer of aqueous reaction systems: The state of the art and environmental applications. *Appl. Catal. B Environ.* **2001**, *29*, 167–176. [CrossRef]
38. Litter, M.; Quici, N. Photochemical Advanced Oxidation Processes for Water and Wastewater Treatment. *Recent Pat. Eng.* **2010**, *4*, 217–241. [CrossRef]
39. Mahamuni, N.N.; Adewuyi, Y.G. Advanced oxidation processes (AOPs) involving ultrasound for waste water treatment: A review with emphasis on cost estimation. *Ultrason. Sonochem.* **2010**, *17*, 990–1003. [CrossRef]
40. Nasseri, S.; Mahvi, A.H.; Seyedsalehi, M.; Yaghmaeian, K.; Nabizadeh, R.; Alimohammadi, M.; Safari, G.H. Degradation kinetics of tetracycline in aqueous solutions using peroxydisulfate activated by ultrasound irradiation: Effect of radical scavenger and water matrix. *J. Mol. Liq.* **2017**, *241*, 704–714. [CrossRef]
41. Jiang, Y.; Pétrier, C.; David Waite, T. Kinetics and mechanisms of ultrasonic degradation of volatile chlorinated aromatics in aqueous solutions. *Ultrason. Sonochem.* **2002**, *9*, 317–323. [CrossRef]
42. Méndez-Arriaga, F.; Torres-Palma, R.A.; Pétrier, C.; Esplugas, S.; Gimenez, J.; Pulgarin, C. Ultrasonic treatment of water contaminated with ibuprofen. *Water Res.* **2008**, *42*, 4243–4248. [CrossRef] [PubMed]
43. Chiha, M.; Merouani, S.; Hamdaoui, O.; Baup, S.; Gondrexon, N.; Pétrier, C. Modeling of ultrasonic degradation of non-volatile organic compounds by Langmuir-type kinetics. *Ultrason. Sonochem.* **2010**, *17*, 773–782. [CrossRef] [PubMed]
44. Pavlovna, I.; Vladimirovna, S.; Mihailovich, I.; Alekseevna, N.; Evgenevna, O.; Olegovna, O. Mechanism of chemiluminescence in Fenton reaction. *J. Biophys. Chem.* **2012**, *3*, 88–100. [CrossRef]
45. Naddeo, V.; Belgiorno, V.; Kassinos, D.; Mantzavinos, D.; Meric, S. Ultrasonic degradation, mineralization and detoxification of diclofenac in water: Optimization of operating parameters. *Ultrason. Sonochem.* **2010**, *17*, 179–185. [CrossRef]
46. Huang, T.; Zhang, G.; Chong, S.; Liu, Y.; Zhang, N.; Fang, S.; Zhu, J. Effects and mechanism of diclofenac degradation in aqueous solution by US/Zn_0. *Ultrason. Sonochem.* **2017**, *37*, 676–685. [CrossRef]
47. Hartmann, J.; Bartels, P.; Mau, U.; Witter, M.; Tümpling, W.V.; Hofmann, J.; Nietzschmann, E. Degradation of the drug diclofenac in water by sonolysis in presence of catalysts. *Chemosphere* **2008**, *70*, 453–461. [CrossRef]
48. Song, W.; Teshiba, T.; Rein, K.; O'Shea, K.E. Ultrasonically Induced Degradation and Detoxification of Microcystin-LR (Cyanobacterial Toxin). *Environ. Sci. Technol.* **2005**, *39*, 6300–6305. [CrossRef]
49. Drijvers, D.; Van Langenhove, H.; Vervaet, K. Sonolysis of chlorobenzene in aqueous solution: Organic intermediates. *Ultrason. Sonochem.* **1998**, *5*, 13–19. [CrossRef]
50. Zhang, G.; Hua, I. Cavitation chemistry of polychlorinated biphenyls: Decomposition mechanisms and rates. *Environ. Sci. Technol.* **2000**, *34*, 1529–1534. [CrossRef]
51. Hoffmann, M.H. Die in Zentraleuropa verwilderten und kultivierten nordamerikanischen Astern. *Feddes Repert.* **1996**, *107*, 163–188. [CrossRef]
52. Kidak, R.; Dogan, S. Degradation of trace concentrations of alachlor by medium frequency ultrasound. *Chem. Eng. Process. Process Intensif.* **2015**, *89*, 19–27. [CrossRef]
53. Adityosulindro, S.; Barthe, L.; González-Labrada, K.; Jáuregui, U.J.; Delmas, H.; Julcour, C. Sonolysis and sono-Fenton oxidation for removal of ibuprofen in (waste) water. *Ultrason. Sonochem.* **2017**, *39*, 889–896. [CrossRef] [PubMed]
54. Villaroel, E.; Silva-Agredo, J.; Petrier, C.; Taborda, G.; Torres-Palma, R.A. Ultrasonic degradation of acetaminophen in water: Effect of sonochemical parameters and water matrix. *Ultrason. Sonochem.* **2014**, *21*, 1763–1769. [CrossRef] [PubMed]

55. Petrier, C.; Jeunet, A.; Luche, J.L.; Reverdy, G. Unexpected frequency effects on the rate of oxidative processes induced by ultrasound. *J. Am. Chem. Soc.* **1992**, *114*, 3148–3150. [CrossRef]
56. Pétrier, C.; Francony, A. Ultrasonic waste-water treatment: Incidence of ultrasonic frequency on the rate of phenol and carbon tetrachloride degradation. *Ultrason. Sonochem.* **1997**, *4*, 295–300. [CrossRef]
57. Beckett, M.A.; Hua, I. Impact of Ultrasonic Frequency on Aqueous Sonoluminescence and Sonochemistry. *J. Phys. Chem. A* **2001**, *105*, 3796–3802. [CrossRef]
58. Petrier, C.; David, B.; Laguian, S. Ultrasonic degradation at 20 kHz and 500 kHz of atrazine and pentachlorophenol in aqueous solution: Preliminary results. *Chemosphere* **1996**, *32*, 1709–1718. [CrossRef]
59. Ziylan, A.; Koltypin, Y.; Gedanken, A.; Ince, N.H. More on sonolytic and sonocatalytic decomposition of Diclofenac using zero-valent iron. *Ultrason. Sonochem.* **2013**, *20*, 580–586. [CrossRef]
60. Gogate, P.R.; Sutkar, V.S.; Pandit, A.B. Sonochemical reactors: Important design and scale up considerations with a special emphasis on heterogeneous systems. *Chem. Eng. J.* **2011**, *166*, 1066–1082. [CrossRef]
61. Madhavan, J.; Kumar, P.S.S.; Anandan, S.; Zhou, M.; Grieser, F.; Ashokkumar, M. Ultrasound assisted photocatalytic degradation of diclofenac in an aqueous environment. *Chemosphere* **2010**, *80*, 747–752. [CrossRef]
62. Ince, N.H. Ultrasound-assisted advanced oxidation processes for water decontaminaration. *Ultrason. Sonochem.* **2018**, *40*, 97–103. [CrossRef] [PubMed]
63. Grebel, J.E.; Pignatello, J.J.; Mitch, W.A. Effect of Halide Ions and Carbonates on Organic Contaminant Degradation by Hydroxyl Radical-Based Advanced Oxidation Processes in Saline Waters. *Environ. Sci. Technol.* **2010**, *44*, 6822–6828. [CrossRef] [PubMed]
64. Yang, Y.; Pignatello, J.J.; Ma, J.; Mitch, W.A. Comparison of Halide Impacts on the Efficiency of Contaminant Degradation by Sulfate and Hydroxyl Radical-Based Advanced Oxidation Processes (AOPs). *Environ. Sci. Technol.* **2014**, *48*, 2344–2351. [CrossRef] [PubMed]
65. Golash, N.; Gogate, P.R. Degradation of dichlorvos containing wastewaters using sonochemical reactors. *Ultrason. Sonochem.* **2012**, *19*, 1051–1060. [CrossRef] [PubMed]
66. Rayaroth, M.P.; Aravind, U.K.; Aravindakumar, C.T. Effect of inorganic ions on the ultrasound initiated degradation and product formation of triphenylmethane dyes. *Ultrason. Sonochem.* **2018**, *48*, 482–491. [CrossRef]
67. Kask, M.; Krichevskaya, M.; Bolobajev, J. Sonolytic degradation of pesticide metazachlor in water: The role of dissolved oxygen and ferric sludge in the process intensification. *J. Environ. Chem. Eng.* **2019**, *7*, 103095. [CrossRef]
68. Serna-Galvis, E.A.; Silva-Agredo, J.; Giraldo-Aguirre, A.L.; Flórez-Acosta, O.A.; Torres-Palma, R.A. High frequency ultrasound as a selective advanced oxidation process to remove penicillinic antibiotics and eliminate its antimicrobial activity from water. *Ultrason. Sonochem.* **2016**, *31*, 276–283. [CrossRef]
69. Cetinkaya, S.G.; Morcali, M.H.; Akarsu, S.; Ziba, C.A.; Dolaz, M. Comparison of classic Fenton with ultrasound Fenton processes on industrial textile wastewater. *Sustain. Environ. Res.* **2018**, *28*, 165–170. [CrossRef]
70. Vilardi, G.; Rodríguez-Rodríguez, J.; Ochando-Pulido, J.M.; Verdone, N.; Martinez-Ferez, A.; Di Palma, L. Large Laboratory-Plant application for the treatment of a Tannery wastewater by Fenton oxidation: Fe(II) and nZVI catalysts comparison and kinetic modelling. *Process Saf. Environ. Prot.* **2018**, *117*, 629–638. [CrossRef]
71. Serna-Galvis, E.A.; Botero-Coy, A.M.; Martínez-Pachón, D.; Moncayo-Lasso, A.; Ibáñez, M.; Hernández, F.; Torres-Palma, R.A. Degradation of seventeen contaminants of emerging concern in municipal wastewater effluents by sonochemical advanced oxidation processes. *Water Res.* **2019**, *154*, 349–360. [CrossRef]
72. Bagal, M.V.; Gogate, P.R. Degradation of diclofenac sodium using combined processes based on hydrodynamic cavitation and heterogeneous photocatalysis. *Ultrason. Sonochem.* **2014**, *21*, 1035–1043. [CrossRef] [PubMed]
73. Liang, J.; Komarov, S.; Hayashi, N.; Kasai, E. Improvement in sonochemical degradation of 4-chlorophenol by combined use of Fenton-like reagents. *Ultrason. Sonochem.* **2007**, *14*, 201–207. [CrossRef] [PubMed]
74. Neppolian, B.; Jung, H.; Choi, H.; Lee, J.H.; Kang, J.W. Sonolytic degradation of methyl tert-butyl ether: The role of coupled fenton process and persulphate ion. *Water Res.* **2002**, *36*, 4699–4708. [CrossRef]
75. Lin, J.G.; Ma, Y.S. Oxidation of 2-Chlorophenol in Water by Ultrasound/Fenton Method. *J. Environ. Eng.* **2000**, *126*, 130–137. [CrossRef]
76. Birkin, P.R.; Silva-Martinez, S. A study on the effect of ultrasound on electrochemical phenomena. *Ultrason. Sonochem.* **1997**, *4*, 121–122. [CrossRef]

77. Klima, J.; Bernard, C.; Degrand, C. Sonoelectrochemistry: Transient cavitation in acetonitrile in the neighbourhood of a polarized electrode. *J. Electroanal. Chem.* **1995**, *399*, 147–155. [CrossRef]
78. Macounova, K.; Klima, J.; Bernard, C.; Degrand, C. Ultrasound-assisted anodic oxidation of diuron. *J. Electroanal. Chem.* **1998**, *457*, 141–147. [CrossRef]
79. Mohapatra, D.P.; Brar, S.K.; Tyagi, R.D.; Picard, P.; Surampalli, R.Y. A comparative study of ultrasonication, Fenton's oxidation and ferro-sonication treatment for degradation of carbamazepine from wastewater and toxicity test by Yeast Estrogen Screen (YES) assay. *Sci. Total Environ.* **2013**, *447*, 280–285. [CrossRef]
80. Vilardi, G.; Ochando-Pulido, J.M.; Stoller, M.; Verdone, N.; Di Palma, L. Fenton oxidation and chromium recovery from tannery wastewater by means of iron-based coated biomass as heterogeneous catalyst in fixed-bed columns. *Chem. Eng. J.* **2018**, *351*, 1–11. [CrossRef]
81. Rubio-Clemente, A.; Chica, E.; Peñuela, G.A. Photovoltaic array for powering advanced oxidation processes: Sizing, application and investment costs for the degradation of a mixture of anthracene and benzo[a]pyrene in natural water by the UV/H_2O_2 system. *J. Environ. Chem. Eng.* **2018**, *6*, 2751–2761. [CrossRef]

Article

Degradation of Losartan in Fresh Urine by Sonochemical and Photochemical Advanced Oxidation Processes

John F. Guateque-Londoño [1,2], Efraím A. Serna-Galvis [1], Yenny Ávila-Torres [3,*] and Ricardo A. Torres-Palma [1,*]

1 Grupo de Investigación en Remediación Ambiental y Biocatálisis (GIRAB), Instituto de Química, Facultad de Ciencias Exactas y Naturales, Universidad de Antioquia UdeA, Calle 70 No. 52-21, Calle Nueva, 050014 Medellín, Colombia; jfguateque@utp.edu.co (J.F.G.-L.); efrain.serna@udea.edu.co (E.A.S.-G.)

2 Maestría en Ciencias Químicas, Facultad de Tecnología, Universidad Tecnológica de Pereira, 660001 Pereira, Colombia

3 Grupo de Investigación QUIBIO, Facultad de Ciencias Básicas, Universidad Santiago de Cali, Santiago de Cali, 760035 Pampalinda, Colombia

* Correspondence: yennytorres@usc.edu.co (Y.Á.-T.); ricardo.torres@udea.edu.co (R.A.T.-P.)

Received: 30 October 2020; Accepted: 29 November 2020; Published: 3 December 2020

Abstract: In this work, the degradation of the pharmaceutical losartan, in simulated fresh urine (which was considered because urine is the main excretion route for this compound) by sonochemistry and UVC/H_2O_2 individually, was studied. Initially, special attention was paid to the degrading action of the processes. Then, theoretical analyses on Fukui function indices, to determine electron-rich regions on the pharmaceutical susceptible to attacks by the hydroxyl radical, were performed. Afterward, the ability of the processes to mineralize losartan and remove the phyto-toxicity was tested. It was found that in the sonochemical treatment, hydroxyl radicals played the main degrading role. In turn, in UVC/H_2O_2, both the light and hydroxyl radical eliminated the target contaminant. The sonochemical system showed the lowest interference for the elimination of losartan in the fresh urine. It was established that atoms in the imidazole of the contaminant were the moieties most prone to primary transformations by radicals. This was coincident with the initial degradation products coming from the processes action. Although both processes exhibited low mineralizing ability toward losartan, the sonochemical treatment converted losartan into nonphytotoxic products. This research presents relevant results on the elimination of a representative pharmaceutical in fresh urine by two advanced oxidation processes.

Keywords: advanced oxidation process; elimination routes; fresh urine; pharmaceutical degradation; processes selectivity; theoretical analysis

---

1. Introduction

Losartan was the first commercialized angiotensin II antagonist pharmaceutical. This is an antihypertensive consumed widely around the world [1]. Urine is the main route of excretion of losartan from the human body, ≈35% of the oral dose is expelled without alterations [2], reaching the wastewater systems. In fact, losartan has been determined in ranges of 0.0197–2.76 $\mu g\ L^{-1}$ in wastewater treatment plants influent (WWTP) [3,4]. This indicates that losartan is not effectively removed by the conventional systems in WWTP.

In the aquatic environment, losartan can promote noxious effects on organisms, and it can be transformed into more toxic and persistent substances [5–7]. The recalcitrance to conventional treatment systems, negative environmental impact, and high excretion of losartan in urine lead to consider alternative options to eliminate this pharmaceutical from aqueous media. Particularly, the application of degradation processes should be focused on primary contamination sources, such as human fresh urine.

Advanced oxidation processes (AOPs, which are based on the production and utilization of radical species to attack pollutants) are interesting options for losartan elimination from urine, to avoid entering into the wastewater systems. Indeed, AOPs such as UVC/H_2O_2 and sonochemistry have been successfully applied for the elimination of different pharmaceuticals in diverse aqueous matrices [8].

In the UVC/H_2O_2 process, UVC light (e.g., photons of 254 nm) promotes the homolysis of hydrogen peroxide, generating hydroxyl radicals (Equation (1)) available to degrade organic contaminants (Equation (2)) [9].

$$H_2O_2 + hv_{(<290\ nm)} \rightarrow 2HO^{\bullet} \quad (1)$$

$$HO^{\bullet} + \text{Pollutant} \rightarrow \text{degradation products} \quad (2)$$

Meanwhile, the sonochemical process, which uses high-frequency ultrasound waves ")))", produces hydroxyl radicals from the breaking of water molecules and dissolved oxygen (Equations (3)–(5)) [10].

$$H_2O +))) \rightarrow HO^{\bullet} + H^{\bullet} \quad (3)$$

$$O_2 +))) \rightarrow 2O^{\bullet} \quad (4)$$

$$H_2O + O^{\bullet}\ ))) \rightarrow 2HO^{\bullet} \quad (5)$$

It should be mentioned that some previous works have evidenced the high potentiality of AOPs to eliminate pollutants in urine [11–20]. However, until now, the treatment of losartan in fresh urine, considering the intrinsic degradation abilities of UVC/H_2O_2 and sonochemistry has not been reported. Moreover, computational analyses about the reactivity of this pharmaceutical toward hydroxyl radical species or phytotoxicity tests of the treated water have not been considered. Thereby, the present research was focused on the losartan treatment in fresh urine by UVC/H_2O_2 and ultrasound individually. The selectivity of the processes toward the pollutant degradation in the urine matrix was established. Firstly, special attention was paid to the action routes of the processes involved in the elimination of losartan. Besides, computational analyses using DFT/Fukui functionals were performed to determine the most regions on losartan reactive to hydroxyl radicals, and these theoretical results were related to primary degradation products coming from the processes action. Additionally, considering the possible reuse of treated urine for water irrigation extra analyses such as mineralization and phytotoxicity were carried out.

2. Materials and Methods

2.1. Reagents

Losartan tablets (50 mg each) were purchased from La Santé S.A. Acetonitrile (HPLC grade), ammonium heptamolybdate (>99.3%), methanol (HPLC grade), potassium iodide (>99.5%), potassium perchlorate (>99.5%), sodium acetate (>99%), sodium chloride (99.9%), sodium dihydrogen phosphate (>99.0%), sodium hydroxide (>99.0%), sodium sulfate (>99.0%), sulfuric acid (95–97%), and urea (>99.0%) were provided by Merck. Ammonium chloride (>99.8%), calcium chloride dihydrate (>99.0%), ferrous sulfate heptahydrate (>99.0%), formic acid (99.0%), hydrogen peroxide (30% w/v), and magnesium chloride hexahydrate (>99.0%) were provided by PanReac. All the reagents were used as received.

The solutions were prepared using distilled water. In all cases, the initial losartan concentration was 43.38 μM (i.e., 20 mg L^{-1}, which is a plausible amount of the antihypertensive excreted in human urine [21]). The fresh urine used for the tests was prepared according to Table 1. The fresh urine was used immediately after its preparation and the pH was adjusted to 6.1.

Table 1. Composition of fresh urine [1].

Compound	Concentration (M)
Urea	0.2664
$NaCH_3COO$	0.1250
Na_2SO_4	0.01619
NH_4Cl	0.03365
NaH_2PO_4	0.02417
KCl	0.05634
$MgCl_2$	0.003886
$CaCl_2$	0.004595
NaOH	0.00300
pH: 6.1	

[1] Composition taken from Amstutz et al. [22].

2.2. Reaction Systems

For the UVC/H_2O_2 process, a homemade aluminum reflective reactor box equipped with UVC lamps (OSRAM HNS®, with the main emission peak at 254 nm, 60 W) was used (Figure 1a). Losartan solutions (50 mL) were placed in beakers under constant stirring. Meanwhile, the sonochemical treatments were performed in a Meinhardt cylindrical glass reactor containing 250 mL of losartan solution. Ultrasonic waves of 375 kHz and 106.3 W L^{-1} (actual ultrasound power density determined by the calorimetric method) were emitted from a transducer at the bottom of the reactor (Figure 1b). For both processes, the experimental conditions (i.e., reagents concentrations, ultrasonic frequency, light power) were selected based on previous works [23,24].

2.3. Analyses

2.3.1. Chromatographic Analyses

Losartan evolution was followed by using a UHPLC Thermo Scientific Dionex UltiMate 3000 instrument equipped with an Acclaim™ 120 RP C18 column (5 µm, 4.6 × 150 mm) and a diode array detector (operated at 230 and 254 nm). The mobile phase was methanol (10% v/v), acetonitrile (44% v/v), and formic acid (46% v/v, 10 mM, and pH 3.0) at a flow of 0.6 mL min^{-1}. Primary transformation products were elucidated by HPLC–MS analyses in our previous work [24]. For the chromatographic analyses, samples of 0.5 mL were periodically taken from the reaction systems (the total taken volume was always lower than 10% of the initial volume in each system). All experiments were performed at least in duplicate.

2.3.2. Oxidizing Species Accumulation

Accumulation of sonogenerated hydrogen peroxide was estimated by iodometry [25]. An aliquot of 600 µL from the reactors was added to a quartz cell containing 1350 µL of potassium iodide (0.1 M) and 50 µL of ammonium heptamolybdate (0.01 M). After 5 min, the absorbance at 350 nm was measured using a Mettler Toledo UV5 spectrophotometer.

2.3.3. Mineralization Determinations

Mineralization degree was established as removal of total organic carbon (TOC). TOC content of the samples was measured using a Shimadzu LCSH TOC analyzer (previously calibrated), according to Standard Methods 5310B (high-temperature combustion method), in which the water sample is homogenized and injected into a heated reaction chamber packed with an oxidative catalyst (platinum spheres). The water is vaporized, and the organic carbon is oxidized to CO_2. The CO_2 from oxidation is transported by a carrier gas stream and is then measured using an IR detector. The TOC analyzer performed the catalytic combustion at 680 °C using high-purity oxygen gas at a flow rate of 190 mL min^{-1}. The apparatus had a nondispersive infrared detector. For the TOC analyses, samples of

7.0 mL were taken from the reaction systems, and for the TOC analyses, the experiments were carried out independently from the initial tests of degradation (to avoid retire amounts higher than 10% of the initial volume in each system).

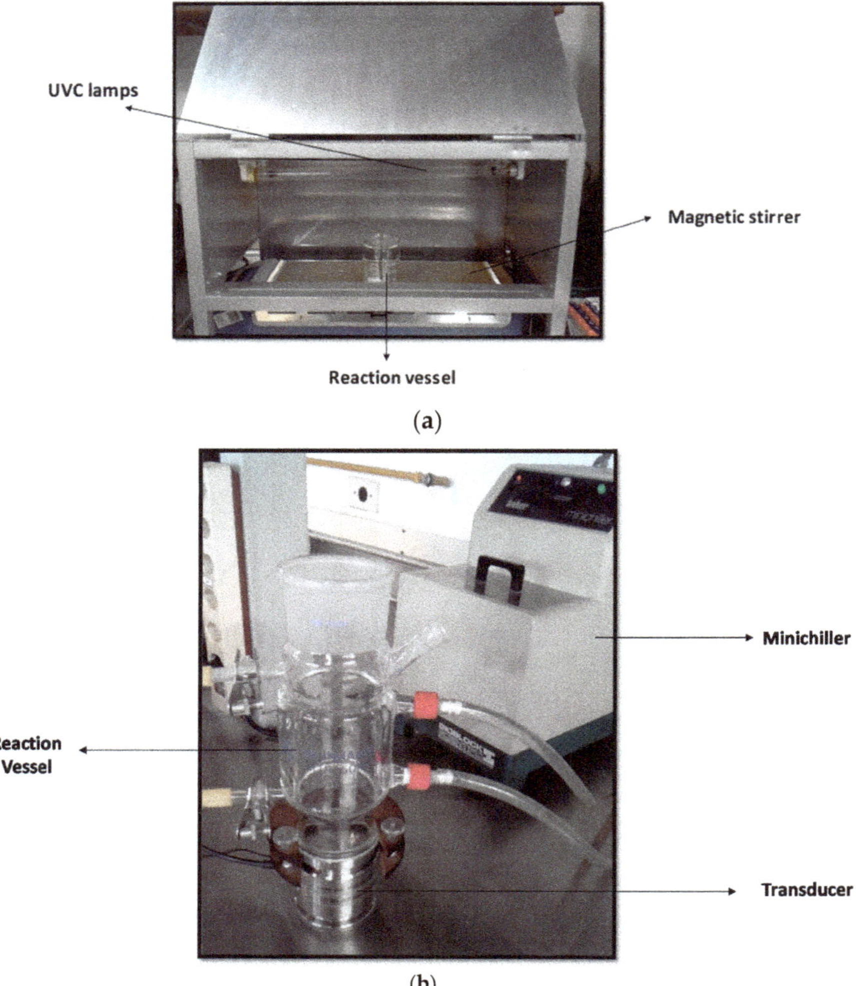

Figure 1. Reactors used in the degradation of losartan. (**a**) UVC/H_2O_2 process; (**b**) sonochemical treatment.

2.3.4. Phytotoxicity Tests

Toxicity against radish seeds (*Raphanus sativus*) was considered. For such purpose, the ratio of seeds germinated (RSG, Equation (6)) and the ratio of root length (RRG, Equation (7)) were determined. As a phytotoxicity parameter, the germination index (GI, Equation (8)) was assessed according to N.J. Hoekstra et al. [26]. For the phytotoxicity tests, samples of 5.0 mL were taken.

$$\text{RSG } (\%) = \frac{\text{Number of seeds germinated in sample}}{\text{Number of seeds germinated in control}} \times 100 \quad (6)$$

$$\text{RRG } (\%) = \frac{\text{mean root length in sample}}{\text{mean root length in control}} \times 100 \quad (7)$$

$$GI(\%) = \frac{RSG \times RRG}{100} \tag{8}$$

2.3.5. Computational Analyses

For the determination of regions on losartan most susceptible to the attack of radical species and electrophilic oxidants, computational analyses were performed by applying the framework of functional density theory (DFT). The antihypertensive structure was optimized with the B3LYP hybrid functional density [27], 6-311++G(2d,2p) method [28] using the dielectric constant for water to simulate the aqueous environment. Thus, f^+ and f^- (i.e., nucleophilic and electrophilic Fukui function indices) values and the average between such values ($\mathbf{f_{ave}}$) were calculated.

3. Results

3.1. *Treatment of Fresh Urine Loaded with Losartan*

The two processes were individually applied to degrade losartan in the simulated fresh urine (FU, whose composition is presented in Table 1). In addition to degradation in urine, losartan was also treated in distilled water (DW). The degradations followed pseudo-first-order kinetics, and their respective rate constants (k) in both matrices were established (see Figure S1 in Supplementary material). Then, the ratio between the degradation rate constants (Rk:k_{FU}/k_{DW}) was calculated. This Rk parameter is an indicator of both the selectivity of processes toward the antihypertensive degradation in the complex matrix and the inhibitory effect of losartan elimination caused by the fresh urine components. Table 2 contains the k and Rk values for each process.

Table 2. Kinetic constants (in min^{-1}) determined in the degradation of losartan in fresh urine (k_{FU}) and distilled water (k_{DW}) for each advanced oxidation processes [1].

AOP	k_{DW} (R^2)	k_{FU} (R^2)	Rk = k_{FU}/k_{DW}
Sonochemistry	0.0549 (0.9972)	0.0437 (0.9975)	0.796
UVC/H_2O_2	0.0532 (0.9987)	0.0245 (0.9981)	0.461

[1] Experimental conditions: [Pollutant] = 43.38 μM, pH: 6.1. Sonochemistry: 106.3 W L^{-1} (375 kHz). UVC/H_2O_2: [H_2O_2] = 500 μM, 60 W.

3.2. *Degradation Routes of Losartan (LOS) in Different AOPs*

To elucidate the routes of the processes action, some specific experiments and measures in distilled water were carried out and results were compared to those obtained in the urine to understand the effect of the matrix components. Results for each treatment in distilled water are detailed in the following subsections.

3.2.1. Action Routes of the UVC/H_2O_2 Process

The UVC/H_2O_2 process may include the action of light of 254 nm, hydrogen peroxide, and radicals. To identify the routes involved in the process, control tests for the individual effects of UVC and H_2O_2 were carried out. Figure 2 compares the degrading effect of individual components of the process in distilled water, plus the losartan elimination in both distilled water and fresh urine (FU) by UVC/H_2O_2.

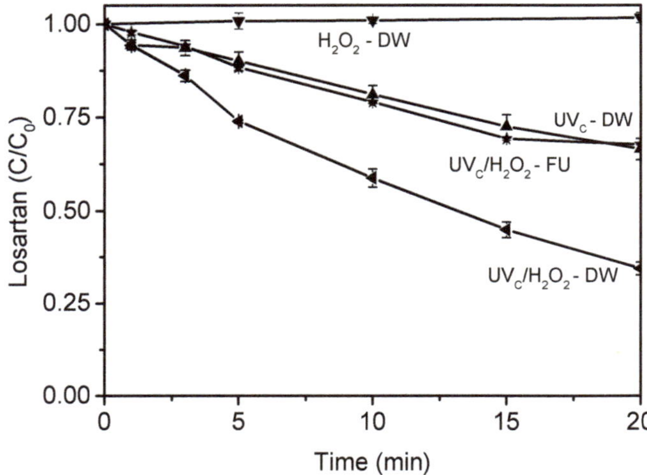

Figure 2. Determination of action routes of the UVC/H_2O_2 process on losartan degradation in distilled water (DW) and fresh urine (FU). Conditions: [LOS] = 43.38 μM, [H_2O_2] = 500 μM, UVC light: 60 W, pH: 6.1.

3.2.2. Degradation Routes Involved in the Sonochemical Treatment

Figure 3 depicts the degradation of LOS in both distilled water and fresh urine (FU) by ultrasound. To determine the degradation route in the sonochemical process, the accumulation of sonogenerated hydrogen peroxide in the presence and absence of the pollutant was also measured (results also presented in Figure 3).

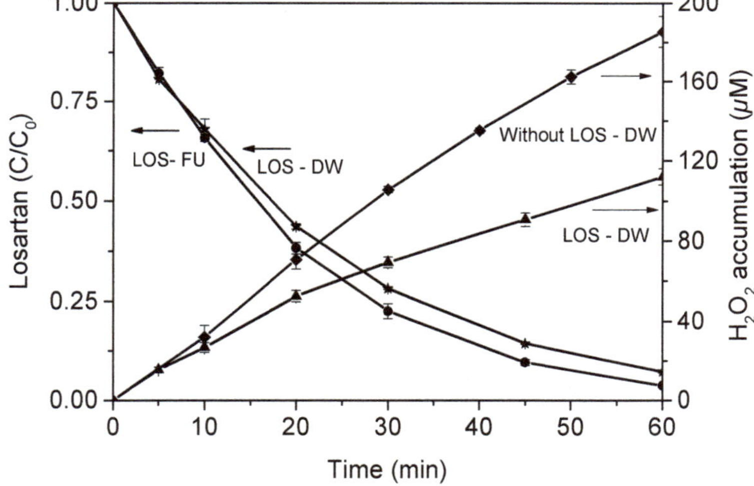

Figure 3. Determination of action routes of the sonochemical treatment on losartan in distilled water (DW) and fresh urine (FU). Conditions: [LOS] = 43.38 μM, 106.3 W L^{-1} (375 kHz), pH = 6.1.

3.3. Analysis of Losartan Susceptibility to Attacks by Radical Species

To establish electron-rich regions on losartan susceptible to attacks of radicals, computational analyses were performed [29–31], and the results from the theoretical calculations were used to better understand the formation of the degradation intermediaries. Table 3 depicts the moieties on losartan having more electron density according to Fukui function indices. In addition to these indices, other related quantities such as local softness and global hardness were determined, the values of which were 17.513 and 0.0571 eV, respectively. Moreover, a donor–acceptor diagram (DAM), to show the donor capability of the pharmaceutical concerning hydroxyl radical (HO·), hydroperoxyl radical (HOO·), and superoxide anion radical ($\cdot O_2^-$), was elaborated (Figure S2).

Table 3. Results of computational analysis for losartan [1].

Structure and Numeration	Atoms	Fukui Function Indices		
		f^-	f^+	$f\,ave$
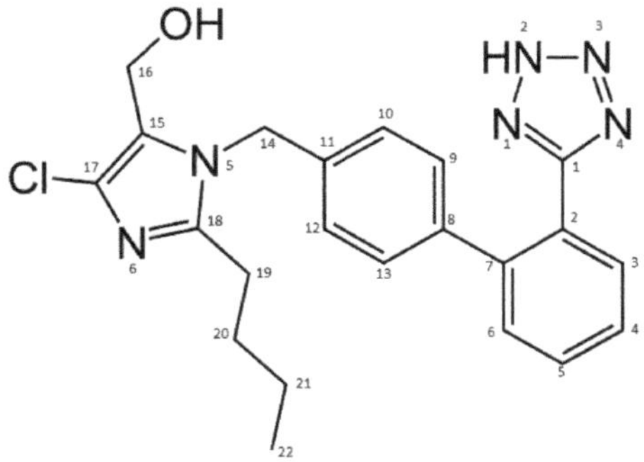	1 C	0.045	0.054	0.049
	2 C	−0.027	0.005	−0.011
	3 C	0.066	−0.022	0.022
	4 C	0.006	0.004	0.005
	5 C	0.055	−0.015	0.020
	6 C	0.004	−0.007	−0.002
	7 C	−0.103	−0.012	−0.058
	8 C	−0.100	−0.031	−0.066
	10 C	−0.112	0.160	0.024
	11 C	−0.094	−0.088	−0.091
	12 C	0.160	−0.104	0.028
	13 C	−0.050	−0.045	−0.048
	14 C	−0.660	0.117	−0.272
	15 C	0.402	0.126	0.264
	16 C	0.007	0.009	0.008
	17 C	0.000	−0.044	−0.022
	18 C	1.841	1.119	1.480
	19 C	−0.661	−0.739	−0.700
	21 C	0.216	−0.116	0.050
	22 C	0.008	−0.141	−0.067
	1 N	0.053	−0.005	0.024
	2 N	−0.043	0.003	−0.020
	3 N	0.016	0.005	0.011
	4 N	0.022	0.000	0.011
	5 N	−0.258	0.044	−0.107
	6 N	−0.066	0.179	0.057
	Cl	0.061	0.058	0.060
	O	0.061	0.000	0.031

[1] Boxes in gray color contains atoms having high values for the Fukui function indices. It should be mentioned that the computational calculations were done for LOS in water.

3.4. *Mineralization and Toxicity Evolution in Distilled Water*

The ability of the two processes to mineralize losartan was analyzed. The experiments were carried out in distilled water to avoid interfering effects of matrix and understand the fundamental aspects of the mineralizing action of the processes. We can mention that if mineralization is carried out in the fresh urine matrix, the urea that has a higher concentration masks the contribution of losartan, making it difficult to evaluate the mineralization of the contaminant under the oxidation processes. The TOC removal, at different treatment times normalized concerning the time necessary to completely degrade losartan in distilled water, was evaluated. Two different treatment times were considered: T (when losartan is 100% degraded) and 2T (the double of time required to 100% remove the antihypertensive). Results for mineralization are presented in Figure 4A.

On the other hand, toxicity modifications exerted by treatment with ultrasound and UVC/H_2O_2 to the distilled water loaded with losartan were tested. Radish seeds (*Raphanus sativus*) were used as probe organisms. The growth index (GI) was used as the toxicity measure (phytotoxicity). Phytotoxicity was established at 2T of treatment for both processes (Figure 4B).

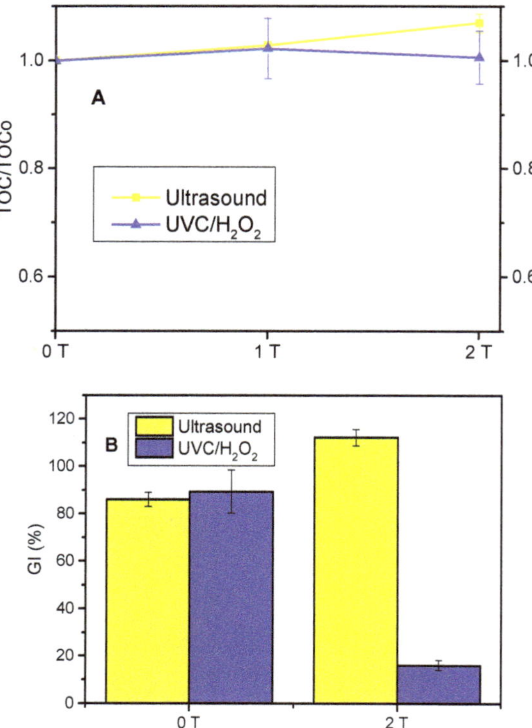

Figure 4. Extent of advanced oxidation treatments in distilled water. (**A**) Mineralization of losartan during the application of different processes; (**B**) evolution of the toxicity of losartan treated solutions against radish seeds. Note: the time was normalized concerning the time necessary to completely degrade losartan. Then, T is the time when losartan is 100% degraded, and 2T means the double of time required to 100% remove the antihypertensive. Experimental conditions as described in Figures 2 and 3.

4. Discussion

4.1. Treatment of Fresh Urine Loaded with Losartan

The Rk values for the ultrasound and UVC/H_2O_2 were 0.79 and 0.46, respectively (Table 2). It can be noted that ultrasound had the highest value for Rk; indicating that the losartan degradation through such process is affected at a low extent (21%) by urine matrix components. Meanwhile, for UVC/H_2O_2, the urine matrix presented a moderate inhibition (54%) of the antihypertensive elimination. These results suggest that the matrix components decreased the efficiency of the processes, which can be related to modifications of degradation routes. The explanations are presented in detail in the next subsections.

4.2. Degradation Routes of Losartan (LOS) in the Different AOPs

4.2.1. UVC/H_2O_2 Process

After the application of the individual components of the UVC/H_2O_2 system to LOS, it was found that hydrogen peroxide (even at 500 μM) did not induce significant removal of losartan (less than 5% elimination after 20 min of treatment). On the contrary, the treatment with the UVC light degraded ≈33.5% of the antihypertensive at 20 min of irradiation. The ultraviolet spectrum of losartan shows light absorption at 254 nm (Figure S3), which suggests that this molecule can be transformed by the UVC light. This is corroborated with the relative high photodegradation coefficient for losartan at UVC light (Cp, 123–190 L Einstein^{-1} cm^{-1} [32]). In fact, organic compounds having Cp values higher than 40 L Einstein^{-1} cm^{-1} can experience direct photolysis [33], which is currently related to the presence of aromatics rings, π-conjugated systems, and heteroatoms [34], as contained in the losartan structure (e.g., biphenyl, imidazole, and tetrazole). These aspects explain the losartan degradation by the UVC light. When losartan was treated by the complete UVC/H_2O_2 system, 65.7% of removal after 20 min was observed (Figure 2). The significant improvement of losartan elimination with the combination of hydrogen peroxide and UVC suggests the participation of radical species in the pollutant degradation. Indeed, as indicated earlier, the UVC/H_2O_2 process generates hydroxyl radical by homolytic rupture of peroxide by UVC light (Equation (1)). Hence, it can be indicated that in this process, the main action routes are the UVC photolysis and the attacks of hydroxyl radicals.

4.2.2. Ultrasound Process

The sonochemical system has three reaction zones: the inner part of cavitation bubbles, where volatile molecules are pyrolyzed by high temperatures and pressures [35–37]; the interfacial region, where hydrophobic substances can react with the sonogenerated hydroxyl radical [38]; the solution bulk, where a small number of hydroxyl radical can react with hydrophilic compounds [39].

When losartan in distilled water was treated by high-frequency ultrasound (375 kHz and 106.3 W L^{-1}), this process led to 97% of pollutant concentration reduction after 60 min of treatment (Figure 3). Since losartan is a nonvolatile compound, degradation by pyrolysis is negligible. Thus, the antihypertensive elimination would be associated with the attack of hydroxyl radicals. On the other hand, it is well-known that during the sonochemical process, hydrogen peroxide is formed by the combination of hydroxyl radicals (Equation (9)). Indeed, H_2O_2 production is an indicator of pollutant interaction with sonochemically formed $HO^{\bullet}$ [40]. Thereby, to prove the participation of sonogenerated $HO^{\bullet}$ in the pollutant degradation, the accumulation of H_2O_2 was determined. Figure 3 shows that the accumulations of H_2O_2 after 60 min of sonication in the absence and the presence of losartan were ≈180 and ≈110 µM, respectively.

$$HO^{\bullet} + HO^{\bullet} \rightarrow H_2O_2 \tag{9}$$

The oxidation of losartan by the accumulated hydrogen peroxide was discarded because pollutant removal by H_2O_2 even at 500 µM was not observed (Figure 2). Then, the lower accumulation of hydrogen peroxide in the presence of losartan is an indicator of the reaction between the $HO^{\bullet}$ and losartan. Moreover, due to the hydrophobic character of losartan (denoted by its high Log K_{OW} value, which is >4.0) [24]), its degradation is expected to occur in the interfacial zone of the system [41] by the sonogenerated hydroxyl radicals.

4.2.3. Understanding the Interference of Urine Matrix

Based on the degradation routes previously established, the interference of the urine matrix on the pharmaceutical degradation by the considered processes can be rationalized. During the application of UVC/H_2O_2 (which has both radical and photolytic routes), the antihypertensive removal was inhibited (by ≈55%) by the urine matrix components (see Rk value in Table 2). This was related to two aspects: the shielding of UVC light and scavenging of hydroxyl radicals. The shielding effect of the urine matrix was demonstrated through the evaluation of the only action of UVC light on losartan in both matrices (i.e., urine and distilled water), which showed a Rk value of 0.8 (Figure S4).

In turn, it is recognized that the inorganic anions such as chloride or bicarbonate, and organic substances like urea and acetate, present in the fresh urine, have relatively high rate constants with hydroxyl radicals (see Table 4), and as a result, they also affect the losartan degradation. It can be remarked the significant contribution of UVC photolysis to the degradation of losartan, as well as to the relative low interference of urine components for the light absorption (see Rk value for photolysis in Figure S4). Considering these findings, a scheme of losartan degradation by UVC/H_2O_2 was proposed in Figure 5a. It can be mentioned that the action of the photogenerated hydroxyl radicals induces transformations to losartan (such topic discussed below in Section 4.3), which is also schematized in this figure.

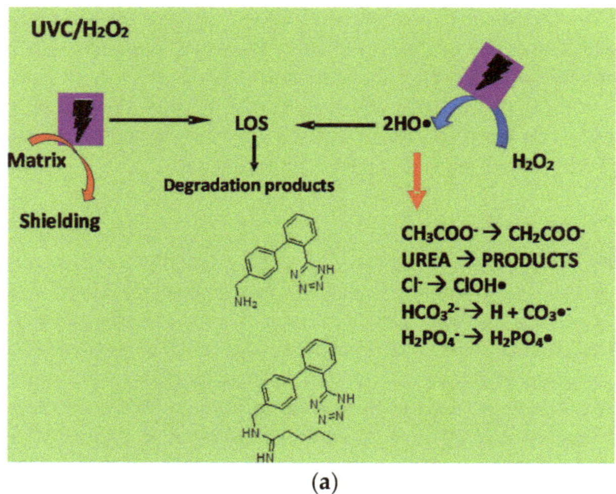

(a)

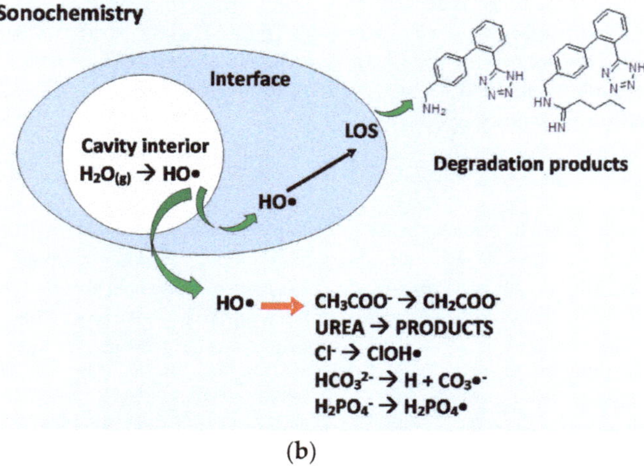

(b)

Figure 5. Scheme of degradation routes and interfering action of the urine components on the tested processes and generation of primary degradation products. (**a**) UVC/H_2O_2; (**b**) sonochemical treatment. Note: black arrows mean degradation routes and red arrows represent interfering action of the urine components.

Table 4. Rate constants of the reactions between hydroxyl radical and the diverse components of fresh urine.

Reaction	Second-Order Rate Constant (k^{2nd}, $M^{-1}\,s^{-1}$)	References
$HO^{\bullet} + Cl^{-} \rightarrow ClOH^{\bullet -}$	4.3×10^{9}	[42]
$HO^{\bullet} + H_2PO_4^{-} \rightarrow HO^{-} + H_2PO_4^{\bullet}$	$\approx 2 \times 10^{4}$	[43]
$HO^{\bullet} + CH_3COO^{-} \rightarrow H_2O + CH_2COO^{\bullet -}$	7.0×10^{7}	[44]
$HO^{\bullet} + H_2NCONH_2 \rightarrow$ products	7.9×10^{5}	[43]
$HO^{\bullet} + HCO_3^{-} \rightarrow CO_3^{\bullet -} + H_2O$	8.5×10^{6}	[45]

In the case of the sonochemical process, for the rationalization of the low inhibitory effect of the urine matrix for the degradation of losartan (Figure 3), we must consider both the degradation route of losartan and the hydrophobic/hydrophilic nature of the substances in the matrix. The urine components are very hydrophilic, as evidenced by their Log Kow values (which are close to zero or negative, see Table S1). Thus, such components are mainly placed in the bulk of the solution and losartan is in the interfacial zone (where there is a high concentration of the sonogenerated $HO^{\bullet}$). Consequently, this pharmaceutical is slightly affected by the ions and/or organic compounds of the urine matrix (as schematized in Figure 5b). It should be indicated that the action of the sonogenerated hydroxyl radicals modifies the structure of losartan (such topic discussed below in Section 4.3), which is also schematized in this figure.

4.3. *Analysis of Losartan Susceptibility to Attacks by Radical Species*

The values of the hardness and softness for losartan indicate its high donor capacity. This is advantageous for attacks of the radicals to the pharmaceutical. Such behavior was also observed in the DAM (Figure S2), which shows that the losartan molecule has a better donor capacity concerning hydroxyl radical, hydroperoxyl radical, and superoxide anion radical. Besides, the computational analyses revealed that atoms on the imidazole moiety (15C, 18C, and 6N), aromatic rings (3C, 5C, 10C, and 12C), tetrazole (1C, 1N, 3N, and 4N), alcohol (O), and alkyl chain (21C) on losartan have the highest values for $\mathbf{f_{ave}}$ (this suggests that such regions on losartan are the most susceptible to transformations by radicals such as $HO^{\bullet}$). Indeed, we can mention that the atom with the highest Fukui function indices is more reactive to hydroxyl radical (the main degrading radical species in the tested AOPs). In the case of losartan, its C18 atom presents a f_{ave} of 1.480, the highest value concerning all the atoms in the entire molecule. This behavior can be associated with the stabilization by resonance among the imidazole ring for the radical generated (Figure S5). In contrast, the attack of hydroxyl radical on the C1 atom in the tetrazole ring for the hydroxyl radical does not lead to such stabilization (Figure S6). In fact, the Fukui function indices for the tetrazole system are smaller than for the imidazole ring. Additionally, in a previous work from our research team, it was reported that for losartan molecule, the HOMO is located in the imidazole ring, whereas LUMO is on the tetrazole ring [46].

The primary products of losartan degradation in distilled water present a good agreement to the computational analysis on reactive regions of losartan (see Table S2 and Table 3). In the sonochemical treatment, three transformation products coming from imidazole ring rupture (TP1, TP2, and TP3), several isomers of biphenyl hydroxylation (TP4a-f), and one product of alcohol moiety oxidation (TP5) have been observed. Additionally, products of hydroxylation/oxidation of the alkyl chain on the antihypertensive have been found (TP6 and TP7, Table S3). Furthermore, analogous primary transformations of losartan induced by UVC/H_2O_2 and photo-Fenton were recently reported. Kaur and Dulova also found the formation of TP2 TP3, TP4, TP5, TP6, and TP7, in addition to TP8 (product of hydroxylation at the imidazole ring) and TP9 (transformation coming from a chlorine removal of the imidazole structure, see Table S3) [4]. In this sense, the region attackable by the hydroxyl radical, indicated by theoretical results correlates with the reported primary transformation products. This highlights the usefulness of computational analysis as a tool to establish the regions on losartan susceptible to degradation by the radicals from the AOPs.

4.4. *Mineralization and Toxicity Evolution*

The ability of the two processes to mineralize losartan in distilled water was tested, showing that none of these processes transformed losartan into carbon dioxide, water, and inorganic ions even at longer treatment times (2T) (Figure 4A). These results can be understood based on the degradation routes involved in each process. In the case of ultrasound, the attack of sonogenerated radicals in the interfacial zone (main route above described) led to hydroxylations/oxidations and rupture of pollutant molecules (see Table S3), which typically generates products more hydrophilic than the parent compound [47]. Hence, due to the hydrophilic nature of losartan degradation products, they are placed far away from the cavitation bubble, and consequently far away from the sonogenerated $HO^{\bullet}$. Thereby, the mineralization of losartan by ultrasound is not observed.

In the case of losartan elimination by the UVC/H_2O_2 process, it was noted the high participation of light (Section 4.2.1). Although UVC has a strong degrading ability through isomerizations or carbon-heteroatoms bond cleavages, its mineralizing power is very low [48]. On the other hand, although the mineralizing ability of $HO^{\bullet}$ is widely recognized, under the tested conditions (moderate H_2O_2 concentration; i.e., 500 μM), the formed amount of such species seems to be not enough to reach some mineralization of losartan. Due to the nonmineralizing ability of ultrasound and UVC/H_2O_2 toward LOS, it was necessary to test the toxicity. To establish the potential reuse of the treated urine for irrigating crops; toxicity tests against radish seeds (*Raphanus sativus*) were performed (Figure 4B). It should be noted that the UVC/H_2O_2 process inhibits the germination of the seeds, this is associated with noxious substances generated in this system. In fact, recent research on losartan degradation by UVC/H_2O_2 process also evidenced that toxicity of solutions against *Daphnia magna* and *Desmodesmus subspicatus* augmented after the treatment [49].

Unlike UVC/H_2O_2, in the sonochemical process, the growth of the radish seeds increased with treatment (see 2T in Figure 4B). This suggests that the losartan by-products generated at large treatment periods of the sono-treatment are beneficial/less toxic for the indicator organism than the parent compound. Such results are coincident with several studies, which reported that the treatment of polluting substances using ultrasonic irradiation reduces the toxicity of solutions [50]. It must be indicated that although both UVC/H_2O_2 and sonochemistry can generate similar primary transformation products by hydroxyl radical attacks to losartan in distilled water (Section 4.3) at long treatment periods they may differ. Additionally, it must be considered that in the sonochemical process mainly acts hydroxyl radicals, whereas in the UVC/H_2O_2 both the radicals and UVC light are responsible for pollutant degradation (Section 4.2.2). Then, the observed differences in toxicity between both processes would be associated with their degradation mode. In the UVC/H_2O_2, the noxious substances could come from the action of UVC light on losartan or its primary degradation products (indeed, a previous work about the treatment of other emerging concern pollutants by UVC also reported the generation of toxic products for some organisms produced by this irradiation [32]).

5. Conclusions

It can be concluded that this research provides relevant information to understand the elimination of a representative pharmaceutical in fresh urine by two advanced oxidation processes having different nature (a photochemical treatment and other sonochemical system). The application of ultrasound and UVC/H_2O_2 individually, for the removal of the model pharmaceutical (antihypertensive losartan) in simulated fresh urine, showed that the sonochemical process was little affected by the urine matrix, exhibiting a high selectivity (Rk = 0.79) for the removal of losartan, which was related to degradation of the pharmaceutical at the interface of the cavitation bubble by the action of $HO^{\bullet}$. Meanwhile, the UVC/H_2O_2 process experienced moderate impacts of the matrix (Rk = 0.46) on the removal of losartan, because their degradation routes involved both photolysis and radical attacks. In turn, both ultrasound and UVC/H_2O_2 processes showed no mineralization of the pollutant in distilled water. Nevertheless, differently to UVC/H_2O_2, the sonochemical system transformed losartan into nonphytotoxic products (evidencing the potential reuse of sono-treated urine to irrigate crops). This illustrates the positive potentiality of ultrasound for the treatment of pharmaceuticals with hydrophobic characteristics in the simulated fresh urine. On the other hand, the computational analyses indicated that atoms on imidazole moiety on losartan were the most susceptible to transformations by the radical species. Such analysis was in good agreement with primary degradation products coming from UVC/H_2O_2 and sonochemical treatments, evidencing that theoretical methods are a useful tool to predict and rationalize the attacks of degrading species in the considered AOPs. Finally, it must be mentioned that losartan degradation was carried out at a pH value of 6.1; however, urine ranges from 4.5 to 8, and the modification of such parameter may change the results about the degradation of pharmaceuticals by the AOPs. Thus, the effect of the urine pH should be evaluated in more detail in future studies.

Supplementary Materials: The following are available online at http://www.mdpi.com/2073-4441/12/12/3398/s1, Text S1: Determination of pseudo-first-order kinetic constants (k), Figure S1: Determination of the kinetic constants, Figure S2: Donor–acceptor diagram (DAM), Figure S3: Absorption spectra of losartan, Figure S4: Comparison of Rk for UVC/H_2O_2 and UVC alone, Figure S5: Resonance hybrid, Figure S6: Hydroxyl radical attack to the tetrazole ring, Table S1: Log K_{OW} of losartan and the components of urine, Table S2: Primary transformation products of losartan during sonochemical treatment, Table S3: Additional products of losartan transformation by UVC/H_2O_2 and photo-Fenton.

Author Contributions: Conceptualization: R.A.T.-P. and E.A.S.-G.; methodology: J.F.G.-L. and E.A.S.-G.; software: Y.Á.-T.; formal analysis: J.F.G.-L. and E.A.S.-G.; investigation: J.F.G.-L.; resources: R.A.T.-P. and Y.Á.-T.; writing—original draft preparation: J.F.G.-L.; writing—review and editing: E.A.S.-G., R.A.T.-P. and Y.Á.-T.; supervision: Y.Á.-T. and R.A.T.-P.; project administration: Y.Á.-T. and R.A.T.-P.; funding acquisition: Y.Á.-T. and R.A.T.-P. All authors have read and agreed to the published version of the manuscript.

Funding: This research was funded by UNIVERSIDAD SANTIAGO DE CALI, project DGI No. 934 621120-G04; Universidad de Antioquia UdeA; MINCIENCIAS COLOMBIA (before named COLCIENCIAS), grant "Jóvenes investigadores por la Paz 2018- J19-18-1, 5381" and project No. 111577757323.

Acknowledgments: Researchers from Grupo de Investigación en Remediación Ambiental y Biocatálisis (GIRAB) thanks Universidad de Antioquia UdeA for the support provided through "PROGRAMA DE SOSTENIBILIDAD"; E. A. Serna-Galvis thanks MINCIENCIAS COLOMBIA for his PhD fellowship during July 2015–June 2019 (Convocation 647 de 2014).

Conflicts of Interest: The authors declare no conflict of interest.

References

1. Gu, Q.; Burt, V.L.; Dillon, C.F.; Yoon, S. Trends in antihypertensive medication use and blood pressure control among United States adults with hypertension: The National Health And Nutrition Examination Survey, 2001 to 2010. *Circulation* **2012**, *126*, 2105–2114. [CrossRef]
2. Israili, Z.H. Clinical pharmacokinetics of angiotensin II (AT 1) receptor blockers in hypertension. *J. Hum. Hypertens.* **2000**, *14* (Suppl. 1), S73–S87. [CrossRef]
3. Gurke, R.; Rößler, M.; Marx, C.; Diamond, S.; Schubert, S.; Oertel, R.; Fauler, J. Science of the Total Environment Occurrence and removal of frequently prescribed pharmaceuticals and corresponding metabolites in wastewater of a sewage treatment plant. *Sci. Total Environ.* **2015**, *532*, 762–770. [CrossRef] [PubMed]
4. Kaur, B.; Dulova, N. UV-assisted chemical oxidation of antihypertensive losartan in water. *J. Environ. Manag.* **2020**, *261*, 110170. [CrossRef]
5. Osorio, V.; Larrañaga, A.; Aceña, J.; Pérez, S.; Barceló, D. Science of the Total Environment Concentration and risk of pharmaceuticals in freshwater systems are related to the population density and the livestock units in Iberian Rivers. *Sci. Total Environ.* **2016**, *540*, 267–277. [CrossRef]
6. Sanzi, F.; Souza, S.; Lopes, L.; Emanoel, J.; Hermes, F.; Alves, L.; Gonçalves, L.; Rodrigues, C.; Barbosa, B.; Moledo, D.; et al. Ecotoxicological effects of losartan on the brown mussel Perna perna and its occurrence in seawater from Santos Bay (Brazil). *Sci. Total Environ.* **2018**, *637–638*, 1363–1371. [CrossRef]
7. Bayer, A.; Asner, R.; Schüssler, W.; Kopf, W.; Weiß, K.; Sengl, M.; Letzel, M. Behavior of sartans (antihypertensive drugs) in wastewater treatment plants, their occurrence and risk for the aquatic environment. *Environ. Sci. Pollut. Res.* **2014**, *21*, 10830–10839. [CrossRef]
8. Miklos, D.B.; Remy, C.; Jekel, M.; Linden, K.G.; Hübner, U. Evaluation of advanced oxidation processes for water and wastewater treatment—A critical review. *Water Res.* **2018**, *139*, 118–131. [CrossRef]
9. Zhou, C.; Gao, N.; Deng, Y.; Chu, W.; Rong, W.; Zhou, S. Factors affecting ultraviolet irradiation/hydrogen peroxide (UV/H_2O_2) degradation of mixed N-nitrosamines in water. *J. Hazard. Mater.* **2012**, *231–232*, 43–48. [CrossRef]
10. Mahamuni, N.N.; Adewuyi, Y.G. Advanced oxidation processes (AOPs) involving ultrasound for waste water treatment: A review with emphasis on cost estimation. *Ultrason. Sonochem.* **2010**, *17*, 990–1003. [CrossRef]

11. Zhang, R.; Yang, Y.; Huang, C.-H.; Zhao, L.; Sun, P. Kinetics and modeling of sulfonamide antibiotic degradation in wastewater and human urine by UV/H_2O_2 and UV/PDS. *Water Res.* **2016**, *103*, 283–292. [CrossRef] [PubMed]
12. Giannakis, S.; Hendaoui, I.; Jovic, M.; Grandjean, D.; De Alencastro, L.F.; Girault, H.; Pulgarin, C. Solar photo-Fenton and UV/H_2O_2 processes against the antidepressant Venlafaxine in urban wastewaters and human urine. Intermediates formation and biodegradability assessment. *Chem. Eng. J.* **2017**, *308*, 492–504. [CrossRef]
13. Giannakis, S.; Jovic, M.; Gasilova, N.; Pastor Gelabert, M.; Schindelholz, S.; Furbringer, J.-M.M.; Girault, H.; Pulgarin, C. Iohexol degradation in wastewater and urine by UV-based Advanced Oxidation Processes (AOPs): Process modeling and by-products identification. *J. Environ. Manag.* **2017**, *195*, 174–185. [CrossRef]
14. Singla, J.; Verma, A.; Sangal, V.K. Parametric optimization for the treatment of human urine metabolite, creatinine using electro-oxidation. *J. Electroanal. Chem.* **2018**, *809*, 136–146. [CrossRef]
15. Giannakis, S.; Androulaki, B.; Comninellis, C.; Pulgarin, C. Wastewater and urine treatment by UVC-based advanced oxidation processes: Implications from the interactions of bacteria, viruses, and chemical contaminants. *Chem. Eng. J.* **2018**, *343*, 270–282. [CrossRef]
16. Yin, K.; He, Q.; Liu, C.; Deng, Y.; Wei, Y.; Chen, S.; Liu, T.; Luo, S. Prednisolone degradation by UV/chlorine process: Influence factors, transformation products and mechanism. *Chemosphere* **2018**, *212*, 56–66. [CrossRef]
17. Luo, J.; Liu, T.; Zhang, D.; Yin, K.; Wang, D.; Zhang, W.; Liu, C.; Yang, C.; Wei, Y.; Wang, L.; et al. The individual and Co-exposure degradation of benzophenone derivatives by UV/H_2O_2 and UV/PDS in different water matrices. *Water Res.* **2019**, *159*, 102–110. [CrossRef]
18. Montoya-Rodríguez, D.M.; Serna-Galvis, E.A.; Ferraro, F.; Torres-Palma, R.A. Degradation of the emerging concern pollutant ampicillin in aqueous media by sonochemical advanced oxidation processes—Parameters effect, removal of antimicrobial activity and pollutant treatment in hydrolyzed urine. *J. Environ. Manage.* **2020**, *261*, 110224. [CrossRef]
19. Lacasa, E.; Herraiz, M. The role of anode material in the selective oxidation of 2 penicillin G in urine. *Chemelectrochem* **2019**, *6*, 1376–1380. [CrossRef]
20. Cotillas, S.; Lacasa, E.; Sáez, C.; Cañizares, P.; Rodrigo, M.A. Removal of pharmaceuticals from the urine of polymedicated patients: A first approach. *Chem. Eng. J.* **2018**, *331*, 606–614. [CrossRef]
21. Schmidt, B.; Schieffer, B. Angiotensin II AT1 Receptor Antagonists. Clinical Implications of Active Metabolites. *J. Med. Chem.* **2003**, *46*, 2261–2270. [CrossRef]
22. Amstutz, V.; Katsaounis, A.; Kapalka, A.; Comninellis, C.; Udert, K.M. Effects of carbonate on the electrolytic removal of ammonia and urea from urine with thermally prepared IrO2 electrodes. *J. Appl. Electrochem.* **2012**, *42*, 787–795. [CrossRef]
23. Serna-Galvis, E.A.; Silva-Agredo, J.; Giraldo, A.L.; Flórez, O.A.; Torres-Palma, R.A. Comparison of route, mechanism and extent of treatment for the degradation of a β-lactam antibiotic by TiO_2 photocatalysis, sonochemistry, electrochemistry and the photo-Fenton system. *Chem. Eng. J.* **2016**, *284*, 953–962. [CrossRef]
24. Serna-Galvis, E.A.; Isaza-Pineda, L.; Moncayo-Lasso, A.; Hernández, F.; Ibáñez, M.; Torres-Palma, R.A. Comparative degradation of two highly consumed antihypertensives in water by sonochemical process. Determination of the reaction zone, primary degradation products and theoretical calculations on the oxidative process. *Ultrason. Sonochem.* **2019**, *58*, 104635. [CrossRef]
25. Serna-Galvis, E.A.; Silva-Agredo, J.; Giraldo-Aguirre, A.L.; Torres-Palma, R.A. Sonochemical degradation of the pharmaceutical fluoxetine: Effect of parameters, organic and inorganic additives and combination with a biological system. *Sci. Total Environ.* **2015**, *524–525*, 354–360. [CrossRef]
26. Hoekstra, N.J.; Bosker, T.; Lantinga, E.A. Effects of cattle dung from farms with different feeding strategies on germination and initial root growth of cress (Lepidium sativum L.). *Agric. Ecosyst. Environ.* **2002**, *93*, 189–196. [CrossRef]
27. Raghavachari, K. Perspective on "Density functional thermochemistry. III. The role of exact exchange". *Theor. Chem. Acc.* **2000**, *103*, 361–363. [CrossRef]
28. Tomasi, J.; Mennucci, B.; Cammi, R. Quantum Mechanical Continuum Solvation Models. *Chem. Rev.* **2005**, *105*, 2999–3094. [CrossRef]
29. An, T.; Yang, H.; Li, G.; Song, W.; Cooper, W.J.; Nie, X. Kinetics and mechanism of advanced oxidation processes (AOPs) in degradation of ciprofloxacin in water. *Appl. Catal. B Environ.* **2010**, *94*, 288–294. [CrossRef]

30. Gurkan, Y.Y.; Turkten, N.; Hatipoglu, A.; Cinar, Z. Photocatalytic degradation of cefazolin over N-doped TiO_2 under UV and sunlight irradiation: Prediction of the reaction paths via conceptual DFT. *Chem. Eng. J.* **2012**, *184*, 113–124. [CrossRef]
31. Li, L.; Wei, D.; Wei, G.; Du, Y. Transformation of cefazolin during chlorination process: Products, mechanism and genotoxicity assessment. *J. Hazard. Mater.* **2013**, *262*, 48–54. [CrossRef]
32. Clara, M.; Starling, V.M.; Souza, P.P.; Le, A.; Amorim, C.C.; Criquet, J. Intensification of UV-C treatment to remove emerging contaminants by UV-C/H_2O_2 and UV-C/$S_2O_8{}^{2-}$: Susceptibility to photolysis and investigation of acute toxicity UV-C. *Chem. Eng. J.* **2019**, *376*, 120856. [CrossRef]
33. Gerrity, D.; Lee, Y.; Von Gunten, U. Prediction of Trace Organic Contaminant Abatement with UV/H_2O_2: Development and Validation of Semi-Empirical Models for Municipal Wastewater Effluents. *Environ. Sci. Water Res. Technol.* **2016**, *2*, 460–473. [CrossRef]
34. Challis, J.K.; Hanson, M.L.; Friesen, K.J.; Wong, C.S. A critical assessment of the photodegradation of pharmaceuticals in aquatic environments: Defining our current understanding and identifying knowledge gaps. *Environ. Sci. Process. Impacts* **2014**, *16*, 672–696. [CrossRef]
35. Adewuyi, Y.G. Sonochemistry: Environmental Science and Engineering Applications. *Ind. Eng. Chem. Res.* **2001**, *40*, 4681–4715. [CrossRef]
36. Cheng, J.; Vecitis, C.D.; Park, H.; Mader, B.T.; Hoffmann, M.R. Sonochemical degradation of perfluorooctane sulfonate (PFOS) and perfluorooctanoate (PFOA) in landfill groundwater: Environmental matrix effects. *Environ. Sci. Technol. Ronmental.* **2008**, *42*, 8057–8063. [CrossRef]
37. Jiang, Y.; Pétrier, C.; David Waite, T. Kinetics and mechanisms of ultrasonic degradation of volatile chlorinated aromatics in aqueous solutions. *Ultrason. Sonochem.* **2002**, *9*, 317–323. [CrossRef]
38. Fernandez, N.A.; Rodriguez-freire, L.; Keswani, M.; Sierra-alvarez, R. Degradation of perfluoroalkyl and polyfluoroalkyl. *Environ. Sci.* **2016**, 975–983. [CrossRef]
39. Yasman, Y.; Bulatov, V.; Gridin, V.V.; Agur, S.; Galil, N.; Armon, R.; Schechter, I. A new sono-electrochemical method for enhanced detoxification of hydrophilic chloroorganic pollutants in water. *Ultrason. Sonochem.* **2004**, *11*, 365–372. [CrossRef]
40. Serna-Galvis, E.A.; Montoya-Rodríguez, D.M.; Isaza-Pineda, L.; Ibáñez, M.; Hernández, F.; Moncayo-Lasso, A.; Torres-Palma, R.A. Sonochemical degradation of antibiotics from representative classes-Considerations on structural effects, initial transformation products, antimicrobial activity and matrix. *Ultrason. Sonochem.* **2018**, *50*, 157–165. [CrossRef]
41. Tran, N.; Drogui, P.; Brar, S.K. Sonochemical techniques to degrade pharmaceutical organic pollutants. *Environ. Chem. Lett.* **2015**, *13*, 251–268. [CrossRef]
42. Lian, L.; Yao, B.; Hou, S.; Fang, J.; Yan, S.; Song, W. Kinetic Study of Hydroxyl and Sulfate Radical-Mediated Oxidation of Pharmaceuticals in Wastewater Effluents. *Environ. Sci. Technol.* **2017**, *51*, 2954–2962. [CrossRef] [PubMed]
43. Buxton, G.V.; Greenstock, C.L.; Helman, W.P.; Ross, A.B. Critical Review of rate constants for reactions of hydrated electron, hydrogen atoms and hydroxyl radicals (·OH/·O in Aqueous Solution). *J. Phys. Chem. Ref. Data* **1988**, *513*. [CrossRef]
44. Minakata, D.; Song, W.; Crittenden, J. Reactivity of Aqueous Phase Hydroxyl Radical with Halogenated Carboxylate Anions: Experimental and Theoretical Studies. *Environ. Sci. Technol.* **2011**, *45*, 6057–6065. [CrossRef]
45. Toth, J.E.; Rickman, K.A.; Venter, A.R.; Kiddle, J.J.; Mezyk, S.P. Reaction kinetics and efficiencies for the hydroxyl and sulfate radical based oxidation of artificial sweeteners in water. *J. Phys. Chem. A.* **2012**, *116*, 9819–9824. [CrossRef]
46. Guateque-Londoño, J.F.; Serna-Galvis, E.A.; Silva-Agredo, J.; Ávila-Torres, Y.; Torres-Palma, R.A. Dataset on the degradation of losartan by TiO_2-photocatalysis and UVC/persulfate processes. *Data Br.* **2020**, *31*. [CrossRef]
47. Singla, R.; Grieser, F.; Ashokkumar, M. The mechanism of sonochemical degradation of a cationic surfactant in aqueous solution. *Ultrason. Sonochem.* **2011**, *18*, 484–488. [CrossRef]
48. Yang, L.; Yu, L.E.; Ray, M.B. Degradation of paracetamol in aqueous solutions by TiO_2 photocatalysis. *Water Res.* **2008**, *42*, 3480–3488. [CrossRef]

49. Adams, E. *Ecotoxicity and Genotoxicity Evaluation of Losartan Potassium after UVC Photolysis and UV/H_2O_2 Process*; Universidade Tecnológica Federal Do Paraná: Curitiba, Brazil, 2019.
50. Emery, R.J.; Papadaki, M.; Freitas dos Santos, L.M.; Mantzavinos, D. Extent of sonochemical degradation and change of toxicity of a pharmaceutical precursor (triphenylphosphine oxide) in water as a function of treatment conditions. *Environ. Int.* **2005**, *31*, 207–211. [CrossRef]

Article

Towards the Removal of Antibiotics Detected in Wastewaters in the POCTEFA Territory: Occurrence and TiO_2 Photocatalytic Pilot-Scale Plant Performance

Samuel Moles [1,*], Rosa Mosteo [1], Jairo Gómez [2], Joanna Szpunar [3], Sebastiano Gozzo [3], Juan R. Castillo [4] and María P. Ormad [1]

1 Water and Environmental Health Research Group, c/María de Luna 3, 50018 Zaragoza, Spain; mosteo@unizar.es (R.M.); mpormad@unizar.es (M.P.O.)
2 Navarra de Infraestructuras Locales SA, av. Barañain 22, 31008 Pamplona, Spain; jgomez@nilsa.com
3 Institute of Analytical Sciences and Physico-Chemistry for Environment and Materials (IPREM), Centre National de la Recherche Scientifique (CNRS), Universite de Pau et des Pays de l'Adour, CEDEX 9 Pau, France; joanna.szpunar@univ-pau.fr (J.S.); sebastiano.gozzo@univ-pau.fr (S.G.)
4 Analytical Spectroscopy and Sensors Group Analytic Chemistry Department, Science Faculty, Environmental Science Institute, University of Zaragoza, 50009 Zaragoza, Spain; jcastilo@unizar.es
* Correspondence: sma@unizar.es

Received: 25 April 2020; Accepted: 15 May 2020; Published: 20 May 2020

Abstract: This research aims to assess the presence of four antibiotic compounds detected in the influent and effluent of wastewater treatment plants (WWTPs) in the POCTEFA territory (north of Spain and south of France) during the period of 2018–2019, and to relate the removal of antibiotic compounds with the processes used in the WWTPs. The performance of a photocatalytic TiO_2/UV-VIS pilot-scale plant was then evaluated for the degradation of selected antibiotics previously detected in urban treated effluent. The main results reflect that azithromycin had the highest mass loadings (11.3 g/day per 1000 inhabitants) in the influent of one of the selected WWTPs. The results also show considerable differences in the extent of antibiotics removal in WWTPs ranging from 100% for sulfadiazine to practically 0% for trimethoprim. Finally, the photocatalytic TiO_2/UV-VIS pilot-scale plant achieved the removal of the four antibiotics after 240 min of treatment from 78%–80% for trimethoprim and enrofloxacin, up to 100% for amoxicillin, sulfadiazine and azithromycin. The catalyst recovery via mechanical coagulation–flocculation–decantation was almost total. The Ti concentration in the effluent of the TiO_2/UV-VIS pilot-scale plant was lower than 0.1% (w/w), and its release into the environment was subsequently minimized.

Keywords: antibiotics; wastewater; removal efficiency; photocatalysis; slurry reactor

---

Highlights:

- Antibiotics mass loadings range from 11,332 mg/day·1000 inhabitants to undetectable levels.
- Sulfadiazine, amoxicillin and azithromycin can be removed from wastewaters, while 80% of trimethoprim and enrofloxacin removal can be achieved after the photocatalytic treatment.
- The facilities provided with trickling filters proved to be more effective in removing antibiotics from wastewaters.

1. Introduction

The problem of the presence of pharmaceutical compounds in wastewater has recently become a matter for concern, not only in terms of human health, but also for the preservation of the environment [1]. Antibiotics are an important group of medicines suitable for the treatment of human infections and in veterinary medicine. Many of them are not completely metabolized by the body so between 30%–90% are excreted and, as a result, they end up in wastewater [2,3]. The main difference with other organic pollutants is that antibiotics represent a potential risk if they are released into the environment because they a direct biological action on microorganisms, generating antimicrobial-resistant bacteria (ARB). As suggested by other authors [4], ARB of animal origin can also be transmitted to humans.

Several studies have pointed out that conventional wastewater treatment plants (WWTPs) are not designed to remove pharmaceuticals, metabolites or drugs [3–5]. Besides urban plants and hospitals, slaughterhouses also generate wastewaters which are not usually incorporated into sewage systems. As a result, they represent a significant source of antibiotics released into the environment. The European Surveillance of Veterinary Antimicrobial Consumption (ESVAC) collects information on how antimicrobial medicines are used in animals across the European Union (EU). According to their latest report [6], Spain is known to be one of the main consumers of veterinary drugs in the EU. As a result, many studies have monitored the occurrence of the most commonly administered pharmaceuticals in urban wastewater, groundwaters and surface water in Spain. The literature informs that concentrations of antibiotics from ng/L to µg/L [7–10]. sulfonamides [11,12], trimethoprim [13,14], β-lactams [15,16], fluoroquinolones [17,18] and macrolides [19–21] all represent a potential risk for the environment. Consequently, a representative antibiotic from each one these groups was analyzed in this research work: sulfadiazine (veterinary use, sulfonamide), trimethoprim (human and veterinary use, trimethoprim), amoxicillin (human and veterinary use, β-lactam), enrofloxacin (veterinary use, fluoroquinolone) and azithromycin (human use, macrolide).

Among the various water treatment techniques used to eliminate these drugs, advanced oxidation processes (AOPs) are suitable for antibiotic degradation [22–24]. Other techniques, such as activated carbon or reverse osmosis, only transfer the contaminants from one phase to another without degrading them. Nevertheless, photocatalysis has been demonstrated to be effective for wastewater treatment as it is cost-effective and simultaneously oxidizes various organic contaminants into inorganic compounds, water and carbon dioxide, and pathogenic microorganisms [24]. Several semiconductors are used in photocatalysis, such as TiO_2, ZnO, and CdS. Among these, TiO_2 has been widely used because of its strong oxidizing power, availability, nontoxicity and price. The catalyst can be employed either in a colloidal or in an immobilized form. Although immobilizing the catalyst might improve the catalyst recovery, immobilized systems show lower degradation efficiencies compared to the suspended counterpart because of a reduction in the surface area [25,26]. Whenever the nanoparticles are dispersed in an aqueous medium, the depth of penetration of the radiation is limited because of absorption/scattering by the catalyst nanoparticles and the dissolved organic species. These systems also require an additional separation process to prevent Ti emission to the environment, and this stage induces further costs [27].

Pilot-scale plants represent the previous step to industrial scale plant. Literature suggest how to operate at lab-scale photocatalytic systems [28–30]. However, design and operation with a pilot-scale plant are necessary to determine how to deal with possible operational problems and establish the optimal operational parameters for real scale operation. Pilot-scale plants also allow one to determine if the real scale process would be economically feasible. Some studies about the application of TiO_2 photocatalysis in wastewater have been reported [13,14,28–31]. However, these research works do not focus on the simultaneous antibiotic removal by a TiO_2 photocatalysis pilot-scale plant applied to real wastewater.

The aim of this research work is to evaluate the presence of selected antibiotic compounds in the inlet and outlet of four WWTPs for the period of 2018–2019. Another objective is to treat selected antibiotics present in real wastewater in a photocatalytic plant by applying TiO_2 in suspension. Finally, the Ti concentration in the final effluent was controlled, to prevent Ti emission to the environment.

2. Materials and Methods

2.1. Site Description and Sample Collection

This research is focused on four WWTPs located in the POCTEFA territory (north of Spain and south of France). These WWTPs are designed to treat urban wastewater of domestic and industrial origins. Table 1 shows the main characteristics of selected WWTPs.

Samples were collected in four sampling campaigns for two years (in the spring and autumn of 2018 and 2019). The inlet and outlet of each WWTP were selected as sampling points to estimate the current removal performance of selected antibiotics in the four WWTPs, aiming to compare the different treatment lines.

Table 1. Main characteristics of each wastewater treatment plants (WWTP).

#WWTP	Population Equivalent	Total Inlet Flow (m^3/day)	Water Treatment Line
1	695,232	129,600	Grit and grease separator/Activated Sludge/Decanter
2	82,500	22,150	Grit and grease separator/Decanter/Trickling filter (first stage) /Decanter/Trickling filter (second stage)/Decanter
3	10,470	10,995	Decanter/Trickling filter (first stage) /Decanter
4	51,336	7500	Grit and grease separator/Decanter/ Moving bed biofilm reactor/Decanter

The sampling was carried out following the EPA method 1694 for the analysis of pharmaceuticals and personal care products in water, soil, sediment, and biosolids by liquid chromatography tandem mass spectrometry (LC/MS/MS) [32]. Amber glass bottles were used to collect 1000 mL samples which were stored under refrigeration at 4 °C. The bottles were fully filled to avoid the presence of air and properly sealed by means of a PTFE seal. According to EPA 1694, the filtration of the samples is necessary in order to remove suspension solids. Two filtration steps were carried out prior to analysis using glass fiber filters (1.6 µm, supplied by GVS) for the first filtration stage and nylon filters (0.45 µm, supplied by GVS) for the second stage, as suggested in other research works [33–35].

Mass loadings of the antibiotics were calculated in each sampling period as the product of the individual concentration of each antibiotic in the samples and the daily flow rate of each WWTP.

Removal efficiencies of the target compounds were determined as the difference between the inlet mass loading and the outlet mass loading divided by the inlet mass loading and expressed as a percentage (Equation (1)).

$$Removalefficiency\ (\%) : \frac{(m_{inf} - m_{eff})}{m_{inf}} \times 100 \tag{1}$$

2.2. Antibiotic Characterization

The quantification of the concentration of antibiotics was carried out via HPLC/MS/MS. Samples were centrifuged for 10 min at 13,000 rpm in Eppendorf tubes and then diluted 40-fold with 0.1% formic acid/MeOH/ACN (80%/10%/10%) before LC-MS/MS analysis. Chromatographic separations were carried out on an Ultra Performance Liquid Chromatography (UPLC) Ultimate 3000 RSLC system (Thermo Fisher Scientific). The column used was an Accucore C18 100 × 2.1 mm, 2.6 µm (Thermo Fisher Scientific). The mobile phases were A H_2O 0.4% formic acid + 5 mM ammonium formate and B MeOH/ACN 1:1 (v/v). A 20µL sample aliquot was injected. Detection was performed on a Q Exactive Plus (Thermo Fisher Scientific) mass spectrometer operated in the targeted single ion monitoring

(SIM) positive mode with a resolution of 70,000. External calibration was used for quantification and validated by standard additions for selected samples; the samples were prepared and analyzed in triplicate. The limits of detection and quantification of each antibiotic are featured in Table 2.

Table 2. Limits of detection and quantification of selected antibiotics.

Antibiotic	LOD (ng/L)	LOQ (ng/L)
Sulfadiazine	0.8	2.5
Trimethoprim	0.8	2.5
Amoxicillin	10	30
Enrofloxacin	1.2	3.7
Azithromycin	2.0	6.5

The four antibiotics investigated in this research work were selected according to their potential risk for the environment and reported occurrence [11–21]. All of them are representative human-use and veterinary-use antibiotics belonging to the main antibiotic groups. The standards were supplied by Sigma-Aldrich. Some characteristics of the selected antibiotic compounds are included in Table 3.

Table 3. Selected antibiotics and chemical information.

Antibiotic	Group	Chemical Abstracts Service Registry Number (CAS Nr.)	MW (g/mol)
Sulfadiazine	Sulfonamide	68-35-9	250
Trimethoprim	Trimethoprim	738-70-5	290
Amoxicillin	β-lactam	26787-78-0	365
Enrofloxacin	Fluoroquinolone	93106-60-6	359
Azithromycin	Macrolide	83905-01-5	749

2.3. *Total Ti Assessment in the Effluent*

The effluent from the pilot-scale plant might contain some Ti which would then be emitted to the environment. The Ti concentration was quantified by an Inductively Coupled Plasma Mass Spectrometry (ICP/MS) ELAN DRC-e, PerkinElmer, Toronto, Canada. A discrete volume sample (100 μL) was injected through a six-way valve, and the carrier was delivered directly to the nebulizer of the spectrometer. A glass concentric slurry nebulizer with a cyclonic spray chamber (Glass Expansion, Melbourne, Australia) was used. Default values were used for the rest of the instrumental parameters. The quantification of TiO_2 was based on monitoring the ICP-MS signal of the isotope ^{49}Ti, using ^{74}Ge as an internal standard. From an on-line calibration with an ionic titanium standard diluted in nitric acid (1%), intensity signals from the ICP-MS for samples were transformed into mass values by integrating the area of the transient signals obtained. All samples were injected in triplicate. The limit of detection of the method was established at 0.81 μg/L and the limit of quantification at 2.70 μg/L.

2.4. *Photocatalytic Oxidation Experiment*

The oxidation assays were carried out in the facility detailed in Figures 1 and 2. First, a 1 m^3 storage tank provided with a stirrer was filled with the water sample. The solution was then pumped (8–16 L/min) to a 0.1 m^3 mixer decanter where the catalyst was stored. The mixer decanter was provided with a stirrer to mix the influent with the catalyst. Subsequently, the mixture was placed in four identical slurry reactors. These reactors are made of aluminium because this material is known to have a high degree of light reflection. The reactors had a volume of 17 L and were provided with a UVA lamp (330–390 nm) of 40W. When the reactors were completely full, the UV/vis lamp in each reactor was turned on and stirring by means of compressed air took place. After treatment in the reactors, the treated water was pumped again to the decanter where a mechanical coagulation–flocculation–decantation treatment (CFD) was applied. Two steps take place in the separation process, coagulation 200 rpm during 5 min, flocculation 40 rpm during 25 min. and 90 min of decantation, resulting in a 120 min total process. Coagulant was not added to the mechanical CFD separation process.

The effluent (clarified phase) was generated and the catalyst remained in the decanter for the next cycle.

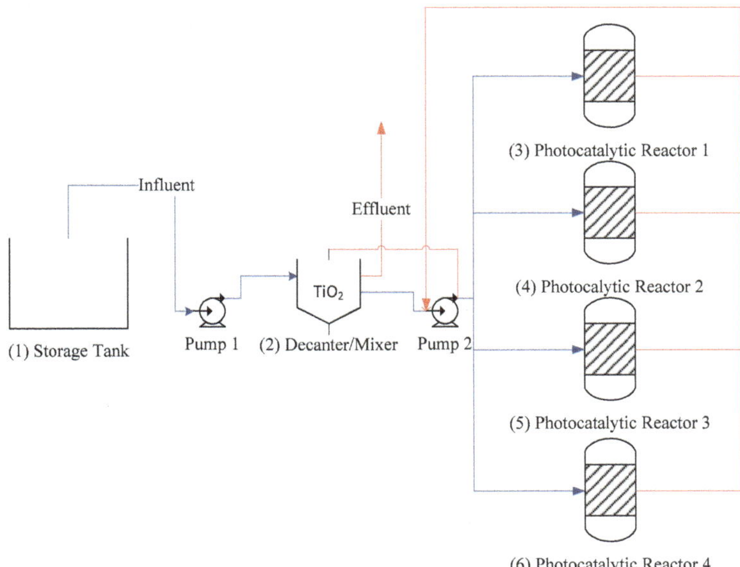

Figure 1. Pilot-scale plant process flow-diagram.

Figure 2. Pilot-scale plant process images.

The sample was prepared by the addition of individual concentrations of 1 mg/L of each antibiotic (amoxicillin, azithromycin, enrofloxacin, trimethoprim and sulfadiazine) simultaneously in the effluent from WWTP2. The physicochemical characteristics of this wastewater were pH = 7.6, DQO = 90 mg/L, Turbidity =11 NTU. The catalyst was applied in suspension in a concentration of 1 g/L of TiO_2 FN2 (supplied by Levenger S.L.). Radiation per unity of volume was 0.3 W/L in each reactor. The temperature ranged from 14 °C to 20 °C during the experiment. The experiments were conducted twice, and their average is represented in the results. The individual antibiotic concentration was quantified via HPLC/MS/MS during 240 min of photocatalytic treatment following the procedure described in Section 2.2.

3. Results

3.1. Occurrence of Target Antibiotics in Urban Wastewaters

Figure 3 shows the mass loading of the selected antibiotics at the four WWTPs during 2018 and 2019. It should be noted that atypical points are not represented. However, they are all available in Tables S1 and S2 of the Supplementary Material. The tables show that the highest mass loading corresponds to azithromycin. This human-use antibiotic presented an average load of around 925 mg/day per 1000 inhabitants and reached a maximum load of 11,332 mg/day per 1000 inhabitants in WWTP4. The azithromycin concentration increased over the four sampling campaigns in all the WWTPs. This fact might be attributed to a major increase in the use of azithromycin [36]. Enrofloxacin (fluoroquinolone group) also followed the same trend: higher loads were detected over the campaigns analyzed in this study. This veterinary-use antibiotic showed median mass loadings of around 200 mg/day per 1000 inhabitants and a maximum of 4329 mg/day per 1000 inhabitants. The enrofloxacin mass load was higher in the spring of both years, while it was rarely found in autumn, and always in lower loads, reflecting a seasonal use of this fluoroquinolone. The seasonal appearance of enrofloxacin has also been recently reported in another research work [37].

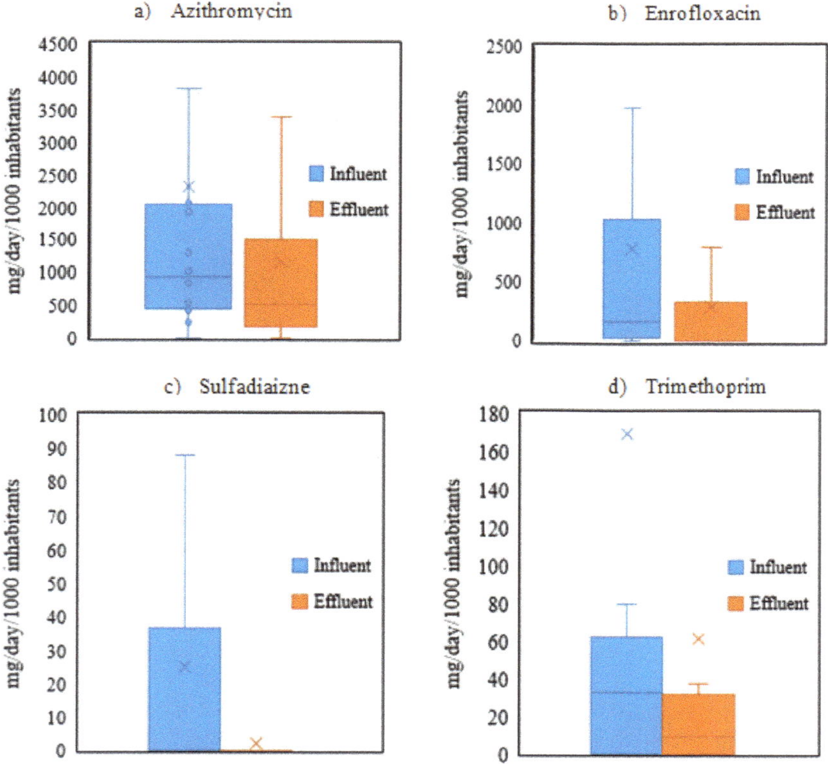

Figure 3. Boxplots of each antibiotic indicating total mass loading values: (**a**) azithromycin, (**b**) enrofloxacin, (**c**) sulfadiazine (**d**) trimethoprim.

By contrast, the average mass loadings of the other two antibiotics (trimethoprim and sulfadiazine) ranged from 35 mg/day per 1000 inhabitants for trimethoprim, to undetectable levels for sulfadiazine. Comparing the median mass loadings of trimethoprim and sulfadiazine with enrofloxacin and azithromycin reveals a difference greater than one order of magnitude.

Amoxicillin was not found in any sample (influent and effluent of the WWTPs). This fact could be attributed to the low stability of amoxicillin and the subsequent generation of degradation products such as amoxicillin penicilloic acid or amoxicillin-diketopiperazine-2′, 5′, as is suggested in the literature [38–40]. These degradation products were found in subsequent campaigns carried out in the same sampling points.

Finally, Tables S1 and S2 show that the reported load of each antibiotic varies significantly, by more than one order of magnitude, between the years and seasons. This trend could be related to the differences in rainfall patterns between the two years: high flows in particular were reported in spring 2018 and in autumn 2019 in both the Ebro River basin and the Cantabrico Occidental River basin. However, the total mass loading of the four antibiotics was relatively higher in the spring: $\sum_{SPRING}$ = 69.4 g/day per 1000 inhabitants versus $\sum_{AUTUMN}$ = 13.6 g/day per 1000 inhabitants.

3.2. Removal Efficiency of Selected WWTPs

The results of the removal efficiency of the antibiotics in each WWTP is shown in Figure 2. These results are also fully detailed in Table S3. Figure 4 shows that the removal efficiencies range from 2%–100%, demonstrating the fact that WWTPs can partially or almost totally remove the target antibiotics.

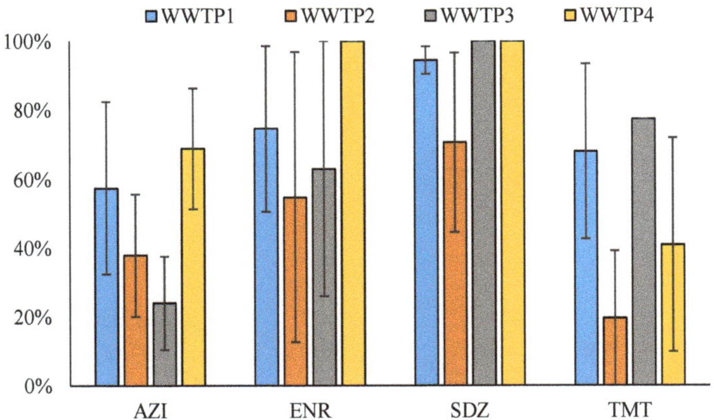

Figure 4. Wastewater treatment plants' removal efficiency for each antibiotic in 2018–2019 (AZI = azithromycin, ENR = enrofloxacin, SDZ = sulfadiazine, TMT = trimethoprim).

The results also suggest a significant variation for each antibiotic in each WWTP, indicating that the removal efficiency strongly depends not only on the specific matrix but also on the season and associated flow. This variation might also be attributed to the different physicochemical characteristics of the antibiotics, such as the degradation rates in the water, organic carbon–water partition coefficients or acid dissociation constants and water solubilities. More precisely, the results suggest that sulfadiazine is the antibiotic with the highest removal efficiency in every case. This trend may be due to the low mass loadings of this antibiotic, which make its removal easier [41]. Enrofloxacin also presents a high removal efficiency after the wastewater treatment, reaching 100% in several samples. By contrast, azithromycin and trimethoprim showed lower removal efficiencies in all the WWTPs. It should be noted that azithromycin had the highest mass loadings, more than one order of magnitude greater than the other antibiotics. It might be also attributed to the fact that azithromycin and trimethoprim have similar carbon–water partition coefficients [42]. Some studies suggest that biological processes, which are present in the four WWTPs, can remove fluoroquinolones effectively, while trimethoprim is more difficult to remove by means of biological treatments [43].

Comparing the different treatments of each WWTP, it can be observed that the facilities provided with a trickling filter (WWTP2 and WWTP3) showed higher removal efficiencies of the target antibiotics. This is consistent with some studies which demonstrate that trickling filters can remove antibiotics and other pharmaceuticals as well as personal care products [44–46]. WWTP4 showed the highest average antibiotic removal in most samples. However, it is important to note that this WWTP is in the Cantabrico River basin where there is higher rainfall than in the Ebro River Basin. As a result, the selected antibiotics had lower mass loadings in WWTP4.

3.3. Photocatalytic Oxidation of Antibiotics

Figure 5 shows the performance of the photocatalytic assays for the simultaneous oxidation of amoxicillin, enrofloxacin, sulfadiazine, trimethoprim and azithromycin during 4 h of treatment. The results show that in only 30 min of treatment, azithromycin and amoxicillin reached a degradation rate of 85% and 75%, respectively. Moreover, after 120 min of treatment, both antibiotics were completely removed from the wastewater. Amoxicillin was previously reported to be easily removed from waters by TiO_2 photocatalysis applied to the isolated compound at lab scale [47]. However, sulfadiazine shows a slower degradation rate, achieving degradation yields of 25% and 100% in 30 min and 240 min, respectively. In contrast, enrofloxacin and trimethoprim were not completely removed from the wastewater after the treatment. The degradation rate of enrofloxacin was relatively high at the beginning of the process (degradation yield of 50% in 30 min of treatment), but complete degradation was not achieved by the end of the treatment. A similar trend has been reported in other research work at lab scale [31]. Finally, trimethoprim showed the slowest initial degradation rate during the first 120 min, as reported in other studies at lab scale [45], while its degradation yield at the end of the treatment was close to 70%.

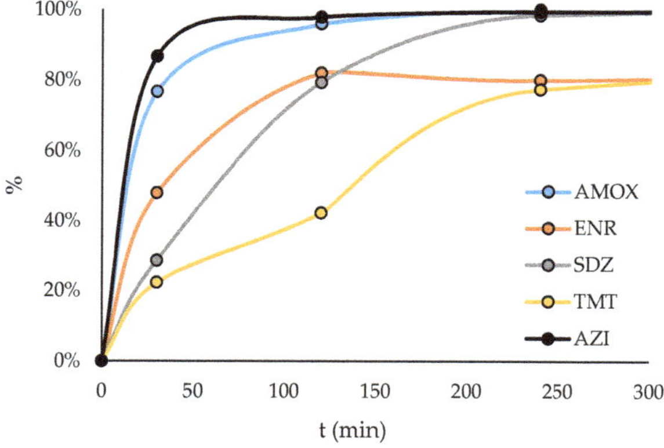

Figure 5. TiO_2 photocatalysis degradation yields of each antibiotic in WWTPE during 240 min of treatment (individual antibiotic initial concentration = 1 mg/L, TiO_2 initial concentration = 1 g/L, radiation per unity of volume = 0.3 W/L). AMX =amoxicillin, AZI = azithromycin, ENR = enrofloxacin, SDZ = sulfadiazine, TMT = trimethoprim.

3.4. Ti Assessment in the Effluent of the Photocatalytic Treatment Plant

In order to determine whether Ti is released into the environment, a quantification of the Ti concentration in the effluent was carried out applying a coagulation–flocculation–decantation treatment. TiO_2 has been demonstrated to be effective after several cycles of photocatalytic treatment for the degradation of pharmaceuticals [48], so it is important to recover it effectively. Fortunately, the results of the Ti assessment suggest that no more than 0.1% of the initial Ti concentration remained in the effluent. This reflects the fact that recovery by means of mechanical coagulation–flocculation–decantation treatment is quite efficient for recovering TiO_2 when used in suspension [48].

4. Conclusions

This research work evaluates the behavior of four antibiotic compounds in four different WWTPs located in the north of Spain. The mass loadings of amoxicillin, enrofloxacin, sulfadiazine, trimethoprim and azithromycin were analyzed in the influent and effluent of the WWTPs. The performance of a TiO_2 photocatalytic treatment plant applied to the simultaneous removal of the antibiotics from real urban treated water was evaluated. This showed that the technology can be used to totally remove some of the selected antibiotics at slightly higher concentrations than those commonly found in wastewaters. The conclusions can be summarized as follows:

1. The mass loadings of the antibiotics ranged from 11,332 mg/day·1000 inhabitants to undetectable levels. Azithromycin had the highest mass loadings, followed by enrofloxacin, trimethoprim, sulfadiazine and amoxicillin.
2. The use of enrofloxacin and azithromycin increased in the locations of the WWTPs during the period of this study.
3. Sulfadiazine, amoxicillin and azithromycin were totally removed from wastewaters in the TiO_2 photocatalytic pilot-scale plant, while 80% removal of trimethoprim and enrofloxacin was achieved by the treatment. Moreover, the facility was able to recover the catalyst after the treatment, minimizing the Ti released into the environment and allowing catalyst reuse.
4. Although WWTPs are not designed to remove antibiotics, they do reduce them. This research shows that biological treatments have a significant influence on antibiotic removal. In particular, the presence of a trickling filter in the water treatment line of the WWTPs has been demonstrated to lead to a higher degree of antibiotic removal. However, the efficiency of the antibiotic removal depends on the physicochemical properties of the antibiotics and on the characteristics of the wastewater.

Supplementary Materials: The following are available online at http://www.mdpi.com/2073-4441/12/5/1453/s1, Table S1: Spring and autumn 2018 Load selected antibiotics in mg/day/1000 inhabitants. Table S2: Spring and autumn 2019 Load selected antibiotics in mg/day/1000 inhabitants. Table S3: Removal efficiencies of the different WWTPs for selected antibiotics.

Author Contributions: Planning photocatalysis experimentation, R.M. and S.M.; general experimental work coordinator, M.P.O.; LC/MS/MS analysis, S.G. (technician) and J.S. (responsible); design and construction of pilot-scale plant, J.G. and S.M.; responsible of nanoparticles detection in the effluent, J.R.C.; experimental photocatalysis technician and principal author, S.M. All authors have read and agreed to the published version of the manuscript.

Funding: This research was funded by DGA_FSE Research Team "Water and Environmental Health" Ref: B43-20R and cofounded by EFA 183/16/OUTBIOTICS.

Acknowledgments: This work was financed by DGA_FSE Research Team "Water and Environmental Health" Ref: B43-20R in the framework of the project EFA 183/16/OUTBIOTICS, Program Interreg-POCTEFA 2014-2020, funded by FEDER.

Conflicts of Interest: The authors declare no conflict of interest.

References

1. Klein, E.Y.; Van Boeckel, T.P.; Martinez, E.M.; Pant, S.; Gandra, S.; evin, S.A.; Goossens, H.; Laxminarayan, R. Global increase and geographic convergence in antibiotic consumption between 2000 and 2015. *Proc. Natl. Acad. Sci. USA* **2018**, *115*, E3463–E3470. [CrossRef] [PubMed]
2. Santos, L.H.; Gros, M.; Rodriguez-Mozaz, S.; Delerue-Matos, C.; Pena, A.; Barceló, D.; Montenegro, M.C. Contribution of hospital effluents to the load of pharmaceuticals in urban wastewaters: Identification of ecologically relevant pharmaceuticals. *Sci. Total Environ.* **2013**, *461–462*, 302–316. [CrossRef] [PubMed]

3. Mceneff, G.; Barron, L.; Kelleher, B.; Paull, B.; Quinn, B. A year-long study of the spatial occurrence and relative distribution of pharmaceutical residues in sewage effluent, receiving marine waters and marine bivalves. *Sci. Total Environ.* **2014**, *476–477*, 317–326. [CrossRef] [PubMed]
4. Kuehn, B.M. Antibiotic-Resistant "Superbugs" May Be Transmitted From Animals to Humans. *Med. News Perspect Futur.* **2007**, *298*, 2125–2126. [CrossRef] [PubMed]
5. Watkinson, A.J.; Murby, E.J.; Costanzo, S.D. Removal of antibiotics in conventional and advanced wastewater treatment: Implications for environmental discharge and wastewater recycling. *Water Res.* **2007**, *41*, 4164–4176. [CrossRef] [PubMed]
6. EMA. Sales of Veterinary Antimicrobial Agents in 31 European Countries in 2017 Trends from 2010 to 2017. Ema/294674/2019. 2019. Available online: https://www.ema.europa.eu/en/documents/report/sales-veterinary-antimicrobial-agents-31-european-countries-2017_en.pdf (accessed on 18 February 2020).
7. García-Galán, M.J.; Garrido, T.; Fraile, J.; Ginebreda, A.; Díaz-Cruz, M.S.; Barceló, D. Simultaneous occurrence of nitrates and sulfonamide antibiotics in two ground water bodies of Catalonia (Spain). *J. Hydrol.* **2010**, *383*, 93–101. [CrossRef]
8. Jurado, A.; Walther, M.; Díaz-Cruz, M.S. Occurrence, fate and environmental risk assessment of the organic microcontaminants included in the Watch Lists set by EU Decisions 2015/495 and 2018/840 in the groundwater of Spain. *Sci. Total Environ.* **2019**, *663*, 285–296. [CrossRef]
9. Boy-Roura, M.; Mas-Pla, J.; Petrovic, M.; Gros, M.; Soler, D.; Brusi, D.; Menció, A. Towards the understanding of antibiotic occurrence and transport in groundwater: Findings from the Baix Fluvià alluvial aquifer (NE Catalonia, Spain). *Sci. Total Environ.* **2018**, *612*, 1387–1406. [CrossRef]
10. García-Gil, A.; Garrido Schneider, E.; Mejías, M.; Barceló, D.; Vázquez-Suñé, E.; Díaz-Cruz, S. Occurrence of pharmaceuticals and personal care products in the urban aquifer of Zaragoza (Spain) and its relationship with intensive shallow geothermal energy exploitation. *J. Hydrol.* **2018**, *566*, 629–642. [CrossRef]
11. Senta, I.; Terzic, S.; Ahel, M. Occurrence and fate of dissolved and particulate antimicrobials in municipal wastewater treatment. *Water Res.* **2013**, *47*, 705–714. [CrossRef]
12. Babić, S.; Ašperger, D.; Mutavdžić, D.; Horvat, A.J.M.; Kaštelan-Macan, M. Solid phase extraction and HPLC determination of veterinary pharmaceuticals in wastewater. *Talanta* **2006**, *70*, 732–738. [CrossRef] [PubMed]
13. Golovko, O.; Kumar, V.; Fedorova, G.; Randak, T.; Grabic, R. Seasonal changes in antibiotics, antidepressants/psychiatric drugs, antihistamines and lipid regulators in a wastewater treatment plant. *Chemosphere* **2014**, *111*, 418–426. [CrossRef] [PubMed]
14. Al Aukidy, M.; Verlicchi, P.; Jelic, A.; Petrovic, M.; Barcelò, D. Monitoring release of pharmaceutical compounds: Occurrence and environmental risk assessment of two WWTP effluents and their receiving bodies in the Po Valley, Italy. *Sci. Total Environ.* **2012**, *438*, 15–25. [CrossRef] [PubMed]
15. Tuc Dinh, Q.; Alliot, F.; Moreau-Guigon, E.; Eurin, J.; Chevreuil, M.; Labadie, P. Measurement of trace levels of antibiotics in river water using on-line enrichment and triple-quadrupole LC-MS/MS. *Talanta* **2011**, *85*, 1238–1245. [CrossRef]
16. Rossmann, J.; Schubert, S.; Gurke, R.; Oertel, R.; Kirch, W. Simultaneous determination of most prescribed antibiotics in multiple urban wastewater by SPE-LC-MS/MS. *J. Chromatogr. B Anal. Technol. Biomed. Life Sci.* **2014**, *969*, 162–170. [CrossRef]
17. Tamtam, F.; Mercier, F.; Le Bot, B.; Eurin, J.; Dinh, Q.T.; Clément, M.; Chevreuil, M. Occurrence and fate of antibiotics in the Seine River in various hydrological conditions. *Sci. Total Environ.* **2008**, *393*, 84–95. [CrossRef]
18. Wagil, M.; Kumirska, J.; Stolte, S.; Puckowski, A.; Maszkowska, J.; Stepnowski, P.; Białk-Bielińska, A. Development of sensitive and reliable LC-MS/MS methods for the determination of three fluoroquinolones in water and fish tissue samples and preliminary environmental risk assessment of their presence in two rivers in northern Poland. *Sci. Total Environ.* **2014**, *493*, 1006–1013. [CrossRef]
19. Dan, A.; Zhang, X.; Dai, Y.; Chen, C.; Yang, Y. Occurrence and removal of quinolone, tetracycline, and macrolide antibiotics from urban wastewater in constructed wetlands. *J. Clean Prod.* **2020**, *252*, 119677. [CrossRef]
20. Samir, A.; Abdel-Moein, K.A.; Zaher, H.M. Emergence of penicillin-macrolide-resistant Streptococcus pyogenes among pet animals: An ongoing public health threat. *Comp. Immunol. Microbiol. Infect. Dis.* **2020**, *68*, 101390. [CrossRef]

21. Milaković, M.; Vestergaard, G.; González-Plaza, J.J.; Petrić, I.; Šimatović, A.; Senta, I.; Kublik, S.; Schloter, M.; Smalla, K.; Udiković-Kolić, N. Pollution from azithromycin-manufacturing promotes macrolide-resistance gene propagation and induces spatial and seasonal bacterial community shifts in receiving river sediments. *Environ. Int.* **2019**, *123*, 501–511. [CrossRef]
22. Yao, N.; Li, C.; Yu, J.; Xu, Q.; Wei, S.; Tian, Z.; Yang, Z.; Yang, W.; Shen, J. Insight into adsorption of combined antibiotic-heavy metal contaminants on graphene oxide in water. *Sep. Purif. Technol.* **2019**, *236*, 116278. [CrossRef]
23. Pan, S.F.; Zhu, M.P.; Chen, J.P.; Yuan, Z.H.; Zhong, L.B.; Zheng, Y.M. Separation of tetracycline from wastewater using forward osmosis process with thin film composite membrane—Implications for antibiotics recovery. *Sep. Purif. Technol.* **2015**, *153*, 76–83. [CrossRef]
24. Moles, S.; Valero, P.; Escuadra, S.; Mosteo, R.; Gómez, J.; Ormad, M.P. Performance comparison of commercial TiO2: Separation and reuse for bacterial photo-inactivation and emerging pollutants photo-degradation. *Environ. Sci. Pollut. Res.* **2020**, 1–15. [CrossRef] [PubMed]
25. van Grieken, R.; Marugán, J.; Pablos, C.; Furones, L.; López, A. Comparison between the photocatalytic inactivation of Gram-positive E. faecalis and Gram-negative E. coli faecal contamination indicator microorganisms. *Appl. Catal. B Environ.* **2010**, *100*, 212–220. [CrossRef]
26. Gumy, D.; Rincon, A.G.; Hajdu, R.; Pulgarin, C. Solar photocatalysis for detoxification and disinfection of water: Different types of suspended and fixed TiO2 catalysts study. *Sol. Energy* **2006**, *80*, 1376–1381. [CrossRef]
27. Malato, S.; Fernández-Ibáñez, P.; Maldonado, M.I.; Blanco, J.; Gernjak, W. Decontamination and disinfection of water by solar photocatalysis: Recent overview and trends. *Catal. Today* **2009**, *147*, 1–59. [CrossRef]
28. Bernabeu, A.; Vercher, R.F.; Santos-Juanes, L.; Simón, P.J.; Lardín, C.; Martínez, M.A.; Vicente, J.A.; González, R.; Llosá, C.; Arques, A.; et al. Solar photocatalysis as a tertiary treatment to remove emerging pollutants from wastewater treatment plant effluents. *Catal. Today* **2011**, *161*, 235–240. [CrossRef]
29. Malesic-Eleftheriadou, N.; Evgenidou, E.; Kyzas, G.Z.; Bikiaris, D.N.; Lambropoulou, D.A. Removal of antibiotics in aqueous media by using new synthesized bio-based poly(ethylene terephthalate)-TiO2 photocatalysts. *Chemosphere* **2019**, *234*, 746–755. [CrossRef]
30. Biancullo, F.; Moreira, N.F.; Ribeiro, A.R.; Manaia, C.M.; Faria, J.L.; Nunes, O.C.; Castro-Silva, S.M.; Silva, A.M. Heterogeneous photocatalysis using UVA-LEDs for the removal of antibiotics and antibiotic resistant bacteria from urban wastewater treatment plant effluents. *Chem. Eng. J.* **2019**, *36*, 304–313. [CrossRef]
31. Cai, Q.; Hu, J. Decomposition of sulfamethoxazole and trimethoprim by continuous UVA/LED/TiO2 photocatalysis: Decomposition pathways, residual antibacterial activity and toxicity. *J. Hazard. Mater.* **2017**, *323*, 527–536. [CrossRef]
32. Englert, B. *Method 1694: Pharmaceuticals and Personal Care Products in Water, Soil, Sediment, and Biosolids by HPLC/MS/MS*; US Environmental Protection Agency (EPA): Washington, DC, USA, 2007.
33. Mirzaei, R.; Yunesian, M.; Nasseri, S.; Gholami, M.; Jalilzadeh, E.; Shoeibi, S.; Bidshahi, H.S.; Mesdaghinia, A. An optimized SPE-LC-MS/MS method for antibiotics residue analysis in ground, surface and treated water samples by response surface methodology- central composite design. *J. Environ. Health Sci. Eng.* **2017**, *15*, 1–16. [CrossRef] [PubMed]
34. Díaz-Bao, M.; Barreiro, R.; Miranda, J.M.; Cepeda, A.; Regal, P. Fast HPLC-MS/MS method for determining penicillin antibiotics in infant formulas using molecularly imprinted solid-phase extraction. *J. Anal. Methods Chem.* **2015**, *2015*, 959675. [CrossRef] [PubMed]
35. Gros, M.; Petrovié, M.; Barceló, D. Multi-residue analytical methods using LC-tandem MS for the determination of pharmaceuticals in environmental and wastewater samples: A review. *Anal. Bioanal. Chem.* **2006**, *386*, 941–952. [CrossRef]
36. European Centre for Disease Prevention and Control. *Antimicrobial Consumption. ECDC. Annual Epidemiological Report for 2017*; ECDC: Stockholm, Sweden, 2018.
37. Lei, K.; Zhu, Y.; Chen, W.; Pan, H.Y.; Cao, Y.X.; Zhang, X.; Guo, B.B.; Sweetman, A.; Lin, C.Y.; Ouyang, W.; et al. Spatial and seasonal variations of antibiotics in river waters in the Haihe River Catchment in China and ecotoxicological risk assessment. *Environ. Int.* **2019**, *130*, 104919. [CrossRef]
38. Lamm, A.; Gozlan, I.; Rotstein, A.; Avisar, D. Detection of amoxicillin-diketopiperazine-2′, 5′ in wastewater samples. *J. Environ. Sci. Health Part A Toxic/Hazardous Subst Environ. Eng.* **2009**, *44*, 1512–1517. [CrossRef]

39. Gozlan, I.; Rotstein, A.; Avisar, D. Amoxicillin-degradation products formed under controlled environmental conditions: Identification and determination in the aquatic environment. *Chemosphere* **2013**, *91*, 985–992. [CrossRef]
40. Li, L.; Guo, C.; Ai, L.; Dou, C.; Wang, G.; Sun, H. Research on degradation of penicillins in milk by β-lactamase using ultra-performance liquid chromatography coupled with time-of-flight mass spectrometry. *J. Dairy Sci.* **2014**, *97*, 4052–4061. [CrossRef]
41. Xia, S.; Jia, R.; Feng, F.; Xie, K.; Li, H.; Jing, D.; Xu, X. Effect of solids retention time on antibiotics removal performance and microbial communities in an A/O-MBR process. *Bioresour. Technol.* **2012**, *106*, 36–43. [CrossRef]
42. Abegglen, C.; Joss, A.; McArdell, C.S.; Fink, G.; Schlüsener, M.P.; Ternes, T.A.; Siegrist, H. The fate of selected micropollutants in a single-house MBR. *Water Res.* **2009**, *43*, 2036–2046. [CrossRef]
43. Tran, N.H.; Chen, H.; Reinhard, M.; Mao, F.; Gin, K.Y.H. Occurrence and removal of multiple classes of antibiotics and antimicrobial agents in biological wastewater treatment processes. *Water Res.* **2016**, *104*, 461–472. [CrossRef]
44. Thompson, A.; Griffin, P.; Stuetz, R.; Cartmell, E. The Fate and Removal of Triclosan during Wastewater Treatment. *Water Environ Res.* **2005**, *77*, 63–67. [CrossRef] [PubMed]
45. Drewes, J.E.; Heberer, T.; Reddersen, K. Fate of pharmaceuticals during indirect potable reuse. *Water Sci. Technol.* **2002**, *46*, 73–80. [CrossRef] [PubMed]
46. Kasprzyk-Hordern, B.; Dinsdale, R.M.; Guwy, A.J. The removal of pharmaceuticals, personal care products, endocrine disruptors and illicit drugs during wastewater treatment and its impact on the quality of receiving waters. *Water Res.* **2009**, *43*, 363–380. [CrossRef] [PubMed]
47. Awfa, D.; Ateia, M.; Fujii, M.; Johnson, M.S.; Yoshimura, C. Photodegradation of pharmaceuticals and personal care products in water treatment using carbonaceous-TiO2 composites: A critical review of recent literature. *Water Res.* **2018**, *142*, 26–45. [CrossRef] [PubMed]
48. Conde-Cid, M.; Álvarez-Esmorís, C.; Paradelo-Núñez, R.; Nóvoa-Muñoz, J.C.; Arias-Estévez, M.; Álvarez-Rodríguez, E.; Fernández-Sanjurjo, M.J.; Núñez-Delgado, A. Occurrence of tetracyclines and sulfonamides in manures, agricultural soils and crops from different areas in Galicia (NW Spain). *J. Clean Prod.* **2018**, *197*, 491–500. [CrossRef]

Article

Evaluation of TiO_2 and SnO Supported on Graphene Oxide (TiO_2-GO and SnO-GO) Photocatalysts for Treatment of Hospital Wastewater

Laura Rosero Parra [1], Lizeth Guerrero Pantoja [1], Natali Lorena Mena [2], Fiderman Machuca-Martínez [1] and Julian Urresta [2,*]

1 Escuela de Ingeniería Química, Universidad del Valle, Cali 760032, Colombia; laura.rosero@correounivalle.edu.co (L.R.P.); lizeth.jazmin.guerrero@correounivalle.edu.co (L.G.P.); fiderman.machuca@correounivalle.edu.co (F.M.-M.)

2 Department of Chemistry, Universidad del Valle, Cali 760032, Colombia; natali.mena@correounivalle.edu.co

* Correspondence: julian.urresta@correounivalle.edu.co; Tel.: +57-3173320433

Received: 7 February 2020; Accepted: 7 April 2020; Published: 19 May 2020

Abstract: The effectiveness of two photocatalysts, TiO_2 and SnO, supported on graphene oxide (TiO_2-GO and SnO-GO) on the removal of organic matter from hospital wastewater effluent was evaluated at laboratory scale. The results of the experimental design allow us to conclude that variables such as catalyst type and catalyst concentration have a significant effect on the organic matter removal efficiency of the photocatalytic process. The highest levels of removal efficiencies—for chemical oxygen demand, 85%, for phenols, 80%, and for dissolved organic carbon, 94%—were achieved using a TiO_2-GO catalyst with a concentration in the wastewater of 1.5 g/L.

Keywords: heterogeneous photocatalysis; titanium dioxide; tin oxide; graphene oxide; chemical oxygen demand

1. Introduction

Among emerging contaminants, pharmaceuticals are one of the major concerns, due mainly to their toxicity, low biodegradability and extensive use [1]. In fact, hospital wastewater is an important source of pollution, not only due to pharmaceuticals and antibiotics not absorbed by the body, but mainly due to the metabolites excreted by patients. In Colombia, most of the hospitals do not have wastewater treatment systems and their effluents are directly discharged into the sewer system, then into the rivers. Drinking water treatment plants are not designed to completely remove substances from hospitals; therefore, direct consumption of tap water could arise as a major public health concern [2].

In Colombia, before 2015, the disposal of hospital wastewater was not regulated by environmental authorities, creating a serious problem of public health and sanitary risk [3]. Nowadays, local environmental authorities, such as DAGMA (Santiago de Cali Administrative Department of Environmental Management) and local public service provider companies such as EMCALI have the power of imposing sanctions in case of non-fulfillment of environmental regulations [4]. For example, between 2017 and 2018 in Santiago de Cali (Colombia), many rehabilitation centers, institutes of radiology, skin centers and clinics have been financially penalized for not complying with wastewater discharge parameters.

Advanced oxidation processes (AOPs) are technically feasible alternatives for the oxidation of the compounds found in hospital wastewater. These methods are based on physicochemical processes which are capable of modifying the chemical structure of the pollutants using highly reactive transient species such as the hydroxyl radicals [5]. Among the AOPs, the photocatalytic process is one of the most studied due to its high efficiency and low implementation costs at commercial level; furthermore, the catalysts can be designed and tuned to absorb energy in certain regions of the visible spectrum, enhancing the process according to the local environmental conditions.

Hospital wastewater contains a complex mixture of active pharmaceutical ingredients and microorganisms. Often, this wastewater is discharged to municipal wastewater treatment plants (WWTPs) without any pre-treatment. The municipal WWTPs are not designed to remove persistent pharmaceuticals. In addition, hazardous wastewater may spread during flooding and combined sewer overflow events. Internationally, there is increasing focus on the potential environmental effects of pharmaceuticals in water environments. Hospitals have been identified as a key source of pharmaceuticals that can act as potent micropollutants. Painkillers such as diclofenac and hormones, for example, can have fatal effects on fish, crustaceans and algae at very low doses. Nowadays, there are now new technologies for treatment of hospital wastewater; conventional methods have been coupled with advanced oxidation processes to obtain high reductions in organic matter content.

For instance, oil and grease traps, sedimentation, homogenization tanks, filtration beds, Fenton processes and the implementation of new photocatalysts are the technologies used for treatment of hospital wastewater with a high content of drug compounds. The study of tetracycline content in hospital wastewater has been important: the use of boron-doped titanium catalysts which absorb visible and UV light enhances the reduction of organic matter content. On the other hand, processes which include upflow anaerobic sludge blankets, anaerobic filters, aerobic processes, and activated sludge extended aeration with final chlorination are only accepted as suitable and efficient technologies for cities with less than 10,000 inhabitants.

In 2019, Monte and collaborators evaluated a hospital wastewater treatment system consisting of a membrane-coupled bioreactor with advanced oxidation, where they showed that the biological phase of the treatment required approximately 20 h of retention, while the advanced oxidation phase took only 5 h, which ensures that processes involving radical hydroxyls as oxidizing agents decrease the production of sludge and involve short residence times. In advanced oxidation processes, groups such as double-bond C-C, activated aromatic groups and non-protonated amines and those antibiotics with higher electronic density have greater affinity with the ozonation process; that is, they will be oxidized more efficiently and quickly [6].

In the same year, Hoang and collaborators, considering the need to implement efficient processes of the reduction of organic and inorganic load, coupled a traditional process of filtration with gravel and sand followed by a biological process of enzymes and plants like *Scirpus validus*, in order to reduce the organic load attributed mainly by acetaminophen. However, it took 15 days to achieve decreases of more than 60%, which shows that traditional processes are not recommended for hospital-type wastewater treatment. At the end of the study, the authors state that, depending on the antibiotic, the biological environment can be used and, if it can be implemented, the minimum retention time should be 40 days. These conclusions lead to the consideration that advanced oxidation processes should prevail over biological processes for the hospital waters, because the flows managed are greater than 5 L/s [7].

In Colombia, there is little research associated with hospital wastewater treatment. In this field, we can mention the preliminary treatment carried out at the Quindío University Departmental Hospital called a fat and oil trap, whose function is to retain suspended solids by sedimentation and by flotation. The fatty material was subjected to a flow homogenization process, ending with a coagulation—flocculation process, which takes care of the removal and reduction of BOD, DOC and suspended solids. The coupling of these three technologies has an efficiency of 59% in the reduction of percentages of organic load in times of residence longer than 24 h, which evidences that it is not feasible for flows greater than 0.2 L/s [8].

In general, most of the recalcitrant compounds found in hospital sewage are difficult to mineralize, and conventional processes do not guarantee the reduction of organic and phenolic-type loads that are present and are continuously monitored by Colombian standards.

Previous research [9] showed that a combination of different semiconductor materials supported on graphene moves the absorption spectra to the visible region, enhancing photocatalytic activity. Characteristics and applications of titanium dioxide (TiO_2) on carbon-supported materials such as activated carbon, fullerene, graphene, carbon nanotubes, etc., have been studied extensively [10]. However, there is no published research on the use of romarchite phase tin oxide (SnO) supported on graphene for photocatalytic processes. The use of SnO as a catalyst seems to be interesting due its physical properties: high surface area, structure, colour and size; SnO could also exhibit high photocatalytic activity due its properties as a semiconductor [11]. In this work, we evaluated the efficiency of two photocatalysts (TiO_2 and SnO) supported on graphene for the treatment of wastewater from a hospital centre in Cali-Colombia. The effect of the catalyst type and catalyst concentration on the organic matter removal efficiency was studied.

The results of this study are a proof of concept based on punctual sampling of hospital wastewater from the Valle de Lili Foundation, which aims to demonstrate the possible interactions of variables associated with the concentrations of recalcitrant contaminants and synthesized catalysts. These results will allow us to analyse and evaluate the oxidative process of this type of wastewater using advanced oxidation processes.

2. Materials and Methods

2.1. Materials and Reagents

Graphite (99 wt%), sodium nitrate (>99 wt%), sulfuric acid (95–97 wt%), glacial acetic acid (>99 wt%), hydrochloric acid (37 wt%), potassium permanganate (99 wt%), hydrogen peroxide (30 wt%), ammonium hydroxide (28 wt%) and ethanol (>99.9 wt%) were purchased from Merck (Darmstadt, Germany) and were used without further purification. Stannous chloride (>99 wt%) was obtained from Panreac (Barcelona, Spain) and titanium tetrachloride (>99 wt%) from Fisher (Pittsburgh, PA). Industrial grade nitrogen was obtained from Cryogas (Cali, Colombia). Milli-Q type water was obtained after two successive steps of filtration, followed by deionization and distillation.

2.2. Synthesis of Graphene Oxide (GO)

In a typical procedure, 0.5 g of graphite powder and 0.5 g of sodium nitrate were mixed in a beaker, then 23 mL of sulphuric acid (H_2SO_4) was added to the beaker, which was previously immersed in a water bath. A quantity of 3 g of potassium permanganate ($KMnO_4$) was added slowly into the solution and dissolved using 20 min of low-intensity (70 W, 60 Hz) indirect sonication. Next, 100 mL of water and 3 mL of hydrogen peroxide were added into the solution, and finally 40 mL of water was added [12].

The solution was centrifuged using an Ortoalresa centrifuge Digitor 20 C at 1900 rpm for 5 min; subsequently, the supernatant was decanted away, and the residual was washed with a 10% hydrochloric acid solution. This mixture was centrifuged, decanted and washed again. The washed solution was dried using a forced convection oven at 80 °C for 48 h to obtain the graphene oxide powder.

2.3. Synthesis of Romarchite Phase SnO and GO Impregnation

Firstly, an aqueous solution of GO was prepared by dissolving 2.1 g of GO in 210 mL of water. This solution was then mixed with the tin precursor solution, which was prepared by mixing 6.67 g of $SnCl_2{\cdot}2H_2O$ in 100 mL of 20 M acetic acid solution at 80 °C. Then, five drops of HCl fuming were added into the solution, and it was left to stand for 24 h.

The solution was quantified by titration with ammonium hydroxide (28%) up to pH 8.0 and was allowed to settle for 24 h at ambient temperature. Then, the solution was dried using a forced convection oven at 100 °C for 3 h. Finally, the dry solid was calcined using a continuous flow of N_2 in an oven with a temperature program as follows: initially, 40 °C for 1 min; then, a first ramp up to

100 °C at a rate of 4 °C/min; 100 °C was then maintained over 25 min; then, a second ramp up to 150 °C at a rate of 1 °C/min; and, finally, 150 °C was maintained over 2.5 h [13].

2.4. Synthesis of TiO_2 and GO Impregnation

A solution was prepared by dissolving 0.0905 g of titanium tetrachloride $TiCl_4$ into 20 mL of anhydrous methanol, then 3 mL of HCl fuming and 20 mL of water were added to the solution. A GO solution prepared by adding 0.1 g of GO in 10 mL of water was added dropwise to the titanium oxide solution by the wet impregnation method [13]. The resulting solution was introduced into a forced convection oven at 100 °C for three hours to obtain a dry powder. Finally, the dry solid was calcined using a continuous flow of N_2 in an oven with a temperature program as follows: initially, 40 °C for 1 min; then, a first ramp up to 100 °C at a rate of 4 °C/min; 100 °C was maintained over 1 h; then, the same rate was used to ramp up to 200 °C; this temperature was maintained for 20 min; then, a third ramp up to 400 °C at a rate of 4 °C/min; and, finally, 400 °C was maintained over 1 h [13].

2.5. Characterization

2.5.1. Photocatalysts

A FT/IR spectrometer (JASCO FT/IR-4100, USA) equipped with an ATR cell was used to characterize the functional groups of the support and the photocatalysts. Diffuse reflectance measurements were carried out with a spectrometer (Ocean Optics, USB 4000, Orlando, FL, USA) using a broadband halogen fiber optic illuminator as a light source (Nikon Inc., NI-30, Orlando, FL, USA). A scanning electron microscope (SEM JEOL, JSM 6490 LV, North Billerica, MA, USA) was used to determine the shape and size distribution of crystalline particles. For the analysis, each sample was coated with gold (Denton Vacum, Desk, North Billerica, MA, USA) and then analyzed at the microscope in a back-scattered mode with an acceleration voltage of 20 kV. Micrographs were taken at 500×.

2.5.2. Samples Analysis

Dissolved organic carbon measurements were carried out in a TOC analyser (Shimadzu TOC, VCPH, Japan) according to Standard Method 5310B. Chemical oxygen demand determinations were made following the Standard Method 5220B, samples were digested for 2 h under closed-reflux at 150 °C, and then a spectrophotometer (Shimadzu, UV 1800, Switzerland) at 610 nm was used to measure the light absorbed by the sample. Phenolic compound measurements were carried out using the Hanna Instruments HI3864 kit.

2.6. Experimental Procedure

Hospital wastewater for the photocatalytic experiments was obtained from a hospital of Cali, Colombia and was prepared by mixing the samples collected from different sewer pipes: oncology department, clinical laboratory, hospitalization rooms, and laundry. Samples were stored in glass recipients and refrigerated at 4 °C. Additionally, for adequate preservation and for COD and phenol characterization, samples were mixed with sulphuric acid. Prior to the experiments, each sample was well stirred to ensure homogeneity of the aliquots.

For this research, hospital wastewater samples were collected specifically to assess the proof of concept. In subsequent stages of this preliminary research, compound sampling periods will be established. The inspection box detected for the sampling of the hospital wastewater involves all the discharges associated with the clinical laboratory, hospitalization, deliveries, oncology and surgery. The wastewater at this sampling point includes fats and oils from coffee shops and laundries. Wastewater samples were taken once monthly during three months in the afternoon.

Quantities of 0.25, 0.75 and 1.5 g of each catalyst were added to 1000 mL volumetric flasks and diluted to the mark with the hospital wastewater. Then, aliquots of 20 mL were exposed to a simulated solar radiation level of 250 W/m^2 for 1 h and under constant stirring (180 rpm).

Photocatalytic experiments were carried out using solar simulator equipment (Atlas, Suntest CPS+, Linsengericht, Germany). After each experiment, the sample was analysed according to Section 2.5.1.

3. Results and Discussion

3.1. Characterization of GO, SnO-GO and TiO_2-GO by Means of FT/IR and SEM

Figure 1 shows the FT/IR spectrum of the GO. The tension peak of the hydroxyl group is seen at 3225 cm^{-1} (peak 1), which represents the water absorbed by the graphene oxide. Peaks 2 and 3 at 1713 cm^{-1} and 1621 cm^{-1}, can be assigned to carbonyl bonds (C=O) and double bonds C=C, respectively. The absorption peak 4 at 1041 cm^{-1} is assigned to the C–O stretching vibrations [14]. The presence of C=O and C–O bonds are evidence of the oxidation of graphite [12].

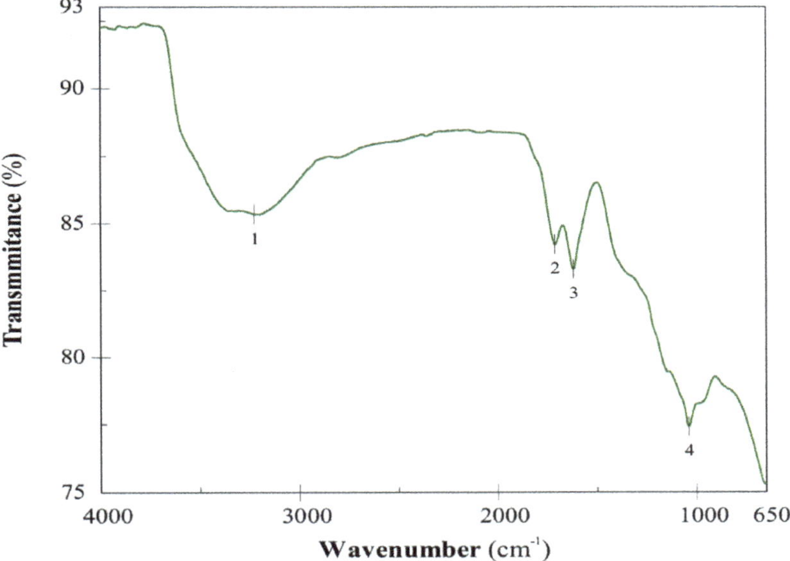

Figure 1. FT/IR spectrum of the graphene oxide (GO).

The characteristic FT/IR spectrum of a SnO-GO photocatalyst is depicted in Figure 2. The absorption peak 1 at 3124 cm^{-1} can be assigned to the water absorbed, which cannot be completely removed during the synthesis at 150 °C. Another characteristic peak (2) appears at 1567 cm^{-1} and can be assigned to C=O stretching of carboxylic and/or carbonyl moiety functional groups. The C–H deformation peak (3) can be noticed at 1398 cm^{-1}. The absorption peak (4) at about 1206 cm^{-1} is assigned to the C–O stretching vibrations. Finally, near to the inorganic region of the spectrum, the characteristic band of the Sn–O–H bond is seen, at 560.2 cm^{-1}. On the other hand, it is important to point out that peaks 2 and 3 can be also assigned to an overtone band of a second vibration of the O–Sn–O bond [15].

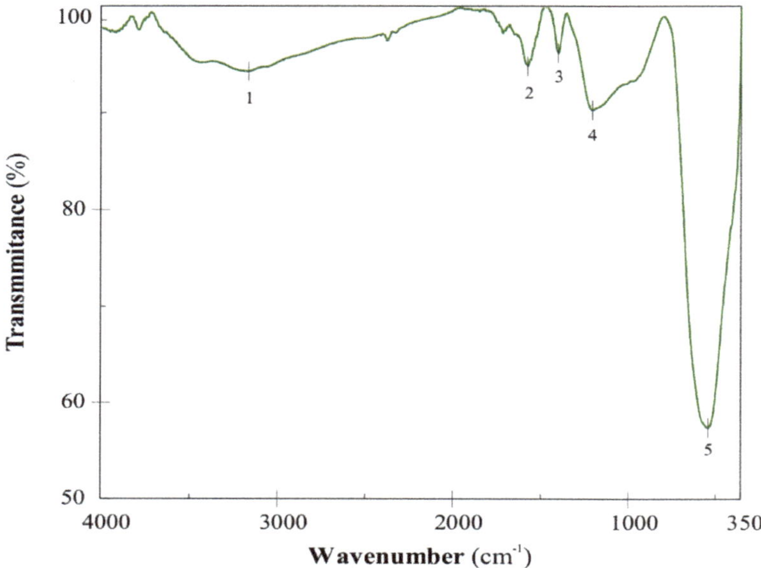

Figure 2. FT/IR spectrum of SnO-GO photocatalyst.

Figure 3 shows the FT/IR spectrum of the TiO_2-GO photocatalyst. Peaks 1 and 2 at 3740 cm^{-1} and 3363 cm^{-1} represent hydroxyl groups. Meanwhile, peaks 3 and 4 at 1707.6 cm^{-1} and 1565.9 cm^{-1} can be assigned to the carbonyl groups and the C=C bonds, respectively. Peaks at 1217.8 cm^{-1} and 1132.9 cm^{-1} appear due the presence of methyl groups in the graphene oxide. In the inorganic region of the spectrum at 541 cm^{-1} the characteristic peak of the Ti-O-Ti bond is seen [14].

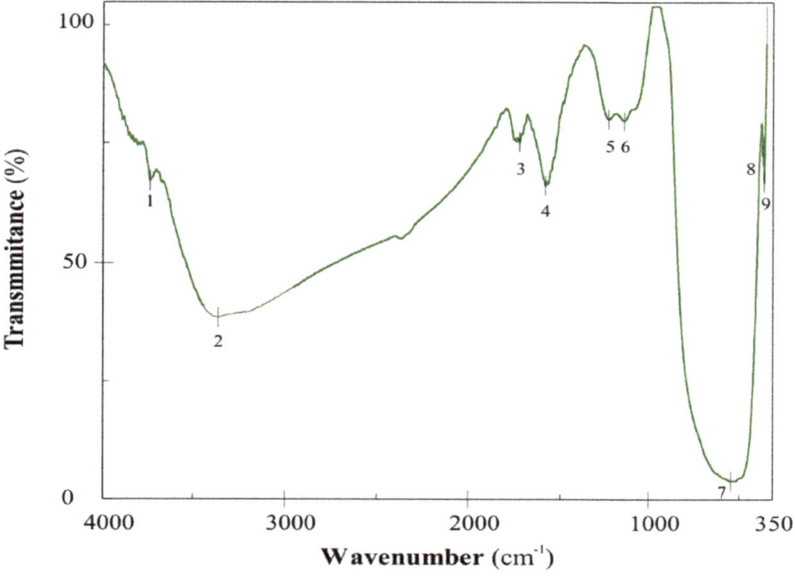

Figure 3. FT/IR spectrum of the TiO_2-GO photocatalyst.

As is mentioned above, peak 1 in the spectrum can be assigned to the hydroxyl groups. It can be noticed that if the drying or calcination temperature increases, the band is decreased, due mainly to the fact that water absorbed decreases as the temperature increases. Meanwhile, the increase of peak 3 intensity in Figure 3 can be explained due the TiO-H interaction, which is more marked due to the high calcination temperature, greater than 300 °C [15].

Comparing the spectrum of GO with the spectrums of SnO-GO and TiO_2-GO, it can be observed that the wavelengths are shifted to shorter wavelengths, towards the inorganic spectrum. This can be assumed to be due to titanium and tin interactions with oxygen and carbon, and the other compounds' interactions.

Figure 4 shows the SEM micrograph of the GO. The micrograph shows an irregular and wrinkled appearance and a close surface morphology, which can be attributed to the exfoliation of the GO layers during the ultrasonic process [16]. Figure 5 shows the SEM micrograph of the SnO-GO; it can be seen that the romarchite impregnated in the GO presents a thin micrometric cluster, approximating a spherical shape. The surface irregularity can be attributed to the acetic acid [17]. Finally, Figure 6 shows the SEM micrograph of the TiO_2-GO. Clusters of crystalline particles with irregular surface can be noticed [18].

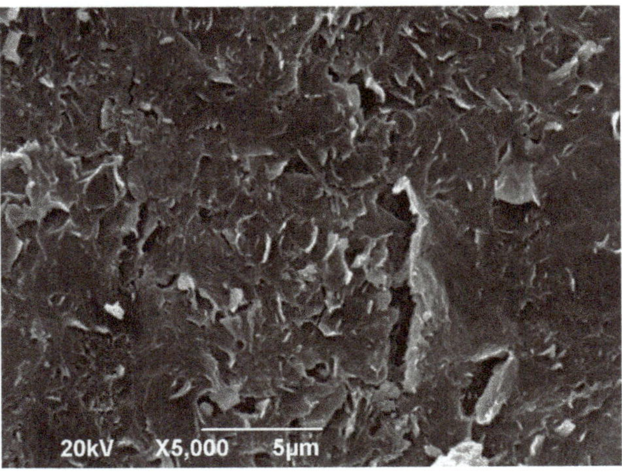

Figure 4. SEM micrograph of the GO.

Figure 5. SEM micrograph of the SnO-GO.

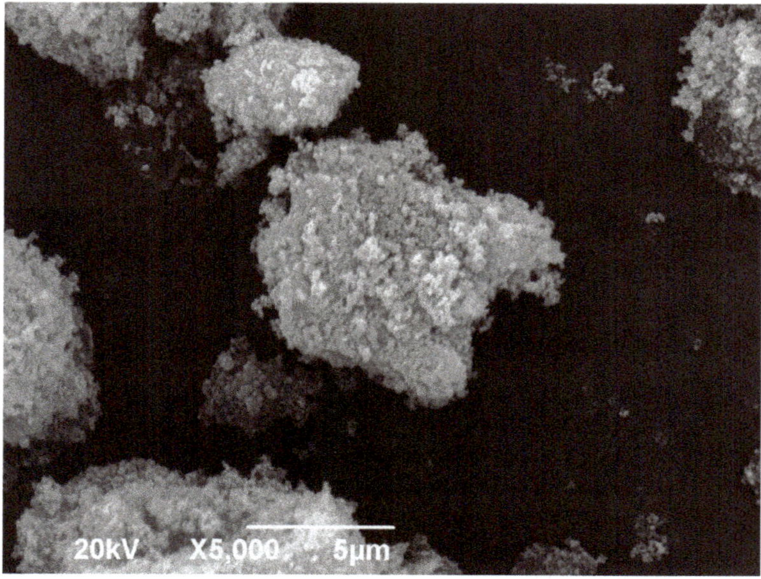

Figure 6. SEM micrograph of the TiO_2-GO.

SEM characterization shows the presence of chlorine in the SnO-GO catalyst (around 3.5% w/w) mainly due to the use of $SnCl_2 \cdot 2H_2O$ as the precursor.

Diffuse reflectance measurements allow us to conclude that all photocatalysts supported on graphene absorb around 80% of the electromagnetic radiation in the visible region (see Figures S1 and S2 Supplementary Information).

3.2. Evaluation of the Photocatalytic Process

Table 1 shows the raw hospital wastewater characterization parameters and the limit values applicable to discharge of wastewater into the public sewer in Colombia [7]. Characterization parameters were evaluated according to Section 2.5.1.

Table 1. Raw wastewater characterization parameters and limit values for discharge.

Parameter	Concentration (ppm)	Limit Value (ppm)
Chemical oxygen demand (COD)	650 ± 6	300
Phenols	5.0 ± 0.1	0.2
Dissolved organic carbon (DOC)	638.8 ± 0.4	N.D

In Colombia, unfortunately, environmental resolution 0631 of 2015 does not include dissolved organic carbon (DOC) measurement as a physicochemical control parameter; therefore, only the detection limit of the method used in the measurement is reported.

A GC-MS analysis of organic compounds in the hospital wastewater allows us to determine the functional groups of pollutants present in the wastewater, in order to know which of the photocatalysts has greater efficiency in the treatment of hospital waters. These were found: sulphonamides, phenols (3,5-dioctoxyphenol) and aromatic compounds. Moreover, we found several compounds with different functional groups, such as esters, ketones, aromatics and amines.

Figure 7 shows the removal efficiencies obtained after the photocatalytic process with Sn-GO and SnO catalysts. The highest organic matter removal efficiency with the Sn-GO photocatalyst was obtained using a concentration of 1.5 g/L; conversely, the lowest efficiency was obtained using a concentration of 0.25 g/L. Similar trends were obtained with the SnO catalyst. It is important to point out that phenol concentrations in the experiments in which we used 0.25 and 0.75 g/L of catalyst remained unaltered, which indicates that there was no phenol conversion. Moreover, a comparison between the catalysts shows that the highest efficiencies were achieved when the TiO_2-GO catalysts were used.

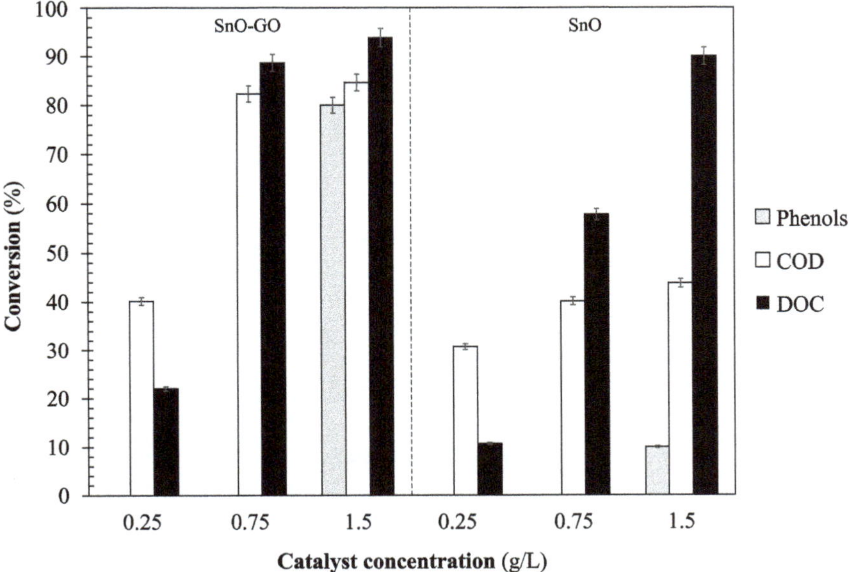

Figure 7. Removal efficiencies of chemical oxygen demand, phenols and dissolved organic matter after the photocatalytic process with SnO-GO and SnO.

Figure 8 shows the removal efficiencies obtained after the photocatalytic process with TiO_2-GO and TiO_2 catalyst. Similar trends are obtained comparing these results with the ones obtained with the SnO-GO catalyst; lower efficiencies were achieved at low catalyst concentrations. As can be noticed, at low concentrations of catalysts (0.25 g/L and 0.75 g/L), concentrations of phenols remain almost unaltered. At the highest catalyst concentrations (0.75 g/L) removal efficiencies of 80%, 85% and 94% for phenols, chemical oxygen demand and dissolved organic matter were obtained, respectively. The use of TiO_2-GO significantly increases the removal efficiencies compared with the TiO_2 catalyst.

The trends obtained suggest that the generation of hydroxyl radicals increases with increasing photocatalyst concentrations, which enhances the organic matter removal efficiency. However, if the concentration of catalyst is too high, it can produce a "shielding" effect or dimming effect, due to the turbidity created by the suspended catalyst.

As was mentioned above, low concentrations of chemical oxygen demand were obtained at the end of the photocatalytic process due to the degradation of organic matter. However, specific compounds such as phenols were not completely mineralized, achieving conversions lower than 91%.

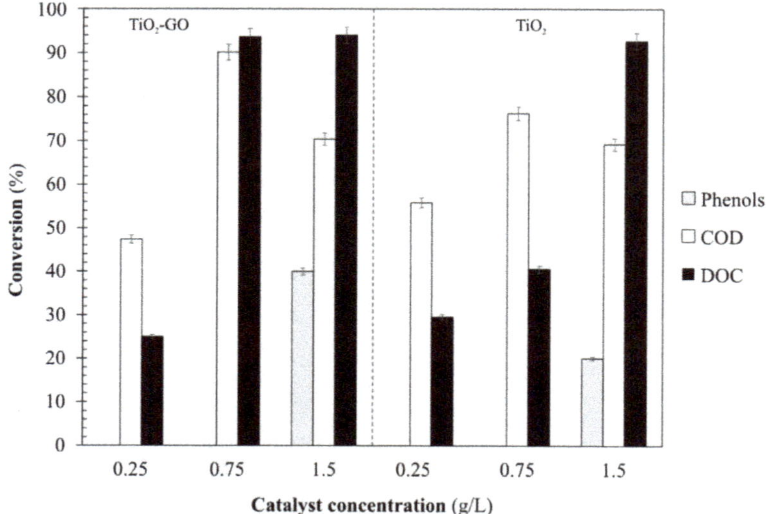

Figure 8. Removal efficiencies of chemical oxygen demand, phenols and dissolved organic matter after the photocatalytic process with TiO_2-GO and TiO_2.

pH of wastewater does not seem to affect the photocatalytic treatment; high removal efficiencies were obtained in a wide range of pH values. Nevertheless, some publications report that pH affects some properties of the catalysts, such as the particle size, the surface charge, and the maximum and minimum of TiO_2 band values due to its amphoteric nature [19].

The concentration of phenol in hospital wastewater is high. In this work, we report a phenol removal efficiency up to 80%, a higher value compared to the values obtained with other methods such as wet air oxidation [20] and adsorption [21], in which conversions of 14% and 10% were respectively reported. The high removal efficiencies obtained with the photocatalytic process can be assigned to the presence of chlorine, to the high amount of absorbed energy and to the properties of the GO-supported catalysts, which enhance the photocatalytic activity.

In advanced oxidation processes that involve hydroxyl radicals, phenol firstly oxidizes to hydroquinone and then to p-benzoquinone. Then, ring opening reactions take place to form maleic acid, which is the main product of the process. Maleic acid can be oxidized in malonic acid, oxalic acid and formic acid [22]. Malonic acid oxidizes to acetic acid; oxalic acid can be converted into formic acid and vice versa. Finally, most of the carboxylic acids can be directly oxidized to CO_2 and H_2O [22].

Figure 9 shows the removal efficiencies obtained in the photocatalytic process with SnO-GO and TiO_2-GO. As can be noticed, the highest organic matter removal efficiencies were obtained when the SnO-GO photocatalyst was used.

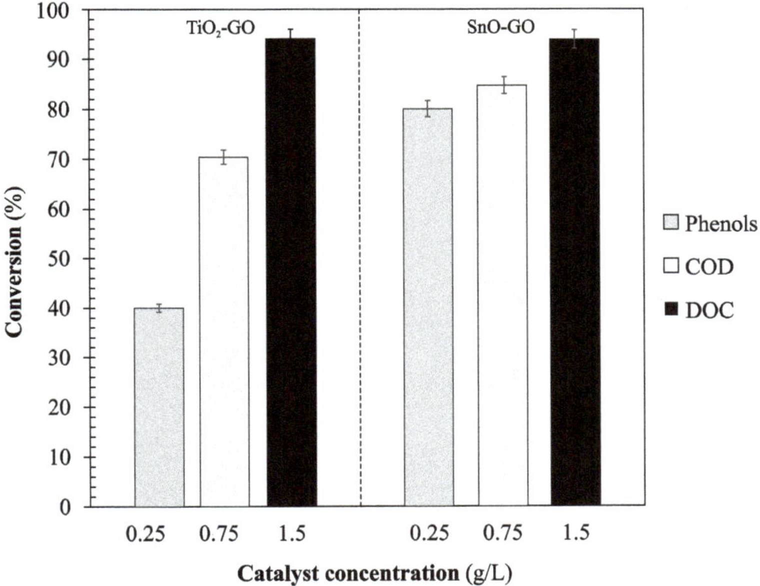

Figure 9. Removal efficiencies obtained in the photocatalytic process with SnO-GO and TiO_2-GO.

3.3. Statistical Analysis

Figure 10 shows the concentration of phenols in the effluent after the photocatalytic treatment with each one of the catalysts synthesized. As it can be seen, the pattern of concentration is identical for all catalysts; low removal efficiencies were achieved in all cases, except when TiO_2-GO with a concentration of 1.5 g/L was used.

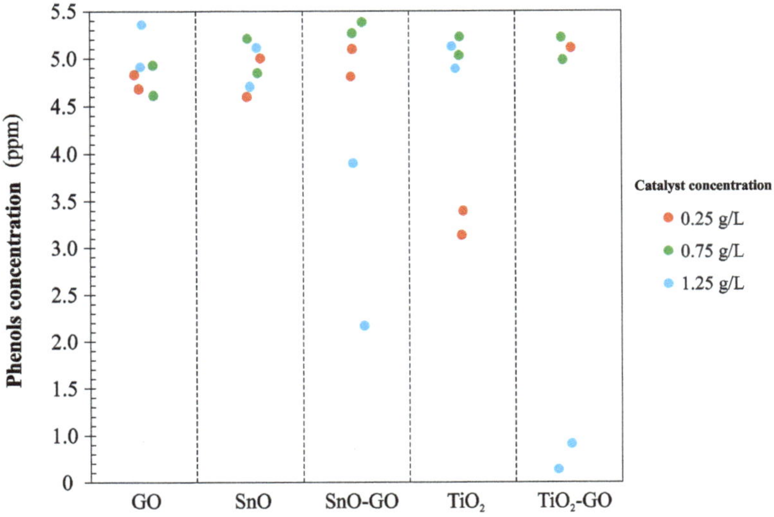

Figure 10. Concentration of phenols in the effluent after the photocatalytic treatment with each one of the catalysts synthesized.

Figure 11 shows the concentrations of COD in the effluent after the photocatalytic treatment with each one of the catalysts synthesized. As can be noticed, the concentrations of catalysts used in the treatment have different effects on the COD removal efficiency; for instance, the lowest COD concentration was achieved using the SnO-GO catalyst with a concentration of 0.75 g/L, while at the same concentration but using the SnO catalyst, the highest COD concentration was obtained.

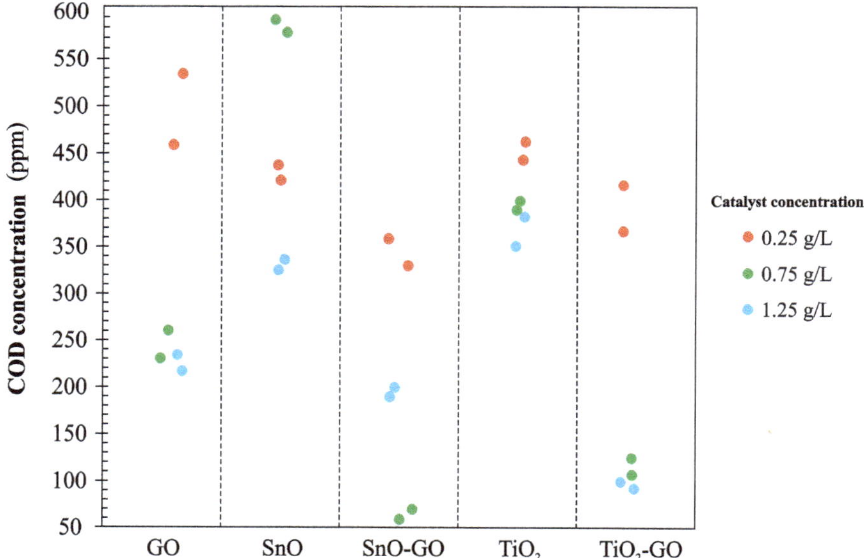

Figure 11. Concentration of COD in the effluent after photocatalytic treatment with each one of the catalysts synthesized.

Figure 12 shows the concentrations of DOC in the effluent after photocatalytic treatment with each one of the catalysts synthesized. In this case, the effect of catalyst concentration on the DOC removal efficiency is proportional; efficiency increases when catalyst concentration increases. The highest DOC removal efficiency was obtained using the SnO-GO catalyst.

A statistical analysis was carried out using the free distribution software R version 3.5.1 (R Development Core Team, 2018). Differences between the photocatalytic treatments at a significance level of 10% were determined. From the F-test, it can be concluded that the interaction (concentration level and catalyst type) has a significant effect on the COD and DOC removal efficiencies. Taking into account the descriptive analysis summarized in Figures 11 and 12 and the results from the F-test, we performed post-ANOVA Duncan tests. In these tests, we compared the interactions at which the highest COD removal efficiencies were obtained: SnO-GO with a concentration of 0.75 g/L and TiO_2-GO with concentrations of 0.75 and 1.5 g/L. We found statistically significant differences between the comparing groups; from this, and considering the phenol removal efficiencies, we can suggest that the TiO_2-GO photocatalyst with a concentration of 1.5 g/L can be used to obtain the highest COD, DOC and phenol removal efficiencies.

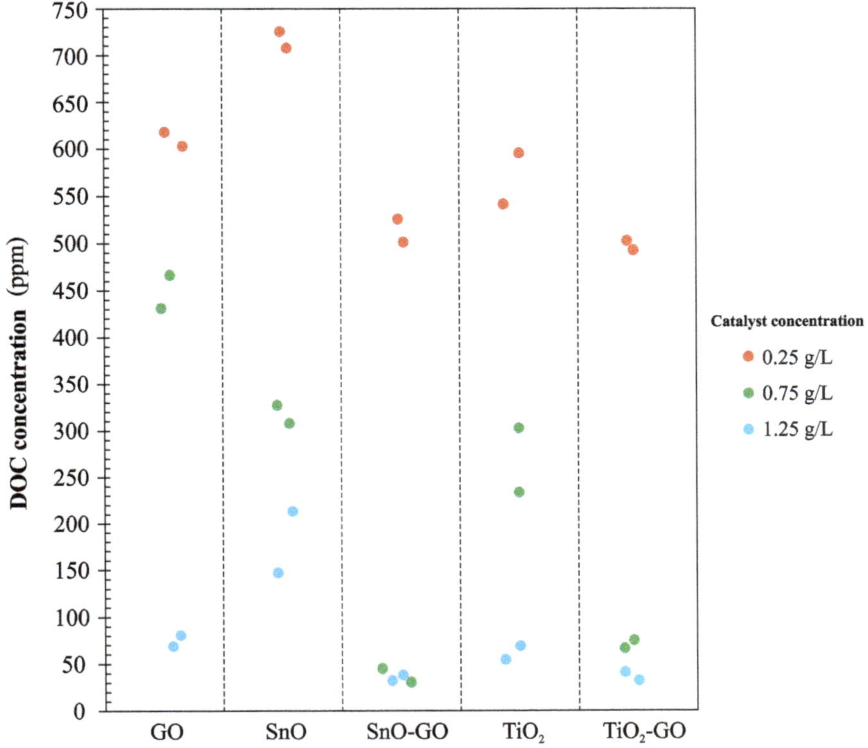

Figure 12. Concentration of DOC in the effluent after the photocatalytic treatment with each one of the catalysts synthesized.

4. Conclusions

In this work, we investigated the use of a heterogeneous photocatalytic process to treat hospital wastewater using two catalysts supported on graphene—SnO-GO and TiO_2-GO—and tested their performance by varying the amount of catalyst loaded in the wastewater. The highest removal efficiencies for chemical oxygen demand (85%), dissolved organic carbon (94%) and phenols (80%) were achieved using the TiO_2-GO catalyst with a concentration of 1.5 g/L. Moreover, we obtained an effluent which met the local environmental regulations in terms of chemical oxygen demand (COD < 200 ppm). However, further work would be needed (e.g., increased residence time, use of radical initiators, increased catalyst load, etc.) to achieve a concentration of phenols lower than 0.2 ppm (the discharge limit value). Findings obtained in this research indicate that the heterogeneous photocatalytic process is a viable alternative for degradation of the pollutants in hospital wastewater. The results obtained in this investigation are a proof of concept, based on punctual indicative sampling, and care must be taken before generalizing the results obtained. However, once this first phase is completed, it could be established that the heterogeneous photocatalytic process is a possible alternative for the degradation of pollutants present in hospital wastewater.

Supplementary Materials: The following are available online at http://www.mdpi.com/2073-4441/12/5/1438/s1, Figure S1. Diffuse reflectance spectra of TiO_2 and TiO_2-GO catalysts; Figure S2. Diffuse reflectance spectra of SnO and SnO-GO catalysts.

Author Contributions: L.R.P., and L.G.P. These authors performed the data acquisition and analysis and prepared all figures. N.L.M., F.M.-M. and J.U., These authors jointly supervised this work. L.R.P. and N.L.M. wrote the main manuscript text. All authors have read and agreed to the published version of the manuscript.

Funding: This research was funded by Universidad del Valle (Colombia), grant number 71030.

Acknowledgments: The authors wish to thank to Laboratory of Applied Catalysis and Catalytic Processes (LICAP), to Chemistry Department, and to Universidad del Valle for financial support (Grant 71030). Also, we wish to acknowledge the help provided by Fundación Valle de Lili (Cali, Colombia) for the wastewater samples and the logistic support.

Conflicts of Interest: The authors declare no conflict of interest.

References

1. Damià, L.; López, M.J. Contaminación y calidad química del agua: El problema de los contaminantes emergentes. *Panel Científico Técnico Seguim La Política Del Agua Jorn Present Result* **2008**, 1–27. [CrossRef]
2. García Gómez, C.; Gortáres Moroyoqui, P.; Drogui, P. Contaminantes emergentes: Efectos y tratamientos de remoción. *Rev. Química Viva* **2011**, *2*, 96–105.
3. Duarte, C.; Gutiérrez, F. Tratamiento de agua residual hospitalaria previamente ozonizada utilizando un reactor anaerobio de lecho fijo. 2013.
4. Ministerio del Medio Ambiente y Desarrollo Sostenible. Resolución 0631 De 2015. Colombia; 2015:73. Available online: https://www.minambiente.gov.co (accessed on 18 May 2020).
5. Gogate, P.R.; Pandit, A.B. A review of imperative technologies for wastewater treatment II: Hybrid methods. *Adv. Environ. Res.* **2004**, *8*, 553–597. [CrossRef]
6. Monte, C. Materiales Nanoestructurados de TiO_2-Grafeno-Macrociclos Tetrapirrólicos para Aplicación en Celdas Solares. Ph.D. Thesis, Universidad de Guadalajara, Guadalajara, Mexico, 2016.
7. Gil Suarez, J. Propuesta para el Sistema de Tratamiento de Aguas Residualesen la E.S.E Hospital Departamental Universitario del Quindío San Juan de Dios. Bachelor's Thesis, Fundación Universidad de América, Bogotá, Colombia, 2018.
8. Vo, T.K.; Bui, X.T.; Chen, S.S.; Nguyen, P.D.; Cao, N.D.; Vo, T.D.; Nguyen, T.T.; Nguyen, T.B. Hospital wastewater treatment by sponge membrane bioreactor coupled with ozonation process. *Chemosphere* **2019**, *230*, 377–383. [CrossRef] [PubMed]
9. Hoang, N.P.; Thammarat, K.; Saroj, K.C. Removal and monitoring acetaminophen-contaminated hospital wasterwater by vertical flow constructed wetland and peroxidase enzymes. *J. Environ. Manag.* **2019**, *250*. [CrossRef]
10. Fernández Velasco, L. Fotodegradación Oxidativa de Fenol con Catalizadores TiO_2 -C. Análisis de la Respuesta Fotoquímica de la fase Carbonosa. Ph.D. Thesis, Universidad de Oviedo, Asturias, Spain, 2012.
11. Urresta, J.D.; González, O.; Rodríguez, J.E. Preparación y caracterización del óxido de estaño y su uso como catalizador en la reacción de epoxidación de esteres grasos insaturados. *Ing. Compet.* **2014**, *16*, 49–59. [CrossRef]
12. Esmaeili, A.; Entezari, M.H. Facile and fast synthesis of graphene oxide nanosheets via bath ultrasonic irradiation. *J. Colloid Interface Sci.* **2014**, *432*, 19–25. [CrossRef] [PubMed]
13. Mena, N. Evaluación de Fotocatalizadores Semiconductores Soportados En Oxido de Grafeno Para Procesos Fotocatalíticos. In Proceedings of the Memorias del XXV CongresoIberoamericano de Catálisis. Ponencia, Montevideo, Uruguay, 18–23 September 2016.
14. Zuluaga, F.; Insuasty, B. *Analisis Orgánico Clásico Y Espectral*, 1st ed.; Universidad del Valle: Cali, Colombia, 2000.
15. López, E. Síntesis y Caracterización de Películas Delgadas de Dióxido de Estaño Dopadas con Hierro al 10%. Master's Thesis, Universidad Nacional de Colombia, Bogotá, Colombia, 2012.
16. Camposeco, R. Adsorción del CO en Nanoestructura de Titania. Ph.D. Thesis, Instituto Politécnico Nacional de México, Ciudad de México, México, 2008.
17. Won-Chun, O.; Feng-Jun, Z. Preparation and Characterization of Graphene Oxide Reduced From a Mild Chemical Method. *Asian J. Chem.* **2011**, *23*, 875–879.

18. Mahdjoub, N.; Allen, N.; Kelly, P.; Vishnyakov, V. SEM and Raman study of thermally treated TiO_2 anatase nanopowders: Influence of calcination on photocatalytic activity. *J. Photochem. Photobiol. A Chem.* **2010**, *211*, 59–64. [CrossRef]
19. Giraldo, L.; Franco, E.; Arango, J. La fotocatálisis como alternativa para el tratamiento de aguas residuales. *Rev. Lasallista Investig.* **2004**, *1*, 83–92. [CrossRef]
20. Yang, S.; Cui, Y.; Sun, Y.; Yang, H. Graphene oxide as an effective catalyst for wet air oxidation of phenol. *J. Hazard. Mater.* **2014**, *280*, 55–62. [CrossRef] [PubMed]
21. Wang, J.; Fu, W.; He, X.; Yang, S.; Zhu, W. Catalytic wet air oxidation of phenol with functionalized carbon materials as catalysts: Reaction mechanism and pathway. *J. Environ. Sci.* **2014**, *26*, 1741–1749. [CrossRef] [PubMed]
22. Santos, A.; Yustos, P.; Quintanilla, A.; García-Ochoa, F.; Casas, J.A.; Rodríguez, J.J. Evolution of Toxicity upon Wet Catalytic Oxidation of Phenol. *Environ. Sci. Technol.* **2004**, *38*, 133–138. [CrossRef] [PubMed]

Article

Insights into the Photocatalytic Bacterial Inactivation by Flower-Like Bi_2WO_6 under Solar or Visible Light, Through in Situ Monitoring and Determination of Reactive Oxygen Species (ROS)

Minoo Karbasi [1,2], Fathallah Karimzadeh [1,*], Keyvan Raeissi [1], Sami Rtimi [2], John Kiwi [2], Stefanos Giannakis [3,*] and Cesar Pulgarin [2]

1 Department of Materials Engineering, Isfahan University of Technology, Isfahan 84156-83111, Iran
2 School of Basic Sciences (SB), Institute of Chemical Science and Engineering (ISIC), Group of Advanced Oxidation Processes (GPAO), École Polytechnique Fédérale de Lausanne (EPFL), Station 6, CH-1015 Lausanne, Switzerland
3 Departamento de Ingeniería Civil: Hidráulica, Universidad Politécnica de Madrid (UPM), E.T.S. Ingenieros de Caminos, Canales y Puertos, Energía y Medio Ambiente, Unidad docente Ingeniería Sanitaria, c/ Profesor Aranguren, s/n, ES-28040 Madrid, Spain
* Correspondence: karimzadeh_f@cc.iut.ac.ir (F.K.); stefanos.giannakis@upm.es (S.G.)

Received: 15 March 2020; Accepted: 10 April 2020; Published: 12 April 2020

Abstract: This study addresses the visible light-induced bacterial inactivation kinetics over a Bi_2WO_6 synthesized catalyst. The systematic investigation was undertaken with Bi_2WO_6 prepared by the complexation of Bi with acetic acid (carboxylate) leading to a flower-like morphology. The characterization of the as-prepared Bi_2WO_6 was carried out by X-ray diffraction (XRD), scanning electron microscopy (SEM), X-ray photoelectron spectroscopy (XPS), specific surface area (SSA), and photoluminescence (PL). Under low intensity solar light (<48 mW/cm^2), complete bacterial inactivation was achieved within two hours in the presence of the flower-like Bi_2WO_6, while under visible light, the synthesized catalyst performed better than commercial TiO_2. The in situ interfacial charge transfer and local pH changes between Bi_2WO_6 and bacteria were monitored during the bacterial inactivation. Furthermore, the reactive oxygen species (ROS) were identified during *Escherichia coli* inactivation mediated by appropriate scavengers. The ROS tests alongside the morphological characteristics allowed the proposition of the mechanism for bacterial inactivation. Finally, recycling of the catalyst confirmed the stable nature of the catalyst presented in this study.

Keywords: flower-like Bi_2WO_6; *E.coli* inactivation; reactive oxygen species (ROS); photocatalysis; solar disinfection; water treatment; pollution

1. Introduction

Over the last few decades, environmental contamination has shifted from the exclusive focus of organic and inorganic pollutants [1], towards the inclusion of bacteria and other organisms [2–4]. Therefore, well-organized methods are urgently required to control the spread [5] or eradicate microorganism-related issues [6]. In recent times, beside the traditional bacterial inactivation methods such as UV disinfection and chlorination, a green, efficient, and cost-effective semiconductor photocatalysis has appeared to be a more promising technique [7,8]. TiO_2 has been extensively reported as an effective bactericidal semiconductor photocatalyst due to its high stability, strong redox potential, low cost, and non-toxic nature, but its band-gap of 3.2 eV allows light absorption up to 387 nm which makes up just over 4% of the total solar spectrum [9–11].

Since solar radiation contains more visible light (~47%) than UV, the appropriate use of this fraction becomes necessary through the employment of efficient visible-light photocatalysts [12]. As a promising visible-light-driven photocatalyst with good chemical and thermal stability, Bi_2WO_6, beside its non-toxic and environmentally friendly nature, is a typical n-type semiconductor composed of accumulated layers of alternating $(Bi_2O_2)^{2+}$ layers and $(WO_4)^{2-}$ octahedral sheets [13,14]. The valence band of Bi_2WO_6 consists of O 2p and Bi 6s hybrid orbitals, its narrowed band gap increases visible light absorption capacity, and photoactivity [15,16], while its photocatalytic activity greatly depends on morphology, particle size, surface area, and interface structure [17,18]. Constructing a unique micro/nano hierarchical structure usually shortens the pathways of water pollutants, absorb incidental light more efficiently, because of multiple-scattering increase, and easily separated from wastewater by filtration or sedimentation methods [13,19,20].

However, despite the long presence of this catalyst as a possible solution, most studies on Bi_2WO_6 have focused on the photocatalytic degradation of organic pollutants, with only a few studies investigating the photocatalytic inactivation of microorganisms. Ren et al. [21] reported *Escherichia coli* degradation in a few hours on Bi_2WO_6 nest-like structures in a pseudo-first order process. Helali et al. [22] prepared a 20 m^2/g SSA Bi_2WO_6 leading to *E. coli* inactivation within four to five hours under solar light on a hydrothermally grown mixture of Bi-nitrate and Na-tungstate in a 65–35% ratio while a similar study has been reported by Amano et al. [23]. However, there is a relatively wide gap in literature on effective preparation of robust structures with high specific surface areas in order to promote efficient disinfection, and a gap in interpreting the pathways to bacterial inactivation by this catalyst.

This study aims to assess a facile preparation method for flower-like Bi_2WO_6 photocatalysts destined for disinfection applications. As such, we assess the preparation parameters (aging, temperature, pH) in order to modify the structural (crystalline) and morphological characteristics (flower-like, nanoparticles). These modifications are envisioned to create a series of catalysts, and their activity under low-intensity solar or visible light will be assessed. Furthermore, the robustness of the catalyst in serial reuse cycles will be evaluated for its stability. Last but not least, special focus will be given to the identification of the pathways that lead to bacterial inactivation in an effort to decrypt the mechanistic action mode of the flower-like Bi_2WO_6.

2. Materials and Methods

2.1. Synthesis of Flower-Like Bi_2WO_6 Samples

All chemicals were of analytical grade. They were used as received without any further purification and were purchased from Merck, Germany. All solutions were prepared with Milli-Q water (18.2 MΩ cm). In a typical hydrothermal procedure for the synthesis of flower-like Bi_2WO_6, 0.5 mmol of $Na_2WO_4 \cdot 2H_2O$ was dissolved in an 80 mL solvent containing 16 mL acetic acid and 64 mL Milli-Q water until attaining a clear solution. Then, 1 mmol of $Bi(NO_3)_3 \cdot 5H_2O$ solid was added to the solution, and a white precipitate immediately emerged. Next, the reaction mixture was stirred for 1 h, transferred into a 120 mL Teflon-lined stainless-steel reactor, and heated at 160 °C for 12 h. The as-formed yellow precipitates were collected, washed with distilled water, and dried in vacuum at 70 °C for 10 h. A schematic representation of the synthesis is illustrated in Scheme 1. The influence of the hydrothermal reaction time and temperature has been explored as shown in Table 1. In order to investigate the effect of morphology on photocatalysis, Bi_2WO_6 nanoparticles (BWO6) were prepared applying the same hydrothermal method at 200 °C for 24 h by the regulation of pH to 10.

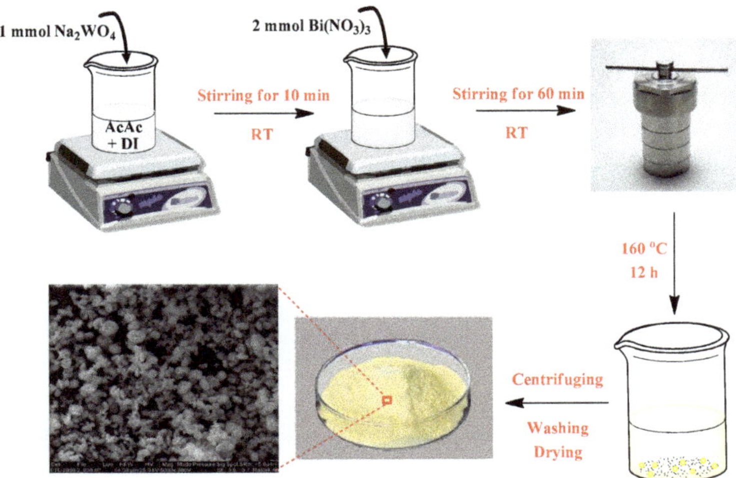

Scheme 1. Schematic illustration of the preparation of flower-like Bi_2WO_6 by hydrothermal method.

Table 1. Bi_2WO_6 obtained at the different synthetic conditions.

Samples	Reaction Time (h)	Reaction Temperature (°C)
BWO1	12	160
BWO2	18	160
BWO3	24	160
BWO4	24	180
BWO5	24	200
BWO6	24	200 (pH = 10)

2.2. Physical Characterization of the Bi_2WO_6 Flakes

The crystallinity and phase identification of the as-prepared samples were determined by powder X-ray diffraction (XRD) using an X'Pert MPD PRO (Panalytical) analyzer, equipped with a ceramic tube (Cu anode, λ = 1.54060 Å), and with a continuous scanning rate in the range of $5° < 2\theta < 80°$. The results were studied with Rietvield refinement by the FullProf program. The morphology developments of the samples were characterized using scanning electron microscopy (SEM, FEI Quanta 200). Before SEM imaging, the samples were coated with a thin layer of gold. The specific surface area and porosity size were obtained using Brunauer–Emmett–Teller (BET) analysis, performed with a BELSORP-mini II analyzer, Japan. The photoluminescence (PL) measurement was carried out using a fluorescence spectrophotometer (Perkin Elmer LS55) equipped with a xenon lamp at an excitation wavelength of λ = 340 nm. The surface atomic percentage of the element in the as-synthesized sample was analyzed using an AXIS NOVA photoelectron spectrometer with a mono-chromatic Al Ka X-ray (hν = 1486.6 eV) source (Kratos Analytical, Manchester, UK). The interfacial in situ voltage and pH variation during the bacterial inactivation was monitored in a pH/mV/Temp meter (Jenco 6230N) equipped with a microprocessor and a RS-232-C IBM interface for data recording.

2.3. Photocatalytic Antibacterial Activity on Bi_2WO_6 and Light Sources

The bacterial strain used was a wild type *E. coli* K12, supplied by the German Collection of Microorganisms and Cell Cultures, DSMZ (No. 498). The master plate and stock solution were prepared according to previous research reported by our laboratory [24,25]. The bacterial concentration of the samples was measured in Colony Forming Units (CFU/mL) and was determined by plating on a non-selective cultivation media, namely, Plate Count Agar (PCA). A total of 1 mL of the sample was withdrawn after each interval and then serial dilutions were made in a sterile 0.8% NaCl/KCl solution. A 100 μL aliquot was pipetted onto a nutrient agar plate and processed using the standard plate count method. The plates were incubated at 37 °C followed by the bacterial evaluation. Experimental results were carried in triplicate runs applying statistical analysis for the calculation of mean and standard deviation (reported in the graphs). Samples were irradiated in the cavity of a SUNTEST solar simulator CPS (Atlas GmbH, Hanau, Germany) with an overall light irradiance of 48 mW/cm^2 (~0.8 × 10^{16} photons/s, Supplementary Figure S1). A cut-off filter was used in the SUNTEST cavity to filter the light <310 nm. A second cut-off filter was also used during bacterial inactivation under visible light with a cut-off blocking the wavelength < 405 nm rendering (Supplementary Figure S2). Finally, after the two filters, the visible light irradiance reaching the sample was 38 mW/cm^2.

3. Results

3.1. Synthesis and Characterization of Bi_2WO_6: X-Ray Diffraction (XRD), Scanning Electron Microscopy (SEM), X-Ray Photoelectron Spectroscopy (XPS), and SSA Determination

Figure 1 depicts the XRD patterns of the as-synthesized Bi_2WO_6 via the hydrothermal method at different reaction times and temperatures. All of the XRD patterns illustrated that characteristic peaks were in good agreement with the orthorhombic phased Bi_2WO_6 in the standard JCPDS card (39-0256) [26]. No other diffraction peaks arising from possible impurities were detected. With the holding time increasing to 24 h, the characteristic peaks became much sharper due to an increase in crystallinity. Understandably, the increment of the temperature with the constant reaction time for 24 h resulted in the same trend because of grain growth. Table 2 illustrates the crystallite size of the samples (using the Scherrer formula based on the half-width of their (113) peak) calculated by the Rietveld method using the FullProf program.

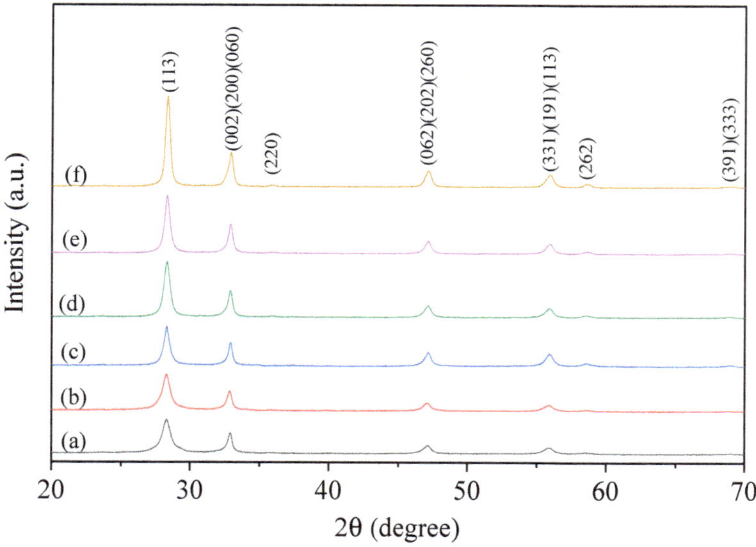

Figure 1. X-ray diffraction patterns of the Bi_2WO_6 samples: (**a**) BWO1, (**b**) BWO2, (**c**) BWO3, (**d**) BWO4, (**e**) BWO5, and (**f**) BWO6. Profiles are shifted in y-scale for clarity.

Table 2. Rietveld structural parameters of the samples.

Samples	Crystallite Size (nm)
BWO1	9
BWO2	10
BWO3	17
BWO4	20
BWO5	22
BWO6	31

The scanning electron microscopy (SEM) images of the Bi_2WO_6 samples prepared under different experimental conditions are shown in Figure 2. Heating at 160 °C for 12 h and 18 h (Figure 2a,b) led to aggregated irregular small Bi_2WO_6 nanoparticles and flower-like microspheres. However, when the heating time was prolonged to 24 h (see Figure 2c), organized hierarchical flower-like Bi_2WO_6 microspheres composed of nanoplates were obtained and the aggregated nanoparticles totally disappeared (Figure 2e,f). The SEM images of as-prepared Bi_2WO_6 nanoparticles are also shown in Figure 2f. The joint effect of nanoparticles assembly followed by the localized ripening mechanism as well as the hierarchical assembly of nanoplates have been also previously reported for the formation mechanism of flower-like microspheres [27,28]. Owing to the absence of discrete nanoplates according to the SEM images at different reaction times (Figure 2), the former mechanism seems to predominate.

Scheme 2 illustrates the proposed formation mechanism of flower-like Bi_2WO_6 microspheres. Nanoparticles initially aggregated, then the self-assembled nanoparticles preferentially grew along <010>. Longer reaction times and higher temperatures result in dissolution of some nanoplates leading concomitantly to re-deposition by Ostwald ripening [27,28].

The relevant reactions leading to the Bi_2WO_6 synthesis in aqueous solutions when working in acetic acid media can be suggested as follows:

$$Bi(NO_3)_3 \rightarrow Bi^{3+} + 3NO_3^{-}, \tag{1}$$

$$Na_2WO_4 \rightarrow 2Na^{+} + WO_4^{2-}, \tag{2}$$

$$CH_3-COOH + H_2O \rightarrow CH_3-COO^{-} + H_3O^{+} \quad pK_a 4.75, \tag{3}$$

$$Bi_3^{+} + WO_4^{2-} + CH_3-COO^{-} \rightarrow CH_3COO^{-} + \left|\begin{matrix} CH_3-COO^{-} \\ \\ CH_3-COO^{-} \end{matrix}\right\rangle M\ , \tag{4}$$

$$\left|\begin{matrix} CH_3-COO^{-} \\ \\ CH_3-COO^{-} \end{matrix}\right\rangle M \rightarrow Bi_2WO_6 \qquad flower-like\ microspheres\ (see\ Figure\ 2)?\ . \tag{5}$$

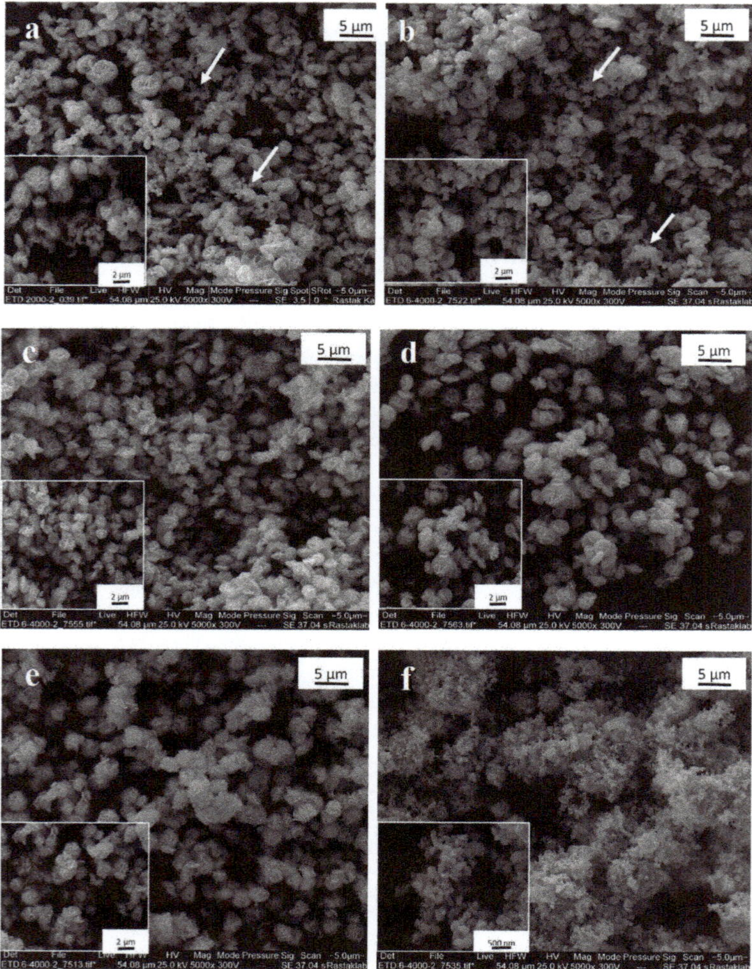

Figure 2. Scanning electron microscopy (SEM) images of Bi_2WO_6 samples prepared under different conditions: (**a**) BWO1, (**b**) BWO2, (**c**) BWO3, (**d**) BWO4, (**e**) BWO5, and (**f**) BWO6.

The initial complex between Bi and acetic acid presents a stability constant of $10^{2.6–2.7}$ [29], which is not in the range found for insoluble complexes/precipitates >$10^{11–12}$ [30–32]. This coordination complex is suggested in Equation (4) (Bi = M). The complex formation which is the precursor of Bi_2WO_6 does not lead to precipitate formation and gradually decomposes releasing Bi^{3+} which reacts with WO_4^{2-}. Therefore, the nanoplate formation leads to aggregates which present inner pores/voids and provide the required contact area for the photocatalytic bacterial inactivation.

In addition to the crystal structure and morphology, the surface chemical composition of the as-synthesized flower-like sample at 200 °C for 24 h was examined by XPS. As shown in the survey XPS spectrum in Figure 3, the Bi, O, W, and C elements were present in the pure Bi_2WO_6. The C element peak can be attributed to adventitious carbon from the sample preparation and/or the XPS instrument itself [33]. The surface atomic concentration ratio of Bi:W:O estimating from XPS peak areas is around 2.0:0.8:5.4, which further confirms its composition of Bi_2WO_6. Furthermore, the peaks centering at 164.7 and 159.4 eV are attributed to the binding energies of Bi $4f_{5/2}$ and Bi $4f_{7/2}$, respectively (inset of Figure 3), confirming Bi^{3+} ions in the crystalline structure [34–36]. The W4f energy region can be designated to be the +6 oxidation state of tungsten in accordance with previous reports [33,36].

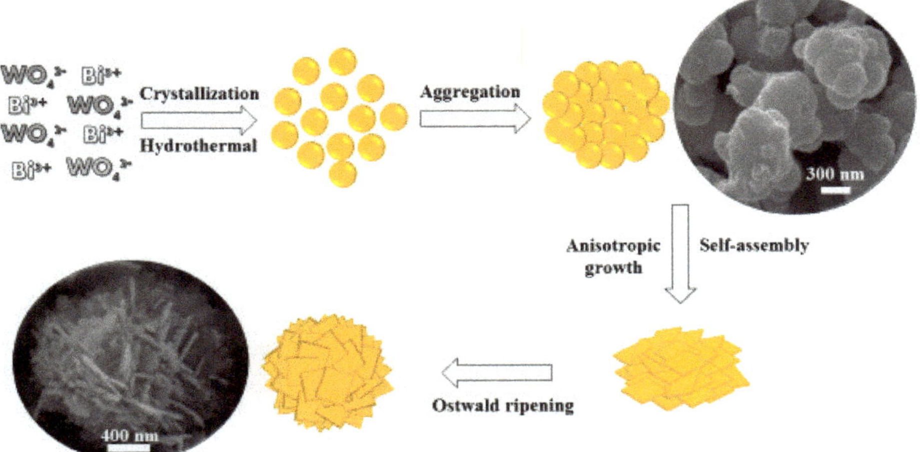

Scheme 2. Schematic illustration of the growth process of the flower-like Bi_2WO_6 microspheres.

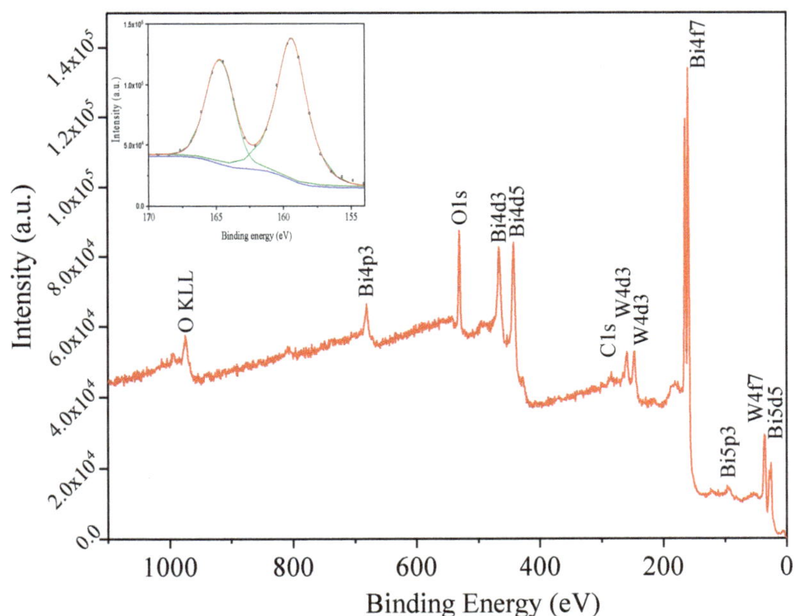

Figure 3. XPS survey spectra of the hydrothermally prepared Bi_2WO_6 sample at 200 °C for 24 h. Inset is the zoom of XPS scans over the $Bi_{4f7/2}$ peak in the 154–170 eV region.

The N_2 adsorption–desorption isotherms of the well-organized flower-like (BWO5) and nanoparticles (BWO6) Bi_2WO_6 are presented in Figure 4. According to IUPAC classification, it can be seen that the isotherm shape for both samples exhibited a typical type IV isotherm with a clear hysteresis loop H3, suggesting the presence of mesopores in the size range of 2–50 nm [37].The insets show the Barrett–Joyner–Halenda (BJH) pore-size distributions and present the evidence for the existence of mesopores (2–50 nm). Table 3 summarizes the BET specific surface areas (SSA) and the pore volumes of BWO5 and BWO6.

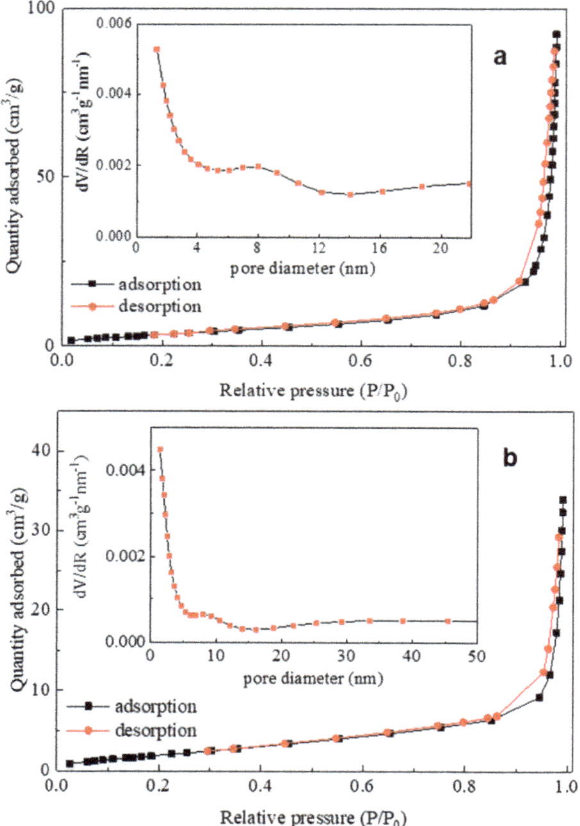

Figure 4. N_2 adsorption–desorption isotherm of the samples: (**a**) flower-like Bi_2WO_6, (**b**) nanoparticle Bi_2WO_6. The insert shows the pore size distribution.

Table 3. Brunauer–Emmett–Teller (BET) parameters of the Bi_2WO_6 samples at various temperatures.

Samples	Surface Areas (m^2 g^{-1})	Total Pore Volumes (cm^3 g^{-1})
BWO5	14.475	0.142
BWO6	6.87	0.0532

3.2. E. Coli Inactivation Kinetics: Effect of the Bacterial Concentration, Amount of Catalyst, Light Dose, and Applied Light Wavelength

Figure 5 shows the complete bacterial inactivation mediated by the BWO5 being faster under low-intensity simulated solar light, compared to the other samples. The *E. coli* inactivation was 95% after 2 h. The effectiveness of a disinfection process resides in the time necessary to inactivate a determined percentage of bacteria. In the Chick–Watson model [38,39], the simplest inactivation model, the inactivation rate shown in Figure 5 is seen to be dependent on the residual bacteria after each specific time during the inactivation process and this allows comparing the effect of the different Bi_2WO_6 samples. Neither irradiation in the absence of Bi_2WO_6 (photolysis) nor runs in the presence of this catalyst in the dark lead to bacterial inactivation of up to 4 h. The latter provides the proof that Bi_2WO_6 is not toxic to *E. coli* and a photocatalytic process is required for their inactivation. As the treatment time increased, the photocatalytic process became more effective, owing to the formation of hierarchical flower-like Bi_2WO_6 microspheres and loss of aggregates and the higher crystallite size. Nevertheless, the nanoparticles (BWO6), which presented lower specific surface area than BWO5, led to lower inactivation rates. The pseudo first-order rates of the Bi_2WO_6 samples during flower-like development (BWO1 and BWO5) compared with Bi_2WO_6 nanoparticles (BWO6) are given in the supplementary material, Figure S3. The pseudo first-order rate constants (k_{app}) of the BWO1, BWO5, and BWO6 were estimated to be 0.0331 min^{-1}, 0.0488 min^{-1}, and 0.0195 min^{-1}, respectively. As can be seen, the photocatalytic inactivation of bacteria mediated by as-developed flower-like Bi_2WO_6 (BWO5) is around 2.5 times faster compared with nanoparticles.

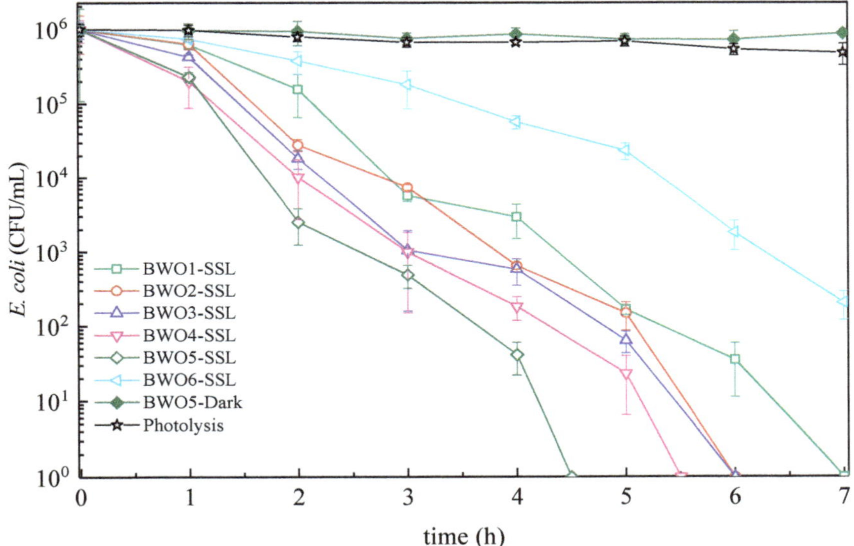

Figure 5. Photocatalytic inactivation of *Escherichia coli* in aqueous dispersions on different Bi_2WO_6 samples in the dark and under simulated solar light (SSL). Experimental conditions: $[Catalyst]_0$ = 0.2 g/L, $[bacteria]_0 = 2 \times 10^6$ Colony Forming Units (CFU)/mL and light intensity: 48 mW/cm^2.

The photoluminescence spectrum of the prepared catalysts was used as a practical method to verify the separation efficiency of photo-generated electron–hole pairs in the semiconductors. Generally, a lower photoluminescence (PL) intensity represents a lower recombination rate of photo-generated charge carriers. The photoluminescence (PL) spectra of the Bi_2WO_6 samples during the flower-like development (BWO1 and BWO5) in comparison with Bi_2WO_6 nanoparticles (BWO6) is shown in Figure 6. The wide absorption-band was observed between 350 nm and 600 nm which is due to the Bi_2WO_6 electron-hole recombination giving rise to the free and bound-exciton luminescence [40]. The PL spectra of the as-synthesized samples through flower-like development (BWO1 and BWO5) exhibited significantly decreased PL intensity related to that of the Bi_2WO_6 nanoparticles. It could be ascribed that the recombination of photo-generated charge carriers is greatly inhibited in the hierarchically flower-like composed of nanosheets. Hence, the efficient separation of photo-generated electron–hole pairs and rapid transfer of electrons to the surface of crystal would be obtained. Moreover, the lower PL-intensity bands shown in BWO5 reflected a higher crystallinity in comparison with BWO1, allowing a lower amount of crystal defects, leading to a higher electron-hole separation and an increased photocatalytic activity [41], a fact that corroborates with the faster inactivation of bacteria (Figure S3).

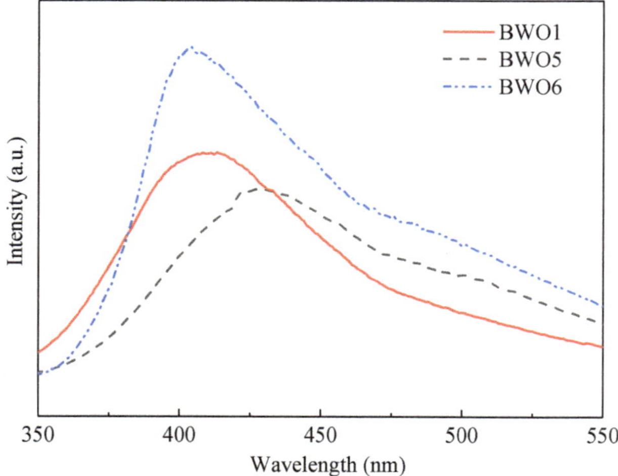

Figure 6. Photoluminescence (PL) spectroscopy of the synthesized samples at different conditions. BWO1: 12 h, 160 °C. BWO5, 24 h, 200 °C. BWO6: 24 h, 200 °C. pH = 10.

Following, the effects of initial catalyst or bacterial concentration were studied, and the results are summarized in Figure 7. The effect of the Bi_2WO_6 concentration on *E. coli* inactivation is shown in Figure 7a. Although increasing Bi_2WO_6 concentration of up to 0.2 mg/mL resulted in higher inactivation rates, increasing the catalyst concentration to 0.4 mg/mL resulted in a slower bacterial inactivation kinetics, most possibly due to a loss in surface area by catalyst agglomeration (particle–particle interactions), as well as a decrease in the penetration of the photon flux by the solution opacity, thereby decreasing the photocatalytic inactivation rate [42]. The effect of the initial concentration on the *E. coli* kinetics mediated by Bi_2WO_6 catalysts is presented in Figure 7b, showing a delay in the time necessary for bacterial inactivation at higher bacterial concentrations. Although this effect can be ascribed to the exhaustion of surface active sites due to opacity in solution [43], we note here that in absolute numbers, the higher the amount of bacteria in solution, the higher the number of available bacteria (for inactivation). Hence, by calculating the amount of cells inactivated in 4 h per mg of catalyst and per minute, we get 2075, 208, and 21 cells min^{-1} mg^{-1} for 10^8, 10^7, and 10^6, respectively. As a result, we report that this catalyst can effectively disinfect higher amounts of microorganisms, albeit in a higher residence time.

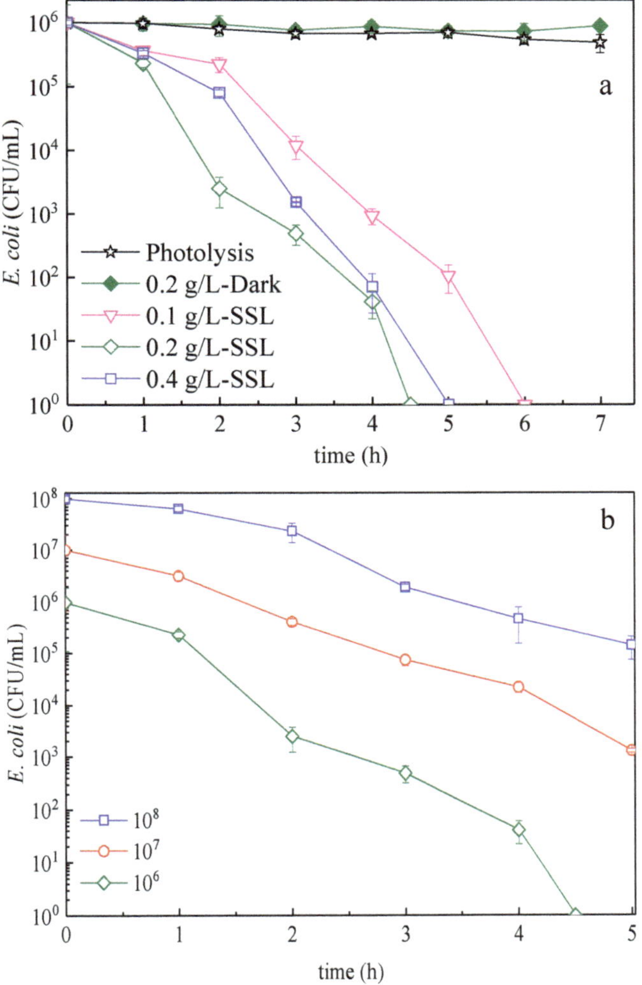

Figure 7. Effect of catalyst and bacterial concentration on inactivation kinetics. (**a**) *E. coli* survival on Bi_2WO_6 samples in the dark and under low intensity solar simulated light. Experimental conditions: $(bacteria)_0 = 2 \times 10^6$ CFU/mL and light intensity: 48 mW/cm^2. (**b**) Initial concentration of *E. coli* (CFU/mL) effects on the bacterial inactivation kinetics mediated by Bi_2WO_6 (200 °C for 4 h) under low intensity solar simulated light. Experimental conditions: $(Catalyst)_0 = 0.2$ g/L and light intensity: 48 mW/cm^2.

Next up in the operational parameters investigation, we assessed the possibility of photonic limitation or saturation of the system. As such, Figure 8a,b shows the effects of the light intensity and composition (UVA–vis or Vis only) on the bacterial degradation kinetics. A higher light dose accelerated the bacterial inactivation because of a higher amount of charges generated in the semiconductor during bacterial disinfection under band-gap irradiation (Figure 8a), since the direct inactivation by light was previously excluded. Figure 8 b illustrates that under visible light, a solution containing 0.2 g/L of Bi_2WO_6 was still efficiently inactivating bacteria and was more effective compared to commercial TiO_2 P25 Degussa (used as reference). These results come from the optical absorption of up to ~450 nm in the visible region by Bi_2WO_6, which is significantly wider than that of TiO_2 P25 Degussa with an absorption of up to 387 nm for the 20 nm particles, making up the bulk of this mixed TiO_2 P25 Degussa rutile–anatase [44].

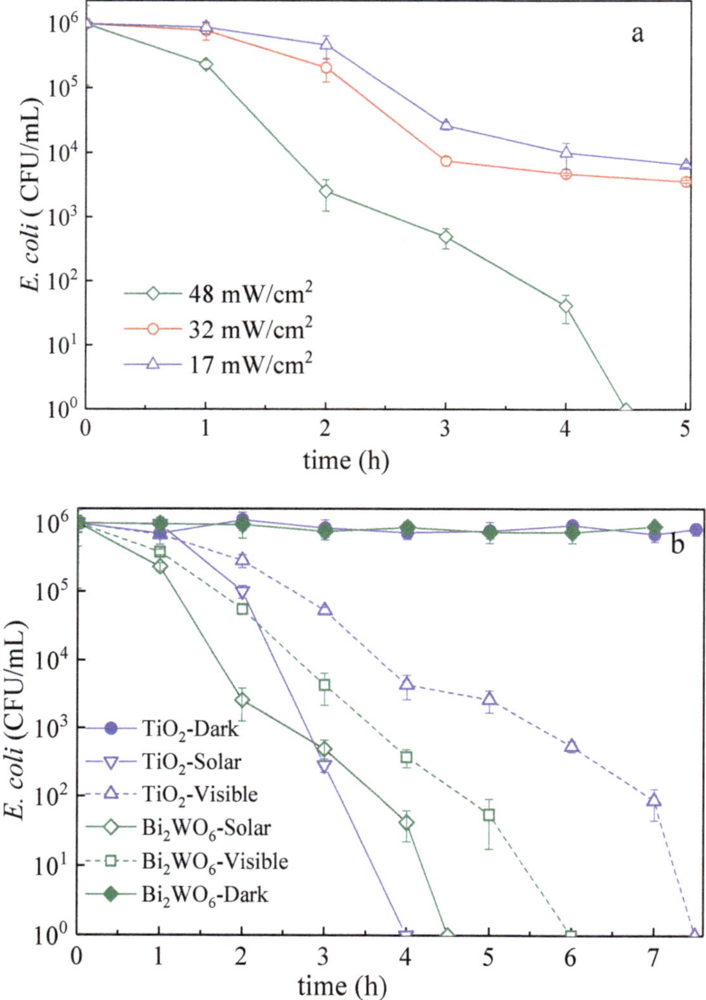

Figure 8. Effect of light irradiance and composition on inactivation kinetics. (**a**) *E. coli* inactivation on Bi_2WO_6 (200 °C for 24 h) under different solar light irradiation intensities. (**b**) *E. coli* inactivation mediated by Bi_2WO_6 (200 °C for 24 h) and TiO_2 under low intensity solar simulated (48 mW/cm^2) and visible light (38 mW/cm^2). Experimental conditions: (Catalyst)$_0$ = 0.2 g/L and (bacteria)$_0$ = 2 × 10^6 CFU/mL.

3.3. Mechanistic Interpretation: ROS-Species Involvement, Interfacial Charge Transfer, and Catalyst Reuse During Bacterial Inactivation

The reactive oxygen species (ROS) such as $^{\cdot}OH$, $O_2^{\cdot -}$, and vb (h^+) play a pivotal role in the photo-degradation of organic pollutants and bacterial inactivation [22,45–47]. To determine the main ROS followed by the photodegradation mechanism, appropriate radical-scavengers such as isopropanol ($^{\cdot}OH$ scavenger), sodium oxalate (a vbh$^+$ hole scavenger), and superoxide dismutase ($O_2^{\cdot -}$ scavenger) were used in the present study. Figure 9 depicts the results of scavenging experiments mediated by the optimized flower-like Bi_2WO_6 (BWO5). The photocatalytic bacterial inactivation could be remarkably suppressed by the addition of isopropanol and sodium oxalate. It is very likely that $^{\cdot}OH$ and h^+ intervene jointly in the bacterial inactivation. Meanwhile, the addition of SOD ($O_2^{\cdot-}$ scavenger) inhibits the bacterial inactivation to a smaller degree compared to vb(h^+) and the $^{\cdot}OH$-radical as shown in Figure 9, traces (a) and (b).

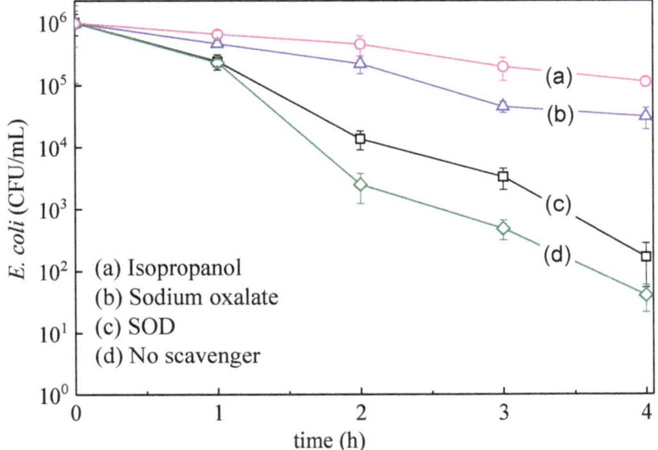

Figure 9. Effect of the scavengers during *E. coli* inactivation on Bi_2WO_6 under solar simulated light for (a) isopropanol as OH-radical scavenger, (b) sodium oxalate a hole vb(h^+) scavenger, (c) superoxide dismutase (SOD) as an $O_2^{\cdot-}$ scavenger, (d) no scavenger. Runs under low intensity solar simulated light (48 mW/cm^2). The solutions contained Bi_2WO_6 (0.2 g/L) and scavenger concentration of 0.1 mM.

The possible reaction mechanism for the inactivation of *E. coli* mediated by Bi_2WO_6 can be proposed as the following, which is shown in Scheme 3. Under visible-light irradiation, the photo-excitation of Bi_2WO_6 implies the transfer of an electron from the valence band (Equation (6)).

$$Bi_2WO_6 + hv \rightarrow e^-(CB) + h^+(VB). \quad (6)$$

As mentioned before, the valence band of Bi_2WO_6 is a hybrid band made up by the O2p and Bi6s orbitals. Under light irradiation, the O2p and Bi6s hybrid orbitals increase the charge transfer in the W5d orbitals of Bi_2WO_6. This moves the valence band (VB) potential to a more positive potential energy narrowing the band-gap and inducing a higher photocatalytic activity [48].

Based on the references, CB and VB potentials of Bi_2WO_6 are 3.08 and 0.36 eV, respectively [49,50]. The redox potential for the dissolved oxygen/superoxide couple (E^0 ($O_2/O_2^{\cdot-}$)), $O_2/HO_2^{\cdot}$, and $OH^-/^{\cdot}OH$ are −0.33 eV, −0.046 eV, and 1.98 eV vs NHE [49], respectively. Comparing the band edge energy level of Bi_2WO_6 with the redox potentials of ROS, it is obvious that the excited holes in the valence band of Bi_2WO_6 were sufficiently more positive than that of $OH^-/^{\cdot}OH$, suggesting that the photogenerated holes on the surface of Bi_2WO_6 could react with OH^-/H_2O to form "non-selective" $^{\cdot}OH$ radicals (Equation (7)). However, the conduction band edge potential of Bi_2WO_6, which is more positive than the standard redox potential of $O_2/O_2^{\cdot-}$ and $O_2/HO_2^{\cdot}$, cannot directly reduce O_2 to $O_2^{\cdot-}$ or $HO_2^{\cdot}$. As shown in Figure 9, the bacterial inactivation is reduced in the presence of SOD-scavengers, which confirms the presence of the $HO_2^{\cdot}$ radicals. Considering the redox potential of O_2/H_2O_2 = +0.682 eV vs NHE [51], H_2O_2 seems to be generated initially (Equation (8)) which is followed by the formation of different species according to the relations 9–10 in the photocatalytic reaction. It is worth noting that the powerful hole can directly attack bacteria cells in the photocatalytic oxidation process, which was also confirmed by the hole scavenger [45,52].

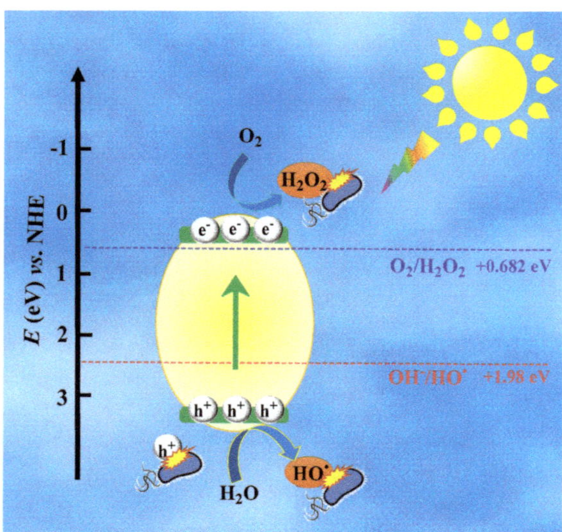

Scheme 3. Schematic diagram showing the photocatalytic inactivation of bacteria on the Bi_2WO_6.

$$h^+(VB) + H_2O \rightarrow HO\bullet + H^+, \tag{7}$$

$$O_2 + 2H^+ + 2e^- \rightarrow H_2O_2, \tag{8}$$

$$H_2O_2 + e^- \rightarrow OH\bullet + OH^-, \tag{9}$$

$$H_2O_2 + h^+ \rightarrow O_2^-\bullet + 2H^+ \ \ or \ \ HO_2^-\bullet + H^+. \tag{10}$$

Figure 10 shows the variation of the interfacial potential and the local pH shift under simulated solar light. At pH ~6, the bacterial inactivation preferentially proceeds via the $O_2^{\cdot-}$ species over $HO_2^{-\cdot}$ as shown in Equation (11) and Figure 9, trace (c).

$$HO_2^{-\cdot} \Leftrightarrow O_2^{\cdot-} + H^+ \qquad pK_a \quad 4.8. \tag{11}$$

The initial pH at time zero in Figure 10 was observed to decrease slightly from 6.0 to 5.9 within four hours of irradiation. The initial pH of 6.0 in this figure is seen to decrease drastically to 5.4 after 8000 s due to the concomitant production of long-lived intermediates carboxylic acids, owing to the degradation of the bacterial membrane. The interface potential is shown to drastically drop within 8000 s (2.2 h) when the bacterial reduction is reduced by 99.90%, which is equivalent to 3 logs as shown in Figure 5. The interface potential recovers to its initial value as shown in Figure 10 after the inactivation of bacteria [52]. The recovery to the initial pH-level occurs when the intermediate acids are mineralized to CO_2 by the photo-Kolbe reaction according to Equation (12) [53,54].

$$RCOO^- + \text{solar light} \rightarrow R + CO^2. \tag{12}$$

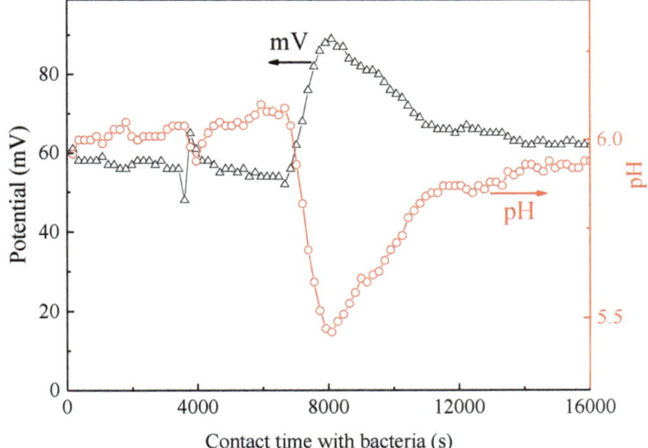

Figure 10. Evolution of the interfacial potential and local pH of an *E. coli* suspension in contact with Bi_2WO_6 under low intensity light irradiation (48 mW/cm^2). Catalyst concentration 0.2 g/L.

Finally, we provide the evidence for synthesizing a stable Bi_2WO_6 flower-like photocatalyst by a repetitive inactivation of a *E. coli* test, which results are shown in Figure 11. In order to evaluate the bacterial inactivation after each cycle, the pseudo first-order rate constants (*kapp*) were calculated and are reported in Table 4. The recycled sample used in Figure 11 was thoroughly washed after each cycle. Practically, no loss of bacterial inactivation was observed. These results show the stable repetitive bacterial inactivation mediated by flower-like Bi_2WO_6 up to five cycles and confirm the potential for the practical application of this photocatalyst in *E. coli* inactivation.

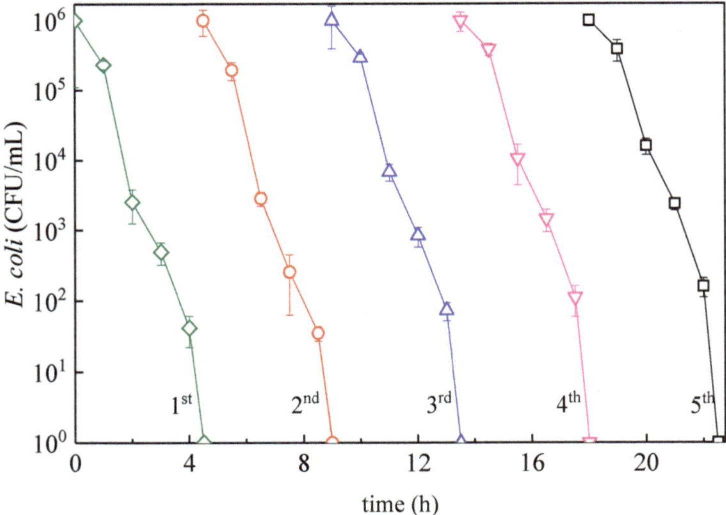

Figure 11. Reusability of flower-like Bi_2WO_6 under low intensity solar simulated light (48 mW/cm^2). Solution parameters: $(Catalyst)_0 = 0.2$ g/L and $(bacteria)_0 = 2 \times 10^6$ CFU/mL.

Table 4. Pseudo first-order rate constants (k_{app}) for *E. coli* inactivation under different conditions consistent with Figure 11.

Cycle Number	k_{app} (min^{-1})
First	0.0488 ± 0.005
Second	0.0494 ± 0.004
Third	0.0484 ± 0.005
Fourth	0.0480 ± 0.006
Fifth	0.0471 + 0.007

4. Conclusions

In the present study, Bi_2WO_6 flower-like samples were prepared at 200 °C attaining a high crystallinity and led a low amount of crystal by hydrothermal growth in acetic acid media. By SEM, XRD, XPS, and PL analysis, the properties of the flower-like Bi_2WO_6 samples and nanoparticles were investigated. These catalysts resulted in effective bacterial inactivation even under visible light and were faster than TiO_2. In addition to higher SSA of flower-like Bi_2WO_6, its lower PL intensity leads to lower recombination of photo-generated electron–hole pairs as a consequence of more efficient photocatalytic activity. The photocatalytic inactivation of bacteria mediated by as-developed flower-like Bi_2WO_6 (BWO5) is around 2.5 times faster when compared with nanoparticles. The samples under light lead to effective Bi_2WO_6 charge separation and the generation of ROS inducing bacterial inactivation. The intermediate ROS species produced by Bi_2WO_6 were identified by the use of the appropriate scavengers, and the $^{\cdot}$OH-radical was identified to be the dominant inactivation mechanism. Finally, the stable performance of the synthesized catalyst during recycling indicates its robustness and may suggest practical application potential.

Supplementary Materials: The following are available online at http://www.mdpi.com/2073-4441/12/4/1099/s1, **Figure S1.** SUNTEST solar simulator light wavelength emission spectrum (Manufacturer: Atlas, CPS+/CPS Instruments Brochure), **Figure S2.** Transmittance of the polymethylmethacrylate filter used to block UV light, **Figure S3.** Pseudo first-order rates of the Bi_2WO_6 samples during flower-like development (BWO1 and BWO5) compared with Bi2WO6 nanoparticles (BWO6).

Author Contributions: Conceptualization, M.K., S.G., C.P., S.R., and J.K.; methodology, M.K., S.G., S.R., J.K., and C.P.; software, M.K.; validation, M.K., S.G., S.R., and J.K.; investigation, M.K.; resources, C.P., F.K., and K.R.; writing—original draft preparation, M.K., F.K., K.R., S.R., J.K., S.G., and C.P.; writing—review & editing, M.K. and S.G.; visualization, M.K.; supervision, F.K., K.R., S.R., J.K., and C.P.; project administration, C.P., F.K., and K.R.; funding acquisition, S.G., C.P., F.K., and K.R. All authors have read and agreed to the published version of the manuscript.

Funding: Minoo Karbasi obtained exchange scholarship from the Ministry of Science, Research and Technology of Iran and the Isfahan University of Technology, and its contribution is hereby acknowledged. Stefanos Giannakis would like to acknowledge the Spanish Ministry of Science, Innovation and Universities (MICIU) for the Ramón y Cajal Fellowship (RYC2018-024033-I).

Conflicts of Interest: The authors declare no conflict of interest.

References

1. Giannakis, S.; Jovic, M.; Gasilova, N.; Pastor Gelabert, M.; Schindelholz, S.; Furbringer, J.M.; Girault, H.; Pulgarin, C. Iohexol degradation in wastewater and urine by UV-based Advanced Oxidation Processes (AOPs): Process modeling and by-products identification. *J. Environ. Manag.* **2017**, *195*, 174–185. [CrossRef] [PubMed]
2. Chong, M.N.; Jin, B.; Chow, C.W.K.; Saint, C. Recent developments in photocatalytic water treatment technology: A review. *Water Res.* **2010**, *44*, 2997–3027. [CrossRef] [PubMed]
3. Pelaez, M.; Nolan, N.T.; Pillai, S.C.; Seery, M.K.; Falaras, P.; Kontos, A.G.; Dunlop, P.S.M.; Hamilton, J.W.J.; Byrne, J.A.; O'Shea, K.; et al. A review on the visible light active titanium dioxide photocatalysts for environmental applications. *Appl. Catal. B Environ.* **2012**, *125*, 331–349. [CrossRef]
4. Campoccia, D.; Montanaro, L.; Arciola, C.R. A review of the biomaterials technologies for infection-resistant surfaces. *Biomaterials* **2013**, *34*, 8533–8554. [CrossRef]

5. Giannakis, S.; Merino Gamo, A.I.; Darakas, E.; Escalas-Cañellas, A.; Pulgarin, C. Impact of different light intermittence regimes on bacteria during simulated solar treatment of secondary effluent: Implications of the inserted dark periods. *Sol. Energy* **2013**, *98*, 572–581. [CrossRef]
6. Giannakis, S.; Darakas, E.; Escalas-Cañellas, A.; Pulgarin, C. Environmental considerations on solar disinfection of wastewater and the subsequent bacterial (re)growth. *Photochem. Photobiol. Sci.* **2015**, *14*, 618–625. [CrossRef]
7. Byrne, J.A.; Dunlop, P.S.M.; Hamilton, J.W.J.; Fernández-Ibáñez, P.; Polo-López, I.; Sharma, P.K.; Vennard, A.S.M. A Review of Heterogeneous Photocatalysis for Water and Surface Disinfection. *Molecules* **2015**, *20*, 5574–5615. [CrossRef]
8. Rizzo, L.; Della Sala, A.; Fiorentino, A.; Li Puma, G. Disinfection of urban wastewater by solar driven and UV lamp–TiO_2 photocatalysis: Effect on a multi drug resistant Escherichia coli strain. *Water Res.* **2014**, *53*, 145–152. [CrossRef]
9. Carré, G.; Hamon, E.; Ennahar, S.; Estner, M.; Lett, M.C.; Horvatovich, P.; Gies, J.P.; Keller, V.; Keller, N.; Andre, P. TiO_2 photocatalysis damages lipids and proteins in Escherichia coli. *Appl. Environ. Microbiol.* **2014**, *80*, 2573–2581. [CrossRef]
10. Ouyang, K.; Dai, K.; Walker, S.L.; Huang, Q.; Yin, X.; Cai, P. Efficient photocatalytic disinfection of Escherichia coli O157:H7 using C70-TiO_2hybrid under visible light irradiation. *Sci. Rep.* **2016**, *6*, 1–8. [CrossRef]
11. Nakata, K.; Fujishima, A. TiO_2 photocatalysis: Design and applications. *J. Photochem. Photobiol. C Photochem. Rev.* **2012**, *13*, 169. [CrossRef]
12. Zhang, L.; Wang, W.; Chen, Z.; Zhou, L.; Xu, H.; Zhu, W. Fabrication of flower-like Bi_2WO_6superstructures as high performance visible-light driven photocatalysts. *J. Mater. Chem.* **2007**, *17*, 2526–2532. [CrossRef]
13. Zhang, L.; Wang, H.; Chen, Z.; Wong, P.K.; Liu, J. Bi_2WO_6 micro/nano-structures: Synthesis, modifications and visible-light-driven photocatalytic applications. *Appl. Catal. B Environ.* **2011**, *106*, 1–13. [CrossRef]
14. Saison, T.; Gras, P.; Chemin, N.; Chanéac, C.; Durupthy, O.; Brezová, V.; Colbeau-Justin, C.; Jolivet, J.P. New insights into Bi_2WO_6 properties as a visible-light photocatalyst. *J. Phys. Chem. C* **2013**, *117*, 22656–22666. [CrossRef]
15. Zhang, L.; Zhu, Y. A review of controllable synthesis and enhancement of performances of bismuth tungstate visible-light-driven photocatalysts. *Catal. Sci. Technol.* **2012**, *2*, 694–706. [CrossRef]
16. Chen, M. Degradation of Antibiotic Norfloxacin by Solar Light/Visible Light-Assisted Oxidation Processes in Aqueous Phase. Ph.D. Thesis, The Hong Kong Polytechnic University, Hong Kong, China, 2013.
17. Shang, Y.; Cui, Y.; Shi, R.; Yang, P. Effect of acetic acid on morphology of $Bi_2WO_6$with enhanced photocatalytic activity. *Mater. Sci. Semicond. Process.* **2019**, *89*, 240–249. [CrossRef]
18. Zhang, Q.; Chen, J.; Xie, Y.; Wang, M.; Ge, X. Inductive effect of poly (vinyl pyrrolidone) on morphology and photocatalytic performance of Bi_2WO_6. *Appl. Surf. Sci.* **2016**, *368*, 332–340. [CrossRef]
19. Shen, R.; Jiang, C.; Xiang, Q.; Xie, J.; Li, X. Surface and interface engineering of hierarchical photocatalysts. *Appl. Surf. Sci.* **2018**, *471*, 43–87. [CrossRef]
20. Li, C.; Chen, G.; Sun, J.; Rao, J.; Han, Z.; Hu, Y.; Zhou, Y. A Novel Mesoporous Single-Crystal-Like $Bi_2WO_6$with Enhanced Photocatalytic Activity for Pollutants Degradation and Oxygen Production. *ACS Appl. Mater. Interfaces* **2015**, *7*, 25716–25724. [CrossRef]
21. Ren, J.; Wang, W.; Zhang, L.; Chang, J.; Hu, S. Photocatalytic inactivation of bacteria by photocatalyst Bi_2 WO_6 under visible light. *Catal. Commun.* **2009**, *10*, 1940–1943. [CrossRef]
22. Helali, S.; Polo-López, M.I.; Fernández-Ibáñez, P.; Ohtani, B.; Amano, F.; Malato, S.; Guillard, C. Solar photocatalysis: A green technology for E. coli contaminated water disinfection. Effect of concentration and different types of suspended catalyst. *J. Photochem. Photobiol. A Chem.* **2014**, *276*, 31–40. [CrossRef]
23. Amano, F.; Nogami, K.; Ohtani, B. Visible Light-Responsive Bismuth Tungstate Photocatalysts: Effects of Hierarchical Architecture on Photocatalytic Activity. *J. Phys. Chem* **2009**, *113*, 1536–1542. [CrossRef]
24. Porras, J.; Giannakis, S.; Torres-Palma, R.A.; Fernandez, J.J.; Bensimon, M.; Pulgarin, C. Fe and Cu in humic acid extracts modify bacterial inactivation pathways during solar disinfection and photo-Fenton processes in water. *Appl. Catal. B Environ.* **2018**, *235*. [CrossRef]
25. Marjanovic, M.; Giannakis, S.; Grandjean, D.; de Alencastro, L.F.; Pulgarin, C. Effect of MM Fe addition, mild heat and solar UV on sulfate radical-mediated inactivation of bacteria, viruses, and micropollutant degradation in water. *Water Res.* **2018**, *140*, 220–231. [CrossRef]

26. Zhang, C.; Zhu, Y. Synthesis of Square Bi2WO6 Nanoplates as High-Activity Visible-Light-Driven Photocatalysts, Chem. *Chem. Mater.* **2005**, *17*, 3537–3545. [CrossRef]
27. Liang, Y.; Shi, J.; Fang, B. Synthesis and electrochemical performance of bismuth tungsten oxides with different composition and morphology. *Chem. Phys. Lett.* **2019**, *716*, 112–118. [CrossRef]
28. Guan, J.; Liu, L.; Xu, L.; Sun, Z.; Zhang, Y. Nickel flower-like nanostructures composed of nanoplates: One-pot synthesis, stepwise growth mechanism and enhanced ferromagnetic properties. *CrystEngComm* **2011**, *13*, 2636. [CrossRef]
29. Donaldson, J.D.; Knifton, J.F.; Ross, S.D. The fundamental vibrational spectra of the formates of the main group elements. *Spectrochim. Acta* **1964**, *20*, 847–851. [CrossRef]
30. Anderegg, G.; Arnaud-Neu, F.; Delgado, R.; Felcman, J.; Popov, K. Critical evaluation of stability constants of metal complexes of complexones for biomedical and environmental applications (IUPAC Technical Report). *Pure Appl. Chem.* **2005**, *77*, 1445–1495. [CrossRef]
31. Portanova, R.; Lajunen, L.H.J.; Tolazzi, M.; Piispanen, J. Critical evaluation of stability constants for alpha-hydroxycarboxylic acid complexes with protons and metal ions and the accompanying enthalpy changes. Part II. Aliphatic 2-hydroxycarboxylic acids (IUPAC Technical Report). *Pure Appl. Chem.* **2003**, *75*, 495–540. [CrossRef]
32. Briand, G.G.; Burford, N. Coordination Complexes of Bismuth(III) Involving Organic Ligands with Pnictogen or Chalcogen Donors. *Adv. Inorg. Chem.* **2000**, *50*, 285–357.
33. Tang, R.; Su, H.; Sun, Y.; Zhang, X.; Li, L.; Liu, C.; Wang, B.; Zeng, S.; Sun, D. Facile Fabrication of Bi_2WO_6/Ag2S Heterostructure with Enhanced Visible-Light-Driven Photocatalytic Performances. *Nanoscale Res. Lett.* **2016**, *11*, 126. [CrossRef] [PubMed]
34. Xu, X.; Ge, Y.; Wang, H.; Li, B.; Yu, L.; Liang, Y.; Chen, K.; Wang, F. Sol–gel synthesis and enhanced photocatalytic activity of doped bismuth tungsten oxide composite. *Mater. Res. Bull.* **2016**, *73*, 385–393. [CrossRef]
35. Li, W.; Wang, Q.; Huang, L.; Li, Y.; Xu, Y.; Song, Y.; Zhang, Q.; Xu, H.; Li, H. Synthesis and characterization of BN/Bi_2WO_6 composite photocatalysts with enhanced visible-light photocatalytic activity. *RSC Adv.* **2015**, *5*, 88832–88840. [CrossRef]
36. Tang, R.; Su, H.; Sun, Y.; Zhang, X.; Li, L.; Liu, C.; Zeng, S.; Sun, D. Journal of Colloid and Interface Science Enhanced photocatalytic performance in Bi_2WO_6/SnS heterostructures: Facile synthesis, influencing factors and mechanism of the photocatalytic process. *J. Colloid Interface Sci.* **2016**, *466*, 388–399. [CrossRef]
37. Guo, Y.; Zhang, G.; Gan, H.; Zhang, Y. Micro/nano-structured $CaWO_4$/Bi_2WO_6 composite: Synthesis, characterization and photocatalytic properties for degradation of organic contaminants. *Dalton Trans.* **2012**, *41*, 12697. [CrossRef]
38. Chick, H. An Investigation of the Laws of Disinfection. *J. Hyg. (Lond)* **1908**, *8*, 92–158. [CrossRef]
39. Watson, H.E. A Note on the Variation of the Rate of Disinfection with Change in the Concentration of the Disinfectant. *J. Hyg. (Lond)* **1908**, *8*, 536–542. [CrossRef]
40. Sun, X.; Zhang, H.; Wei, J.; Yu, Q.; Yang, P.; Zhang, F. Preparation of point-line Bi_2WO_6@TiO_2 nanowires composite photocatalysts with enhanced UV/visible-light-driven photocatalytic activity. *Mater. Sci. Semicond. Process.* **2016**, *45*, 51–56. [CrossRef]
41. Docampo, P.; Guldin, S.; Steiner, U.; Snaith, H.J. Charge Transport Limitations in Self-Assembled TiO_2 Photoanodes for Dye-Sensitized Solar Cells. *J. Phys. Chem. Lett.* **2013**, *4*, 698–703. [CrossRef]
42. Kh, M.; Reza, A.; Kurny, F.G. Parameters affecting the photocatalytic degradation of dyes using TiO_2: A review. *Appl. Water Sci.* **2017**, *7*, 1569–1578.
43. Adhikari, S.; Banerjee, A.; Eswar, N.K.R.; Sarkar, D.; Madras, G. Photocatalytic inactivation of E. Coli by ZnO-Ag nanoparticles under solar radiation. *RSC Adv.* **2015**, *5*, 51067–51077. [CrossRef]
44. Winkler, J. *Titanium Dioxide: Production, Properties and Effective Usage*; Vincentz Network: Hanover, Germany, 2013; ISBN 9783866308121.
45. Nosaka, Y.; Nosaka, A.Y. Generation and Detection of Reactive Oxygen Species in Photocatalysis. *Chem. Rev.* **2017**, *117*, 11302–11336. [CrossRef] [PubMed]
46. Benabbou, A.K.; Derriche, Z.; Felix, C.; Lejeune, P.; Guillard, C. Photocatalytic inactivation of *Escherischia coli*: Effect of concentration of TiO2 and microorganism, nature, and intensity of UV irradiation. *Appl. Catal. B Environ.* **2007**, *76*, 257–263.

47. Regmi, C.; Joshi, B.; Ray, S.K.; Gyawali, G.; Pandey, R.P. Understanding Mechanism of Photocatalytic Microbial Decontamination of Environmental Wastewater. *Front. Chem.* **2018**, *6*, 33. [CrossRef]
48. Wei, Z.; Zhu, Y.; Kudo, A. *Nanostructured Photocatalysts*; Springer International Publishing: Basel, Switzerland, 2016; ISBN 978-3-319-26077-8.
49. Xu, J.; Wang, W.; Sun, S.; Wang, L. Enhancing visible-light-induced photocatalytic activity by coupling with wide-band-gap semiconductor: A case study on Bi_2WO_6/TiO_2. *Appl. Catal. B Environ.* **2012**, *111–112*, 126–132. [CrossRef]
50. Schneider, J.; Bahnemann, D.; Ye, J.; Puma, G.L.; Dionysiou, D.D. *Photocatalysis*; Royal Society of Chemistry: Cambridge, UK, 2016; ISBN 9781782620419.
51. Wu, D.; Wang, W.; Ng, T.W.; Huang, G.; Xia, D.; Yip, H.Y.; Lee, H.K.; Li, G.; An, T.; Wong, P.K. Visible-light-driven photocatalytic bacterial inactivation and the mechanism of zinc oxysulfide under LED light irradiation. *J. Mater. Chem. A* **2016**, *4*, 1052–1059. [CrossRef]
52. Sherman, I.; Gerchman, Y.; Sasson, Y.; Gnayem, H.; Mamane, H. Disinfection and Mechanistic Insights of Escherichia coli in Water by Bismuth Oxyhalide Photocatalysis. *Photochem. Photobiol.* **2016**, *92*, 826–834. [CrossRef]
53. Kiwi, J.; Nadtochenko, V. New Evidence for TiO_2 Photocatalysis during Bilayer Lipid Peroxidation. *J. Phys. Chem. B* **2004**, *108*, 17675–17684. [CrossRef]
54. Kraeutler, B.; Bard, A.J. Heterogeneous photocatalytic decomposition of saturated carboxylic acids on titanium dioxide powder. Decarboxylative route to alkanes. *J. Am. Chem. Soc.* **1978**, *100*, 5985–5992. [CrossRef]

Article

Degradation of Hexacyanoferrate (III) from Gold Mining Wastewaters via UV-A/LED Photocatalysis Using Modified TiO_2 P25

Augusto Arce-Sarria [1,2], Kevin Mauricio Aldana-Villegas [1], Luis Andres Betancourt-Buitrago [1], Jose Ángel Colina-Márquez [3], Fiderman Machuca-Martínez [1] and Miguel Angel Mueses [3,*]

1 Escuela de Ingeniería Química, Universidad del Valle, Cali 760032, Colombia; augusto.arce@correounivalle.edu.co (A.A.-S.); kevin.aldana@correounivalle.edu.co (K.M.A.-V.); betancourt.luis@correounivalle.edu.co (L.A.B.-B.); fiderman.machuca@correounivalle.edu.co (F.M.-M.)

2 Tecnoparque Nodo Cali, GIDEMP Materials and Products Research Group, ASTIN Center, SENA Regional Valle, Cali 760003, Colombia

3 Modeling & Application of Advanced Oxidation Processes, Photocatalysis & Solar Photoreactors Engineering, Chemical Engineering Program, Universidad de Cartagena, Cartagena 130001, Colombia; jcolinam@unicartagena.edu.co

* Correspondence: mmueses@unicartagena.edu.co

Received: 6 July 2020; Accepted: 6 September 2020; Published: 10 September 2020

Abstract: The photocatalytic degradation of potassium hexacyanoferrate (III) was assessed in a bench-scale compound parabolic collectors (CPC) reactor assisted with a light-emitting diode (LED) UV-A source emitting at 365 nm, and using a modified TiO_2 as a catalyst via the hydrothermal treatment of commercial Aeroxide P25. The experiments were performed under oxic and anoxic conditions in order to observe a possible reduction of the iron. The modified TiO_2 showed a specific surface area 2.5 times greater than the original Aeroxide P25 and its isotherm and hysteresis indicated that the modified catalyst is mesoporous. The bandgap energy (E_g) of the modified TiO_2 increased (3.34 eV) compared to the P25 TiO_2 band gap (3.20 eV). A specific reaction rate constant of 0.1977 min^{-1} and an electrical oxidation efficiency of 7.77 kWh/m^3 were obtained in the photocatalytic degradation. Although the TiO_2 P25 yields a photocatalytic degradation 9.5% higher than that obtained one with the modified catalyst (hydrothermal), this catalyst showed better performance in terms of free cyanide release. This last aspect is a significant benefit since this can help to avoid the pollution of fresh water by reusing the treated wastewater for gold extraction. A photocatalytic degradation of the cyanocomplex of 93% was achieved when the process occurred under oxic conditions, which favored the removal. Summarizing, the hydrothermal method could be a promising treatment to obtain TiO_2-based catalysts with larger specific areas.

Keywords: photocatalysis; UV-LED; TiO_2; hexacyanoferrate; mining; hydrothermal method

1. Introduction

Small and medium industries of gold extraction use the leaching process with sodium cyanide for mining the gold contained in the extracted ore, before precipitation of the metallic gold in the presence of zinc. During the process, the cyanide extracts undesired metals and thus forms several types of cyano complexes. The produced wastewater is rich in metallic complexes that are formed when the free cyanide interacts with the different metals present in the ores such as Ni, Fe, Co, Au, Ag, etc. These cyano complexes are very stable and recalcitrant compounds, which are hard to remove by natural remediation, resulting in the pollution of rivers, lakes and groundwater sources. Besides, solar photolysis releases free cyanide, which is highly harmful to ecosystems [1]. Advanced oxidation processes (AOPs), such as ozone-based treatments, alkaline chlorination, hydrogen peroxide-based processes, biological and photocatalytic processes, can be used as alternative treatment technologies for these mining wastewaters [2].

The heterogeneous photocatalysis is an AOP where a solid semiconductor, assisted by UV radiation, promotes the generation of free hydroxyl radicals (•OH) and the degradation of diverse pollutants. The most commonly used semiconductor is the titanium dioxide (TiO_2), and it can be used as a base oxide for the synthesis of other photoactive catalysts as well. The TiO_2 is preferred because of its low cost, easy handling, and low toxicity. In general, when the photocatalyst is irradiated with photons with energy greater than the bandgap (Eg) of the semiconductor, the excited electrons are promoted from the valence band to the conduction band of the semiconductor, leading to the formation of electron–hole pairs. The strong oxidative potential of the holes (h^+) oxidizes the hydroxyl anions of water for generating •OH, whereas the electrons of the conduction band can react with oxygen for generating superoxide ions ($O_2^{\bullet -}$) or promote other reduction reactions. Those radicals are the main species responsible for the oxidation reactions in the photocatalytic process [3,4].

To improve the semiconductors' •OH-generating performance, several studies have been focused on the preparation of semiconductors with enhanced radiation absorption. Different methods of preparation have been reported, namely hydrothermal [5], sol-gel [6], anodic oxidation, template method, and chemical vapor deposition (CVD) [7,8]. TiO_2 catalysts doped with rare earth and transition metals have been modified to improve their •OH electron transfer properties. Some modifications on their morphology have also been made to produce structures such as nanorods, nanotubes, nanospheres, nanoflowers, among others [9–12].

The hydrothermal method has been widely used for the nanomaterial synthesis of TiO_2 with diverse morphologies. This methodology is controlled by different variables, namely the precursors used, pH, temperature and reaction time [13]. Nowadays, TiO_2-based nanowires, TiO_2 nanotubes [14], carbon nanotubes [15], nanofibers, nanoflowers, and others have been successfully modified by hydrothermal treatment [16]. This method has become a very important tool for obtaining advanced materials due to its advantages, such as low cost, low operating temperatures, energy saving and lower impact to the environment (according to the principles of green chemistry) [10,12,17], in comparison to anodic oxidation and CVD methods. The hydrothermal treatment has been applied to the synthesis of nitrogen and carbon co-doped TiO_2 [18], Sn-doped TiO_2 nanoparticles composites [19], silica-titania combination of sol-gel-hydrothermal TiO_2 nanoparticles [20], and both anatase and rutile TiO_2 [21]. Moreover, several applications of TiO_2 nanoparticles synthesized by the hydrothermal method have been reported such as hydrogen production via CO_2 reduction, degradation of emergent pollutants and selective oxidation [22].

Huang and Chien [23] showed that the degradation of methylene blue increases from 65% to 95% with titania nanotubes compared to the powder. Camposeco et al. [24] compared the degradation between nanotubes and Evonik P25, showing that the catalytic activity was improved from 54 to 93% for methylene blue degradation and from 37% to 60% for the elimination of methylene orange. However, there is a lack of specific information about the use of titania modified via hydrothermal process for treating gold mining wastewater under UV/LED radiation.

In this work, the degradation of potassium hexacyanoferrate, which is a complex occurring as a by-product in gold mining wastewaters, via photocatalysis with hydrothermally treated TiO_2, was studied. The mechanism proposed by Grieken et al. 2005 [25] or the hexacyanoferrate (III) reduction to hexacyanoferrate (II) and the subsequent degradation by heterogeneous photocatalysis is depicted in Figure 1. After the progressive abatement of the CN^- groups in the molecule, the free cyanide can remain stable in solution due to the high pH of treatment or to produce cyanate by photocatalytic degradation, which is less toxic than the free cyanide. Nonetheless, the free cyanide is an advantage if the treated wastewater can be reused for the gold extraction. This would reduce the fresh water and cyanide consumptions and a consequent diminution of cyanide presence in water bodies.

The mechanism of free cyanide release is congruent with the reported literature [26–29]. The oxic conditions were analyzed in order to compare these results with the obtained ones in our previous work [27]. A further contribution respect to the reported literature is the use of the modified P25 via hydrothermal treatment and its potential improvement for the potassium hexacyanoferrate removal.

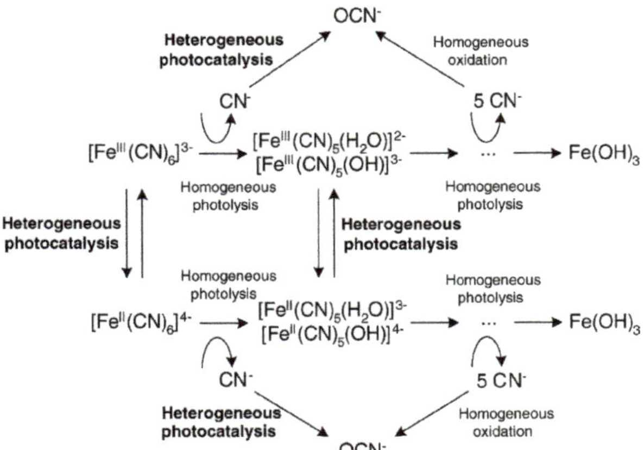

Figure 1. Mechanism of the heterogeneous photocatalytic degradation of Hexacyanoferrate [25]. Reprinted from Applied Catalysis B: Environmental, 55, Rafael van Grieken *, José Aguado, María-José López-Muñoz, Javier Marugán, Photocatalytic degradation of iron–cyanocomplexes by TiO_2 based catalysts, 201–211, Copyright (2005), with permission from Elsevier.

The photocatalytic performance of the obtained titania was evaluated by analyzing the effect of the catalyst load on the overall efficiency of the photodegradation under both oxic and anoxic conditions. In addition, the impact of the variation of the power supplied by the UV source and of the initial concentration of the cyanocomplex, was assessed. All the experiments were carried out in a bench-scale compound parabolic collector (CPC) photoreactor with artificial UV/LED radiation.

2. Materials and Methods

2.1. Catalyst Treatment

The catalyst was modified by using the hydrothermal treatment [30–34]. Six grames of Aeroxide P25 (Evonik®, Essen, Germany) were mixed with 100 mL of a 10-M solution of NaOH (Merck, Darmstadt, Germany). The solution was stirred to avoid the formation of agglomerates and then it was decanted into a 120-mL beaker. Subsequently, it was transferred to a stainless-steel sealed reactor. The reactor temperature increased up to 120 or 180 °C during 24 or 72 h, according to the 2^3 experimental design described in Table 1. The white precipitate was washed with a 0.1-M HCl (Merck, Darmstadt, Germany) solution under stirring. The solid was recovered by centrifugation followed by a series of washing cycles with deionized water until the pH of the supernatant was 7.4. After drying the solid at 100 °C for 24 h, it was calcinated at 400 or 500 °C during four hours, with a heating gradient of 10 °C/min. Figure 2 shows the detailed procedure for the synthesis of photocatalysts.

Table 1 shows the different conditions of reaction time, reaction temperature and calcination temperature used to prepare each of the eight catalysts. For the statistical analysis, an analysis of variance (ANOVA) was carried out, considering a significance level of 0.05.

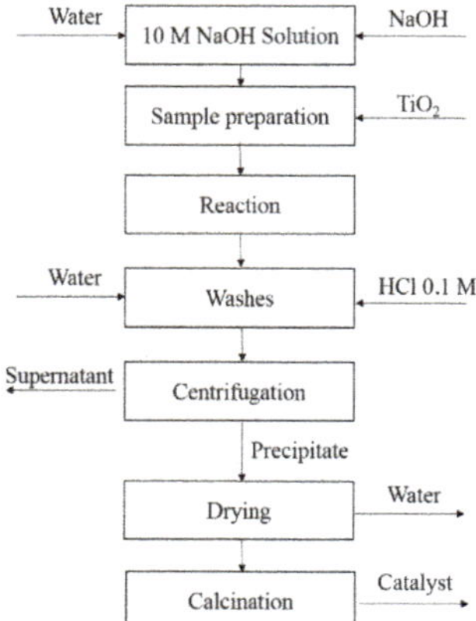

Figure 2. Schematic diagram of the hydrothermal synthesis method used.

Table 1. Experimental runs of the 2^3 factorial design.

Labels	Reaction Temperature (°C)	Reaction Time (h)	Calcination Temperature (°C)
SL400	120 (S)	24 (L)	400
SL500	120 (S)	24 (L)	500
SH400	120 (S)	72 (H)	400
SH500	120 (S)	72 (H)	500
LL400	180 (L)	24 (L)	400
LL500	180 (L)	24 (L)	500
LH400	180 (L)	72 (H)	400
LH500	180 (L)	72 (H)	500

2.2. Catalyst Evaluation

The evaluation of the performance of the modified catalysts was carried out in a bench-scale CPC reactor assisted by a UV/LED radiation source [27,35]. The reactor consisted of four Pyrex tubes with an outside diameter of 2 cm and a length of 11 cm, which were connected to a 750-mL container through a centrifugal pump. The input power of the centrifugal pump was 50 W. The container was sealed at the top with a stopper, which had openings for sampling and oxygen/nitrogen inlet to the gas diffuser [27].

Four 30 W LEDs (TaoYuan Electron Ltd. TY-365 nm, Hong Kong, China) connected in parallel, were used as the artificial light source. The light output was set up with a tilt angle of 115–125° and of 900–1200 mW of radiation intensity per LED [27,35]. Each LED (model GW GPS-3030D, GWINSTEK, Veldhoven, Netherlands) was equipped with a cooling system consisting of heat sinks and a 12-V fan. The UVA radiation intensity was measured with a UV radiometer (DELTA OHM model HD2102.2, Deltha Ohm S.r.l., Padova, Italy) and it was varied by adjusting the current intensity supplied to the LEDs at a constant voltage of 30 V. The reactor had a reactive volume and a total irradiated area of 138.23 cm^3 and 276.4 cm^2, respectively. The ratio of the illuminated volume to the total volume was 0.23. This ratio is useful to characterize the reactive system volume used with respect to those used by other authors and thus be able to compare its performance.

Once the system was loaded with the matrix to be degraded, the LEDs were placed above the tubes at approximately 3 cm of height, whereas the parabolic collectors were placed below the reactor. The use of these reflective surfaces provides a more homogeneous distribution of the radiation reflected to reactor walls since the bottom of the tubes could be illuminated evenly [36].

The hexacyanoferrate III ($K_3[Fe(CN)_6]$, CAS 13746-66-2, (Panreac AppliChem, Darmstadt, Germany) was selected as the model cyanocomplex of the gold mining wastewaters. The control experiments (physical adsorption, i.e., without light; or photolysis, i.e., without catalyst) were carried out with 60 mL of solutions of 100 ppm of the pollutant. For the physical adsorption experiment, the solution was kept under continuous stirring in a 500-mL beaker, under darkness conditions. For the photolysis experiment, the power of the UV-LEDs was set at 30 W that supplies the maximum intensity of UV radiation. For both experiments, an aliquot of 5 mL was taken every 10 min during two hours (time set for the reaction).

The results obtained for the removal were estimated with the Equation (1):

$$\%Degradation = \left(1 - \frac{C}{C_0}\right) \times 100 \tag{1}$$

where C is the final concentration and C_0 the initial concentration

For each optimization step, 500 mL of a solution of 100 ppm of hexacyanoferrate was prepared. For keeping the solution pH above 12, 1 mL of a 10 M solution of NaOH was previously added to 500 mL of hexacyanoferrate solution. After an adsorption stage carried out under darkness conditions for 20 min, the LEDs were turned on to perform the photocatalytic runs. The experiments were carried out at room temperature (20 °C) and 10 mL aliquots (less than 10% of the total volume) were taken at different time intervals. For oxic and anoxic experiments, air or nitrogen was sparged, according to the case, into the solution at a constant flow rate of 0.5 L/min. The optimization study was executed in four stages:

(1) Variation of the catalyst dose (0.1, 0.3, 0.5 and 0.7 g/L) to determine the best performing catalyst dose, at oxic conditions for an hour.
(2) Comparison of reactions (during two hours) under anoxic and oxic conditions, using the best performing catalyst dose selected in the previous stage to select the best conditions for the following experiments: oxic (air) or anoxic (nitrogen).
(3) Variation of the radiation intensity, by testing the power supplied by the LEDs (10, 20 and 30 W) during 3 h of reaction.
(4) Variation of the initial concentration of the contaminant (50–100 ppm) during three hours of reaction.

The hexacyanoferrate (III) concentration was followed by UV-VIS (JASCO V-730 spectrophotometer, Easton, MD, USA) at 303 nm, corresponding to its maximum absorbance wavelength in the UV spectrum. The measurement of total dissolved iron was performed using atomic absorption spectrometry (Thermo Scientific iCE 3000, Waltham, MA, USA) and the measurement of CN^- by titration with $AgNO_3$ according to the Standard Methods 4500 [37].

A kinetic law with a two-step reaction was used to describe the degradation of hexacyanoferrate (III). The first step (faster) corresponds to the adsorption of $Fe(CN)_6^{3-}$ onto the surface of TiO_2 and degradation of the iron modified, whereas the second step (slower) corresponds to the reduction of the iron present in the cyano-metallic complex (that corresponds to the removal of dissolved iron) [38].

For the kinetic analysis of the photo reductive process of the iron cyanocomplex, a pseudo first-order reaction rate equation was proposed (Equations (2) and (3)), as suggested by previous studies [39–41]:

$$-\frac{dC}{dt} = k'C \tag{2}$$

$$ln(C_0/C) = k't \tag{3}$$

where k' is the pseudo first-order rate constant (min^{-1}), C_0 and C are the initial and final concentrations of the iron complex in solution, respectively. The $ln\ (C_0/C)$ was plotted versus time for obtaining the k' value, which is the slope of the equation of the line.

2.3. Characterization

The crystalline phases of the resulting solid from the hydrothermal synthesis were characterized using X-ray diffraction (XRD) on a X'per PRO-PANalytical diffractometer with CuKα radiation

(0.1542 nm) with a 2θ sweep between 0° and 90°. The surface area was determined by the Brunauer–Emmett–Teller method (BET) by adsorption–desorption of nitrogen (N_2) at 77 K and the volume and size of the pore were determined by the Barrett–Joyner–Halenda method (BJH) in a Micromeritics equipment ASAP 2020 V4.01 (Micromeritics, Norcross, GA, USA).

The morphology was analyzed by scanning electron microscopy (SEM) and X-ray energy dispersion spectrometry (EDS) was used for the analysis of elemental composition of the catalyst in a JEOL JSM 6490 LV brand equipment. The semiconductor bandgap (E_g) was estimated by measuring the material transmittance with UV-vis diffuse reflectance spectroscopy (UV DRS) in a Thermo Scientific Evolution 300 PC series EVOP068001 spectrophotometer. Finally, the Fourier-transform infrared spectroscopy (FT-IR) was used to identify the functional groups of the inorganic and organic substances (FT/IR-4100 type-A).

2.4. Estimation of the Electric Oxidation Efficiency (E_{Eo})

The IUPAC has proposed methods to calculate the electrical consumption of an AOP, depending on the type of reactor and the amount of contaminant to be treated. For low concentrations, it is proposed to use the electric energy per order (E_{Eo}). This parameter consists of the electrical energy (kWh) required to remove the pollutant up to 90% of its initial concentration per volume unit. The E_{Eo} can be calculated using the Equation (4), following the methodology proposed by Shirzad-Siboni et al. [41] and Daneshvar et al. [40]:

$$E_{E_o} = \frac{1000\ P\ t}{60\ V\ log\left(C_0/C_f\right)} \tag{4}$$

where P is the power supplied to the system (kW) and it is defined as the product of electric potential and the current intensity (A); V is the total reactive volume (L), and t is time (h). From Equations (3) and (4), the E_{Eo} can be calculated as follows:

$$E_{E_o} = \frac{38.4P}{Vk'} \tag{5}$$

3. Results and Discussion

3.1. Photolysis and Adsorption

The control tests in 3-h experiments showed that the photolysis contributes moderately to the removal of contaminants and the release of free cyanide. A 17% of photolytic removal of hexacyanoferrate and a 12% of cyanide release were achieved, which is in agreement with the results reported in this literature review [26]. On the contrary, the adsorption had a minor effect both in the elimination of contaminants and in the release of cyanide, respectively, 10% and less than 5% after three hours of experimentation. It was observed that 8% of the initial hexacyanoferrate concentration was adsorbed during the first 20 min of the experiment and therefore the dark period for the photocatalytic runs was set at 20 min.

3.2. Evaluation of Synthesized Materials

3.2.1. Catalyst Load

This behavior observed in the Figure 3 is explained by the lower flow of photons into the reactive system resulting from the higher turbidity (catalyst loads higher than 0.5 g/L) of the slurry to treat [42]. This screening effect limits the effectiveness of the treatment by decreasing the local volumetric rate of photon absorption for tubular photoreactors, which has been analyzed by Colina-Marquez et al. in 2010 [43] and Mueses et al. in 2013 [44]. Those studies reported an optimal catalyst load of 0.3 g/L for CPC reactors, approximately. In turn, Osathaphan et al. [45] used catalyst loads between 0.1 and 4 g/L without affecting the reductive treatment considerably. Given the best results when using 0.5 g/L of both SL400 and SL500, both catalysts were promising to degrade the cyanocomplex. To select the best performing catalysts modified, photocatalytic experiments were performed using 0.5 g/L of each catalyst to degrade the pollutant during 2 h of reaction. For the further experiments, 0.5 g/L of SL400 was selected, due to the better performance and also in order to save energy in the calcination process.

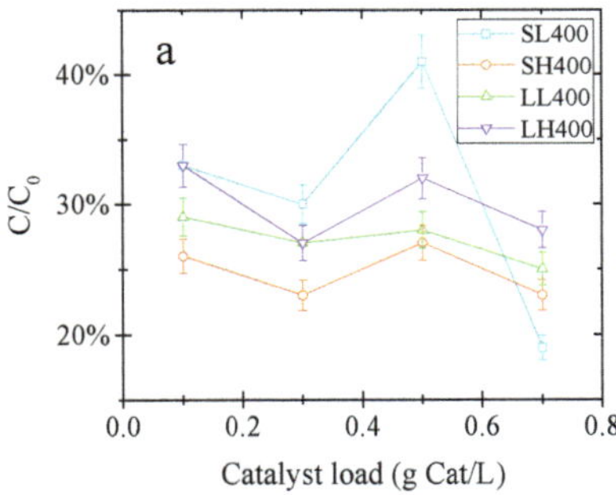

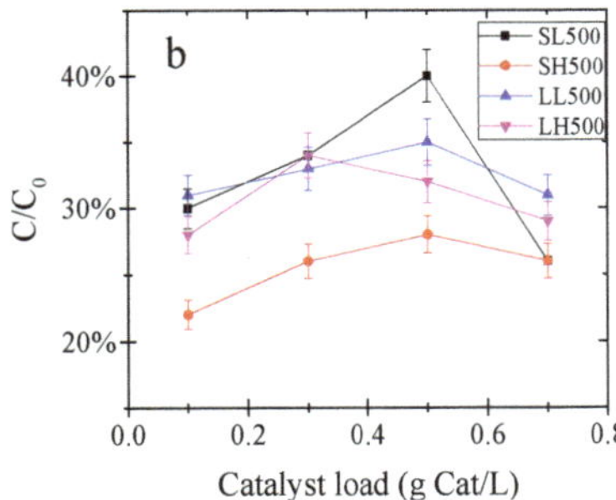

Figure 3. Evaluation of the best catalyst load to remove the cyanocomplex (hexacyanoferrate III) after 1 h of reaction (**a**) Catalysts calcined at 400 °C; (**b**) Catalysts calcined at 500 °C. Operating conditions: Initial pollutant concentration of 100 ppm, 20 min of adsorption, LED power supply of 20 W, air flow of 0.5 L/min.

Table 2 shows the results of the degradation obtained at different synthesis temperatures, calcination temperature and synthesis times.

Table 2. Degradation percentage of $K_3[Fe(CN)_6]$ after 2 h of reaction, using a catalyst load of 0.5 g/L.

Calcination Temperature	**400 °C**				**500 °C**			
Synth. Time / Synthesis Temp.	**24 (L)**		**72 (H)**		**24 (L)**		**72 (H)**	
	SL400		SH400		SL500		SH500	
120 °C (S)	53	51	50	50	49	48	47	49
	LL400		LH400		LL500		LH500	
180 °C (L)	50	48	48	50	43	44	51	50

Operating conditions: initial pollutant concentration of 100 ppm, 20 min of adsorption, LED nominal power of 20 W, air flow of 0.5 L/min. Each experiment was done in duplicate.

A statistical analysis (see Table 3) of the information reported in Table 2 was carried out by using Statgraphics® Centurion XVI (version 16.2.04, Statpoint Technologies Inc., The Plains, VA, USA) and it was found that the calcination temperature was the most significant effect on the response variable within the evaluated intervals (see Figure 4), obtaining better results with 400 °C. The second most significant effect was the synthesis temperature and the best results were obtained at 120 °C; however, it is not statistically significant. Comparing the information of the table with the Pareto chart (Figure 4), it can be observed that the calcination temperature has a negative effect; that means that an increase of this variable represents a degradation decrease. This behavior can be attributed to the reduction of the surface area of the catalyst or material sintering at higher temperatures [6].

Table 3. ANOVA for degradation percentage of $K_3[Fe(CN)_6]$.

Source of Variation	SS	df	MS	F	*p*-Value
A: Calcination Temperature	225.625	1	225.625	10.62	0.0116
B: Synthesis time	50.625	1	50.625	2.38	0.1613
C: Synthesis temperature	105.625	1	105.625	4.97	0.0563
AB	180.625	1	180.625	8.50	0.0194
AC	0.5625	1	0.5625	0.26	0.6208
BC	225.625	1	225.625	10.62	0.0116
Blocks	0.0625	1	0.0625	0.03	0.8681
Total error	17.0	8	2.125		
Total (corr.)	964.375	15			

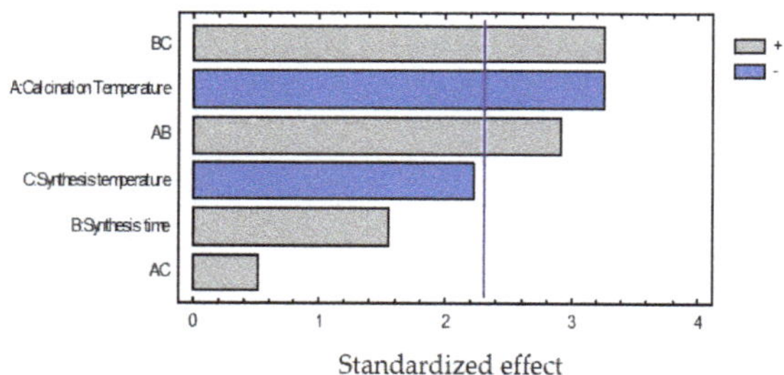

Figure 4. Standardized Pareto Chart for degradation percentage of $K_3[Fe(CN)_6]$.

On the other hand, although the synthesis time was not significant, its interactions with the other variables were meaningful and synergistic. This behavior is interesting because it means that a simultaneous increase of the calcination and synthesis temperatures with the synthesis time represents an improvement on the pollutant removal. In fact, the interaction between the synthesis time and the synthesis temperature (BC) is as significant as the effect of the calcination temperature. In addition, it was found that the best results for the degradation of the cyanocomplex were obtained for the catalyst modified at 24 h—120 °C to 400 °C (SL400). Considering all these facts, the following stages were carried out using SL400.

3.2.2. Tests Under Oxic and Anoxic Conditions

The degradation of the cyanocomplex by photocatalysis using SL400 was evaluated under oxic and anoxic conditions, to evaluate the importance of the presence of oxygen (Figure 5).

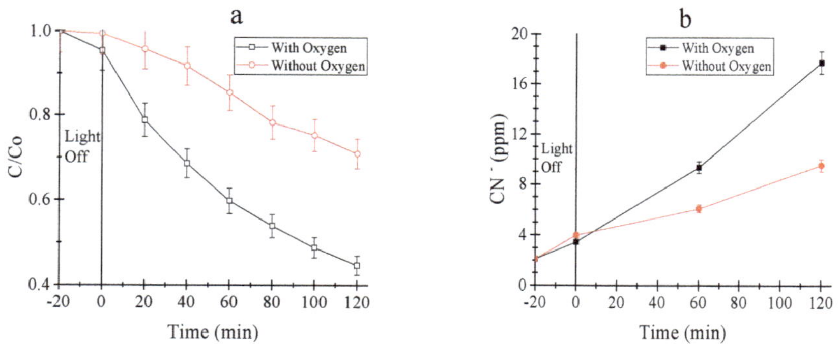

Figure 5. (**a**) Cyanocomplex (hexacyanoferrate III) degradation; and (**b**) release of free cyanide. Operating conditions: Initial concentration of $K_3[Fe(CN)_6]$ of 100 ppm, catalyst load of 0.5 g/L, 20 min of adsorption, Power supply: 20 W, air or nitrogen flow of 0.5 L/min, reaction time of 2 h.

After two hours of reaction, 56% of the cyanocomplex was degraded in the presence of oxygen, whereas it was only 29% when air was replaced by nitrogen. In turn, the cyanide release was two times higher when air containing oxygen was used (18 ppm in the presence of oxygen and 9 ppm using nitrogen). Finally, for the total removal of iron, a removal of 40% was achieved in the presence of oxygen and only 15% under an inert atmosphere. The higher degradation of the cyanocomplex and release of free cyanide in the presence of oxygen can be ascribed to the, electrons directly reducing iron and the oxidation of the complex by holes, hydroxyl radicals and superoxide anions. In contrast to our results that showed that the presence of oxygen during the reaction increases the degradation of the complex, Yang et al. [46] and Ku and Jung [47] reported a better performance of the P25 TiO_2 for the removal of the studied contaminants under anoxic conditions. In these reports, the authors observed that the presence of oxygen did not have a significant effect on the contaminant removal, whereas a higher reduction was showed with nitrogen.

3.2.3. Effect of the Radiation Intensity

The availability of UV photons directly affects the generation of electron–hole pairs. By comparing the results obtained at 10, 20 and 30 W (Figure 6a), it can be observed that the radiation intensity higher effect when increasing from 10 to 20 W than after a further increase to 30 W. Regarding to the degradation of the cyanocomplex, removals of 55, 73 and 79% were obtained with 10, 20 and 30 W, respectively. Additionally, iron removals of 30, 48 and 60% were achieved for 10, 20, and 30 W, respectively. The dissolved iron concentration was analyzed to corroborate its removal from the solution and its deposition onto the catalyst surface (Figure 6b).

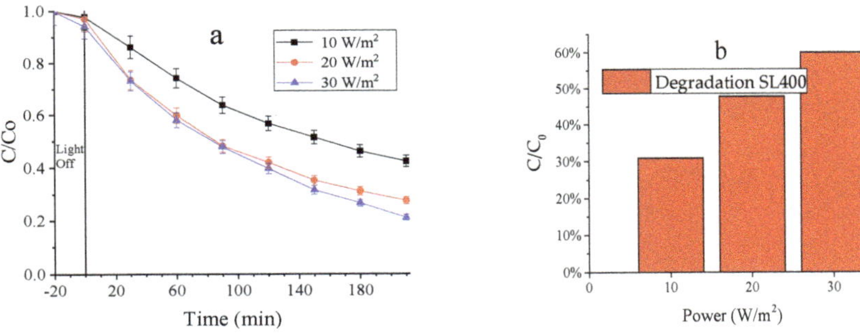

Figure 6. (**a**) Degradation of the cyanocomplex (hexacyanoferrate III) at 10 W, 20 W and 30 W; (**b**) Total removal of dissolved iron. Operating conditions: Initial concentration of $K_3[Fe(CN)_6]$ of 100 ppm, catalyst load of 0.5 g/L, 20 min of adsorption, air flow of 0.5 L/min, reaction time of 3.5 h.

The degradation values obtained with 20 and 30 W exhibited similar behaviors. An energy increase of 33% (20 to 30 W) yielded just an increase of 8.12% for the cyanocomplex degradation. This means that this energy increase is not enough to significantly affect the degradation performance. Therefore, the radiation intensity of 20 W was selected as the best condition due to the less energy consumption. Similar results were obtained by Rodriguez and Ossa [27], reporting a better but not significant performance when working at 30 than 20 W, and thus the selection of an inferior power supply to avoid an additional electrical consumption.

3.2.4. Comparison between Modified TiO_2 and the Raw P25

By comparing the raw and treated TiO_2, the degradation efficiency obtained with SL400 was 70%, whereas TiO_2 P25 led to a photocatalytic removal of 80% (Figure 7a). In turn, 20 ppm of cyanide are released by SL400 and less 10% is observed for TiO_2 P25, with 18 ppm of cyanide released (Figure 7b). Although for the complex degradation, the TiO_2 P25 showed better results; regarding to the free cyanide release, the SL400 showed a performance 10% higher. As the initial concentration of contaminant increases, the degradation decreases, as it was documented in the studies of Yang et al. [46] and Samarghandi et al. [39]. The cyanide release can be beneficial since it can be reused in the mining processes where such cyanide can be returned for the mineral (gold) re-extraction process. This feature would make the use of the synthesized material economically and environmentally attractive and also attenuate its weakness against P25 in terms of degradation of the hexacyanoferrate complex.

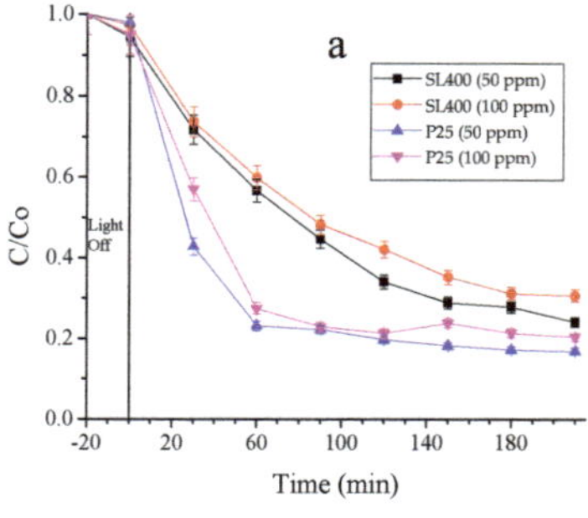

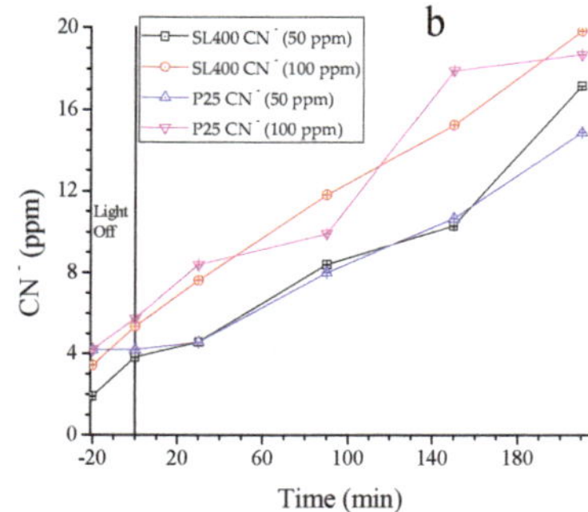

Figure 7. (**a**) Cyanocomplex degradation; (**b**) Free cyanide released. Operating conditions: 0.5 L/min of air bubbled, 0.5 g cat/L solution, 3.5 h of reaction with 30 min of adsorption and power supply of 20 W.

Van Grieken et al. [25] reported that the oxidative degradation of hexacyanoferrate (100 ppm of initial concentration) releases around 20 ppm of CN^- in 240 min of irradiation by using mercury lamps. In this study, the same amount of cyanide ion was released in 210 min by using a UVA/LED photon source.

Table 4 shows the values found for the pseudo first-order speed constant (min^{-1}) for a reaction time of 210 min. As it can be seen, the P25 TiO_2 rate constants are higher than the SL400 ones for both initial concentrations of the pollutant. This can be explained because of the differences in superficial area, particle size distribution, semiconductor purity and other features in electronic properties.

Table 4. Pseudo first order rate constants.

Initial Concentration (ppm)	Catalyst Type	Apparent Reaction Rate Constant k' (min^{-1})
100	P25	0.1924
100	SL400	0.1679
50	P25	0.211
50	SL400	0.1977

3.2.5. Electric Oxidation Efficiency

Table 5 shows the E_{Eo} values obtained for the P25 and the SL400 sample with two different concentrations of hexacyanoferrate.

Table 5. Electrical oxidation efficiency for the catalysts used.

Catalyst Type	Voltage (V)/Amperage (A)	Initial Concentration of Contaminant	E_{Eo} (kWh/m^3)
P25	30/0.8	100	7.98
SL400	30/0.8	100	9.15
P25	30/0.8	50	7.28
SL400	30/0.8	50	7.77

The P25 still exhibits better performance regarding to the energy consumption. This behavior is related to the higher activity of the commercial standard, which was discussed previously. The obtained results are similar to the reported ones by Daneshvar et al. [40], which did not exceed 10 kWh/m^3.

On the other hand, when the value obtained is compared with the study of Rodriguez and Ossa [27], it was found that the E_{Eo} is 40 and 20 times lower, respectively, than the presented ones in Table 5. In these works, it was reported the same concentration of $Fe(CN)_6$ but with the use of different catalysts.

3.3. Characterization of the Photocatalyst

3.3.1. Fourier-Transform Infrared Spectroscopy (FT-IR)

The Figure 8 shows the IR spectra of the SL400 before and after usage in the photocatalytic experiments. Four bands are highlighted that are common in both spectra. As described by Thennarasu et al. [48], the peaks observed around 3300–3400 cm^{-1} correspond to the stretching vibrations (stress) of the •OH and around 1600 cm^{-1} arises from the water bending mode that can be associated with water absorbed by the catalyst due to the presence of moisture in the materials by contact with air. The main bands below 1000 cm^{-1} were attributed to the Ti-O and Ti-O-Ti bending vibrations. The band around 1300 cm^{-1} is attributed to the C-H bending vibrations.

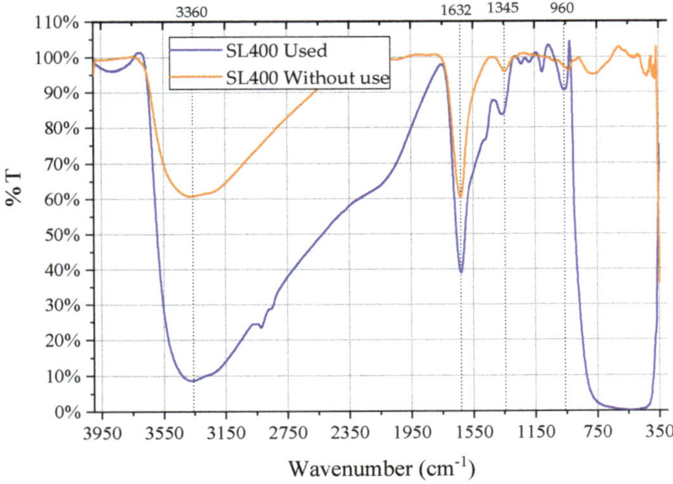

Figure 8. FTIR spectra of SL400 before and after the photocatalytic reaction.

3.3.2. XRD Results

According to Mozia et al. [33], the peaks found at 2θ of 24°, 28° and 48° as those observed for SL400 (Figure 9) correspond to titanates of the form $A_2Ti_2O_5{\cdot}H_2O$ and $A_2Ti_3O_7$. The sodium titanates ($Ti_{12}O_{36}Na_4$ or Ti_3O_9Na) exhibit peaks at 10°, 24°, 28°, 48° and 62°, which evidence the presence of the anatase phase of TiO_2 at 25°, 62°, and 82°. The analysis showed no significant amount of rutile since may be found at calcination temperatures over 600 °C.

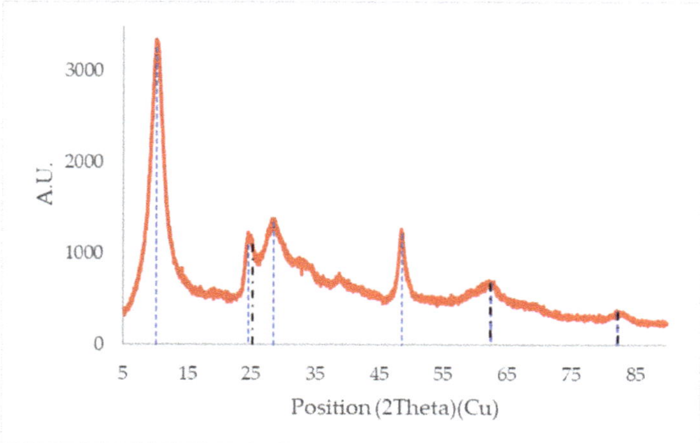

Figure 9. XRD pattern for SL400 without use. The blue dotted lines represent the peaks associated with titanates and the black dotted line represents the peaks associated with the anatase phase of TiO_2.

The most significant difference between the SL400 diffractogram (Figure 9) and that of P25 (Figure 10) without modifications [49], is the sharper peaks obtained by XRD for the commercial P25. This means a more crystalline structure for the unmodified P25 and some amorphous characteristics for the modified material (SL400). This modification affected the overall performance of the modified material regarding to the activity and, therefore, the pollutant removal. In addition, the XRD of SL400 does not have characteristic peaks of rutile phase as P25, which are known to improve the photocatalytic activity thanks to its synergistic effect with the anatase.

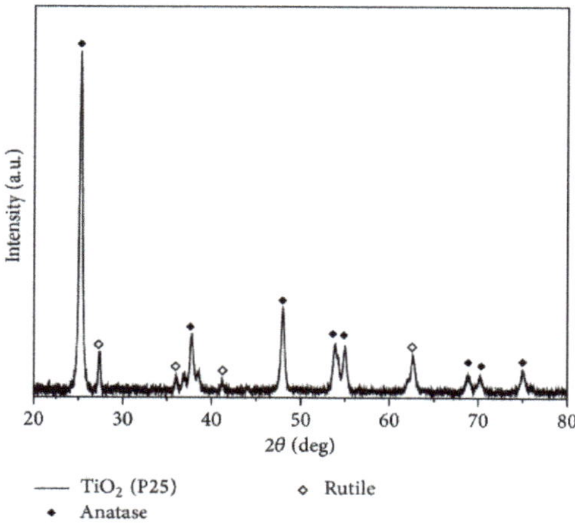

Figure 10. XRD pattern for TiO_2 P25 [49].

3.3.3. EDS Results

The Figure 11 shows a micrograph obtained from SL400. Additionally, an energy-dispersive X-ray spectroscopy analysis (EDS) was performed for elemental detection of the modified catalyst (see Figure 11). This analysis shows the type of elements present in different analyzed areas of the catalyst, where the presence of Carbon (C), Oxygen (O), Sodium (Na) and Titanium (Ti) were exhibited, with their respective composition, as shown in Table 6.

Figure 11. EDS analysis of the modified catalyst SL400.

Table 6. EDS results in % weight in the modified catalyst.

Spectra	C (%)	O (%)	Na (%)	Ti (%)	Total (%)
1	-	34.87	5.13	60.00	100
2	4.62	44.52	7.37	43.49	100
3	4.56	43.86	6.22	45.36	100
4	3.18	47.63	7.69	41.50	100
5	5.67	46.00	7.02	41.31	100
Max	5.67	47.63	7.69	60.00	
Min	3.18	34.87	5.13	41.31	
Average	3.61	43.38	6.69	46.33	

According to the EDS results, the presence of carbon in the material (3–6%), probably from impurities in the precursors used for the synthesis, can affect negatively the photocatalytic performance because of the number of active sites on the semiconductor surface decreases as the carbon occupies them.

3.3.4. Surface Area Results

The surface area was 127.84 m^2/g, which is greater than the surface area of the precursor material (50 m^2/g). The pore volume of the total amount absorbed was 0.197 cm^3/g and the pore size distribution analyzed by the BJH method was approximately 58 Å (5.8 nm) for an average particle size of 469 Å (46.9 nm). An isotherm of type IV was observed (Figure 12) with a hysteresis type III, which suggests that this catalyst is a mesoporous solid (2–50 nm).

Although the sample SL400 has a surface area higher than the P25's one, the number of active sites could not exceed the amount of sites of the TiO_2 P25, since the modified catalyst did not exceed the photocatalytic activity of the precursor. In addition, the absence of rutile phase affects the overall activity of the TiO_2, since this phase in the P25 acts synergistically with the anatase to improve the activity of the catalyst. The surface area is similar to those obtained by Turki et al. [50], Sikhwivhilu et al. [51] and Fen et al. [52]. On the other hand, some studies have obtained values higher than 200 m^2/g as is the case of Thennarasu et al. [48] and Camposeco et al. [53] with important photocatalytic activity.

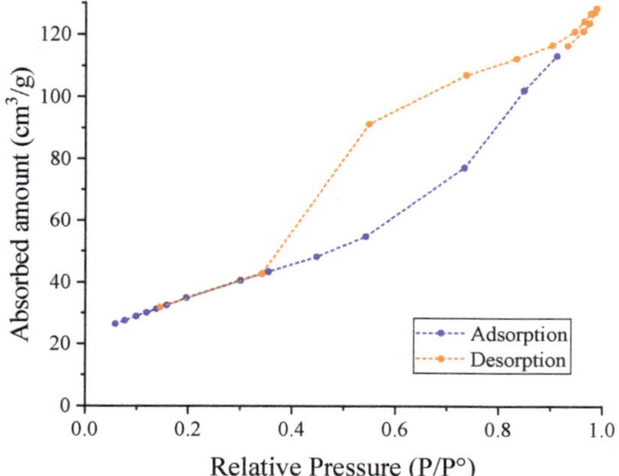

Figure 12. Isotherm absorption-desorption of the modified catalyst SL400.

3.3.5. Bandgap Energy Estimation by DRS

The bandgap energy (E_g) is one of the most important parameters in the photocatalytic activity of TiO_2 since it determines the effective wavelength interval for photon absorption. This parameter was estimated with the Kubelka-Munk theory according to the methodology reported by López and Gómez [54] (see Figure 13). It has to be considered that the crystal size, the particle size, the aggregation state of the particles, and the impurities present in the solid and the method of synthesis, can significantly affect the E_g.

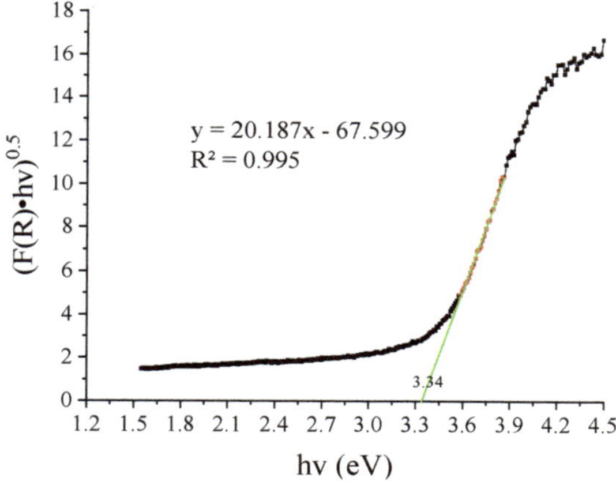

Figure 13. Estimation of bandgap energy (E_g) of the catalyst SL400 using the Kubelka-Munk function, the red points represent those used to obtain the slope of the line and obtain the intercept on the x axis.

The energy of the bandgap obtained was 3.34 eV and the wavelength (estimated with Equation (6)) at which the catalyst is activated is 370 nm.

$$\lambda = \frac{h \times c}{h_v} \tag{6}$$

If these values are compared with those reported for TiO_2 P25 (E_g = 3.20 eV; λ = 385 nm), it is expected that the modified catalyst underperform respect to the commercial standard, regarding the UV photons absorption. This can be a significant drawback when it is intended to use a wide spectrum photons source.

4. Conclusions

The modified TiO_2 P25, via the hydrothermal method, did not improve the $Fe(CN)_6$ removal with respect to the obtained one with the original P25. This could be attributed to the loss of both the rutile phase and the material crystallinity. In addition, the increase of the bandgap energy for the modified P25 is another drawback since it affects the photon absorption by the semiconductor. Although the higher free cyanide release achieved with the modified material can be considered as a shortcoming regarding to the environmental potential of this material, in this particular case, this can be beneficial since this free cyanide could be reused for the gold extraction process and so, obtain a closed cycle for the water use. Furthermore, the increase of the specific surface area can be a promising result, in terms of physical adsorption of the studied pollutant or metallic cations.

While at a first sight the hydrothermal method did not improve the activity of the P25, further studies should be carried out to obtain more information about the structural modifications of the catalyst and potential advantages for photocatalytic applications.

Author Contributions: Conceptualization, A.A.-S. and L.A.B.-B.; methodology, A.A.-S. and L.A.B.-B.; software, L.A.B.-B. and A.A.-S.; validation, A.A.-S., L.A.B.-B., F.M.-M. and J.Á.C.-M.; formal analysis, K.M.A.-V. and A.A.-S.; investigation, K.M.A.-V., A.A.-S. and L.A.B.-B.; resources, F.M.-M.; data curation, K.M.A.-V. and L.A.B.-B.; writing—original draft preparation, K.M.A.-V.; writing—review and editing, M.A.M. and J.Á.C.-M.; visualization, M.A.M.; supervision, A.A.-S.; project administration, A.A.-S.; funding acquisition, F.M.-M. and M.A.M. All authors have read and agreed to the published version of the manuscript.

Funding: This research received external funding from "Departamento Administrativo de Ciencia, Tecnología e Innovación (COLCIENCIAS) through GRANT 1106-669-45250 "Recuperación de oro y tratamiento de aguas residuales cianuradas en la industria aurífera de la región pacífico colombiana" and the Universidad del Valle.

Acknowledgments: The authors are grateful to Universidad del Valle and the COLCIENCIAS for the Ph.D. scholarship 567-2012. Also, the authors thank the Biotechnology and Nanotechnology Laboratory of Tecnoparque (Nodo Cali) for the support with the analytical techniques. Colina-Márquez and Mueses thank the University of Cartagena. All authors send thanks to Ana Rita Lado Ribeiro from the University of Porto for the much-appreciated writing revisions of this manuscript.

Conflicts of Interest: The authors declare no conflict of interest.

References

1. Moran, R. El Cianuro en la Minería: Algunas Observaciones sobre la Química, Toxicidad y Análisis de las Aguas Asociadas con la Minería. *Ecología* **1999**, *99*, 23.
2. Kuyucak, N.; Akcil, A. Cyanide and Removal Options from Effluents in Gold Mining and Metallurgical Processes. *Miner. Eng.* **2013**, *50–51*, 13–29. [CrossRef]
3. López-Vásquez, A.F.; Colina-Márquez, J.A.; Machuca-Martínez, F. Multivariable analysis of 2,4-d herbicide photocatalytic degradation. *DYNA* **2011**, *78*, 119–125.
4. Pichat, P. A brief survey of the practicality of using photocatalysis to purify the ambient air (indoors or outdoors) or air effluents. *Appl. Catal. B* **2019**, *245*, 770–776. [CrossRef]
5. Lozano-Morales, S.; Morales, G.; Lopez Zavala, M.; Arce, A.; Machuca, F. Photocatalytic Treatment of Paracetamol Using TiO_2 Nanotubes: Effect of pH. *Processes* **2019**, *7*, 319. [CrossRef]
6. Arce-Sarria, A.; Machuca-Martínez, F.; Bustillo-Lecompte, C.; Hernández-Ramírez, A.; Colina-Márquez, J. Degradation and Loss of Antibacterial Activity of Commercial Amoxicillin with TiO2/WO3-Assisted Solar Photocatalysis. *Catalysts* **2018**, *8*, 222. [CrossRef]
7. Sulka, G.D.; Brzózka, A.; Liu, L. Fabrication of diameter-modulated and ultrathin porous nanowires in anodic aluminum oxide templates. *Electrochim. Acta* **2011**, *56*, 4972–4979. [CrossRef]
8. Soto Rodriguez, P.E.D.; Olivares, F.; Gómez-Ruiz, S.; Cabrera, G.; Villalonga, R.; Segura del Río, R. Functionalized carbon nanotubes decorated with fluorine-doped titanium dioxide nanoparticles on silicon substrate as template for titanium dioxide film photo-anode grown by chemical vapour deposition. *Thin Solid Film.* **2018**, *656*, 30–36. [CrossRef]
9. Wang, H.; Guo, Z.; Wang, S.; Liu, W. One-dimensional titania nanostructures: Synthesis and applications in dye-sensitized solar cells. *Thin Solid Film.* **2014**, *558*, 1–19. [CrossRef]

10. Byrappa, K.; Adschiri, T. Hydrothermal technology for nanotechnology. *Prog. Cryst. Growth Charact. Mater.* **2007**, *53*, 117–166. [CrossRef]
11. Wang, Y.; He, Y.; Lai, Q.; Fan, M. Review of the progress in preparing nano TiO_2: An important environmental engineering material. *J. Environ. Sci.* **2014**, *26*, 2139–2177. [CrossRef] [PubMed]
12. Pang, Y.L.; Lim, S.; Ong, H.C.; Chong, W.T. A critical review on the recent progress of synthesizing techniques and fabrication of TiO_2-based nanotubes photocatalysts. *Appl. Catal. A* **2014**, *481*, 127–142. [CrossRef]
13. Kasuga, T.; Hiramatsu, M.; Hoson, A.; Sekino, T.; Niihara, K. Titania Nanotubes Prepared by Chemical Processing. *Adv. Mater.* **1999**, *11*, 1307–1311. [CrossRef]
14. Suzuki, Y.; Yoshikawa, S. Synthesis and Thermal Analyses of TiO_2-Derived Nanotubes Prepared by the Hydrothermal Method. *J. Mater. Res.* **2004**, *19*, 982–985. [CrossRef]
15. Gogotsi, Y.; Libera, J.A.; Yoshimura, M. Hydrothermal synthesis of multiwall carbon nanotubes. *J. Mater. Res.* **2000**, *15*, 2591–2594. [CrossRef]
16. Nguyen Phan, T.-D.; Pham, H.-D.; Viet Cuong, T.; Jung Kim, E.; Kim, S.; Woo Shin, E. A simple hydrothermal preparation of TiO_2 nanomaterials using concentrated hydrochloric acid. *J. Cryst. Growth* **2009**, *312*, 79–85. [CrossRef]
17. Yoshimura, M.; Byrappa, K. Hydrothermal Processing of Materials: Past, Present and Future. *J. Mater. Sci.* **2007**, *43*, 2085–2103. [CrossRef]
18. Dolat, D.; Quici, N.; Kusiak-Nejman, E.; Morawski, A.W.; Li Puma, G. One-step, hydrothermal synthesis of nitrogen, carbon co-doped titanium dioxide (N,CTiO_2) photocatalysts. Effect of alcohol degree and chain length as carbon dopant precursors on photocatalytic activity and catalyst deactivation. *Appl. Catal. B* **2012**, *115–116*, 81–89. [CrossRef]
19. Nguyen-Phan, T.-D.; Pham, V.H.; Chung, J.S.; Chhowalla, M.; Asefa, T.; Kim, W.-J.; Shin, E.W. Photocatalytic performance of Sn-doped TiO_2/reduced graphene oxide composite materials. *Appl. Catal. A* **2014**, *473*, 21–30. [CrossRef]
20. Li, Z.; Hou, B.; Xu, Y.; Wu, D.; Sun, Y.; Hu, W.; Deng, F. Comparative study of sol–gel-hydrothermal and sol–gel synthesis of titania–silica composite nanoparticles. *J. Solid State Chem.* **2005**, *178*, 1395–1405. [CrossRef]
21. Andersson, M.; Österlund, L.; Ljungström, S.; Palmqvist, A. Preparation of Nanosize Anatase and Rutile TiO2 by Hydrothermal Treatment of Microemulsions and Their Activity for Photocatalytic Wet Oxidation of Phenol. *J. Phys. Chem. B* **2002**, *106*, 10674–10679. [CrossRef]
22. Liu, N.; Chen, X.; Zhang, J.; Schwank, J.W. A review on TiO_2-based nanotubes synthesized via hydrothermal method: Formation mechanism, structure modification, and photocatalytic applications. *Catal. Today* **2014**, *225*, 34–51. [CrossRef]
23. Huang, K.-C.; Chien, S.-H. Improved visible-light-driven photocatalytic activity of rutile/titania-nanotube composites prepared by microwave-assisted hydrothermal process. *Appl. Catal. B* **2013**, *140–141*, 283–288. [CrossRef]
24. Camposeco, R.; Castillo, S.; Mejía-Centeno, I.; Navarrete, J.; Gomez, R. Effect of the Ti/Na molar ratio on the acidity and the structure of TiO_2 nanostructures: Nanotubes, nanofibers and nanowires. *Mater. Charact.* **2014**, *90*, 113–120. [CrossRef]
25. Van Grieken, R.; Aguado, J.; López-Muñoz, M.-J.; Marugán, J. Photocatalytic degradation of iron–cyanocomplexes by TiO_2 based catalysts. *Appl. Catal. B* **2005**, *55*, 201–211. [CrossRef]
26. Betancourt-Buitrago, L.A.; Hernandez-Ramirez, A.; Colina-Marquez, J.A.; Bustillo-Lecompte, C.F.; Rehmann, L.; Machuca-Martinez, F. Recent Developments in the Photocatalytic Treatment of Cyanide Wastewater: An Approach to Remediation and Recovery of Metals. *Processes* **2019**, *7*, 225. [CrossRef]
27. Betancourt-Buitrago, L.A.; Ossa-Echeverry, O.E.; Rodriguez-Vallejo, J.C.; Barraza, J.M.; Marriaga, N.; Machuca-Martínez, F. Anoxic photocatalytic treatment of synthetic mining wastewater using TiO2 and scavengers for complexed cyanide recovery. *Photochem. Photobiol. Sci.* **2019**, *18*, 853–862. [CrossRef]
28. Caicedo, D.F.; Brum, I.A.S.; Buitrago, L.A.B. Photocatalytic degradation of ferricyanide as synthetic gold mining wastewater using TiO_2 assisted by H_2O_2. *REM-Int. Eng. J.* **2020**, *73*, 99–107. [CrossRef]
29. Kim, S.H.; Lee, S.W.; Lee, G.M.; Lee, B.-T.; Yun, S.-T.; Kim, S.-O. Monitoring of TiO_2-catalytic UV-LED photo-oxidation of cyanide contained in mine wastewater and leachate. *Chemosphere* **2016**, *143*, 106–114. [CrossRef]

30. Liu, C.; Zhang, L.; Liu, R.; Gao, Z.; Yang, X.; Tu, Z.; Yang, F.; Ye, Z.; Cui, L.; Xu, C.; et al. Hydrothermal synthesis of N-doped TiO_2 nanowires and N-doped graphene heterostructures with enhanced photocatalytic properties. *J. Alloy. Compd.* **2016**, *656*, 24–32. [CrossRef]
31. Procek, M.; Stolarczyk, A.; Pustelny, T.; Maciak, E. A Study of a QCM Sensor Based on TiO_2 Nanostructures for the Detection of NO_2 and Explosives Vapours in Air. *Sensors* **2015**, *15*, 9563–9581. [CrossRef] [PubMed]
32. Chang, G.; Cheng, Z.; Warren, R.; Song, G.; Shen, J.; Lin, L. Highly Efficient Photocatalysts for Surface Hybridization of TiO_2 Nanofibers with Carbon Films. *ChemPlusChem* **2015**, *80*, 827–831. [CrossRef] [PubMed]
33. Mozia, S.; Borowiak-Paleń, E.; Przepiórski, J.; Grzmil, B.; Tsumura, T.; Toyoda, M.; Grzechulska-Damszel, J.; Morawski, A.W. Physico-chemical properties and possible photocatalytic applications of titanate nanotubes synthesized via hydrothermal method. *J. Phys. Chem. Solids* **2010**, *71*, 263–272. [CrossRef]
34. Mozia, S. Application of temperature modified titanate nanotubes for removal of an azo dye from water in a hybrid photocatalysis-MD process. *Catal. Today* **2010**, *156*, 198–207. [CrossRef]
35. Betancourt-Buitrago, L.A.; Vásquez, C.; Veitia, L.; Ossa-Echeverry, O.; Rodriguez-Vallejo, J.; Barraza-Burgos, J.; Marriaga-Cabrales, N.; Machuca-Martínez, F. An approach to utilize the artificial high power LED UV-A radiation in photoreactors for the degradation of methylene blue. *Photochem. Photobiol. Sci.* **2017**, *16*, 79–85. [CrossRef]
36. Blanco, J.; Malato, S.; Peral, J.; Sánchez, B.; Cardona, I. Diseño de reactores para fotocatálisis: Evaluación comparativa de las distintas opciones. In *Eliminación de Contaminantes por Fotocatálisis Heterogénea*; Blesa, M., Ed.; CYTED: Buenos Aires, Argentina, 2001; p. 253.
37. Rice, E.W.; Bridgewater, L.; American Public Health; American Water Works; Water Environment. *Standard Methods for the Examination of Water and Wastewater*; American Public Health Association: Washington, DC, USA, 2012.
38. Osathaphan, K.; Ruengruehan, K.; Yngard, R.; Sharma, V. Photocatalytic degradation of Ni(II)-Cyano and Co(III)-cyano complexes. *Water Air Soil Pollut.* **2013**, *224*, 1647. [CrossRef]
39. Samarghandi, M.; Yang, J.-K.; Giahi, O.; Shirzad-Siboni, M. Photocatalytic reduction of hexavalent chromium with illuminated amorphous FeOOH. *Environ. Technol.* **2014**, *36*, 1–30. [CrossRef]
40. Daneshvar, N.; Aleboyeh, A.; Khataee, A.R. The evaluation of electrical energy per order (EEo) for photooxidative decolorization of four textile dye solutions by the kinetic model. *Chemosphere* **2005**, *59*, 761–767. [CrossRef]
41. Shirzad-Siboni, M.; Farrokhi, M.; Darvishi Cheshmeh Soltani, R.; Khataee, A.; Tajassosi, S. Photocatalytic Reduction of Hexavalent Chromium over ZnO Nanorods Immobilized on Kaolin. *Ind. Eng. Chem. Res.* **2014**, *53*, 1079–1087. [CrossRef]
42. Barakat, M.; Chen, Y.T.; Huang, C.P. Removal of toxic cyanide and Cu(II) Ions from water by illuminated TiO. *Appl. Catal. B* **2004**, *53*, 13–20. [CrossRef]
43. Colina-Márquez, J.; Machuca, F.; Li Puma, G. Radiation Absorption and Optimization of Solar Photocatalytic Reactors for Environmental Applications. *Environ. Sci. Technol.* **2010**, *44*, 5112–5120. [CrossRef] [PubMed]
44. Mueses, M.A.; Machuca-Martinez, F.; Li Puma, G. Effective quantum yield and reaction rate model for evaluation of photocatalytic degradation of water contaminants in heterogeneous pilot-scale solar photoreactors. *Chem. Eng. J.* **2013**, *215–216*, 937–947. [CrossRef]
45. Osathaphan, K.; Chucherdwatanasak, B.; Rachdawong, P.; Sharma, V.K. Photocatalytic oxidation of cyanide in aqueous titanium dioxide suspensions: Effect of ethylenediaminetetraacetate. *Sol. Energy* **2008**, *82*, 1031–1036. [CrossRef]
46. Yang, J.-K.; Lee, S.-M.; Farrokhi, M.; Giahi, O.; Shirzad-Siboni, M. Photocatalytic removal of Cr(VI) with illuminated TiO2. *Desalin. Water Treat.* **2012**, *46*, 1–6. [CrossRef]
47. Ku, Y.; Jung, I.-L. Photocatalytic reduction of Cr(VI) in aqueous solutions by UV irradiation with the presence of titanium dioxide. *Water Res.* **2001**, *35*, 135–142. [CrossRef]
48. Thennarasu, S.; Rajasekar, K.; Balkis Ameen, K. Hydrothermal temperature as a morphological control factor: Preparation, characterization and photocatalytic activity of titanate nanotubes and nanoribbons. *J. Mol. Struct.* **2013**, *1049*, 446–457. [CrossRef]
49. White, L.; Koo, Y.; Yun, Y.; Sankar, J. TiO_2 Deposition on AZ31 Magnesium Alloy Using Plasma Electrolytic Oxidation. *J. Nanomater.* **2013**, *2013*, 319437. [CrossRef]

50. Turki, A.; Kochkar, H.; Guillard, C.; Berhault, G.; Ghorbel, A. Effect of Na content and thermal treatment of titanate nanotubes on the photocatalytic degradation of formic acid. *Appl. Catal. B* **2013**, *138–139*, 401–415. [CrossRef]
51. Sikhwivhilu, L.; Sinha Ray, S.; Coville, N. Influence of bases on hydrothermal synthesis of titanate nanostructures. *Appl. Phys. A* **2009**, *94*, 963–973. [CrossRef]
52. Fen, L.B.; Han, T.K.; Nee, N.M.; Ang, B.C.; Johan, M.R. Physico-chemical properties of titania nanotubes synthesized via hydrothermal and annealing treatment. *Appl. Surf. Sci.* **2011**, *258*, 431–435. [CrossRef]
53. Camposeco, R.; Castillo, S.; Navarrete, J.; Gomez, R. Synthesis, characterization and photocatalytic activity of TiO2 nanostructures: Nanotubes, nanofibers, nanowires and nanoparticles. *Catal. Today* **2016**, *266*, 90–101. [CrossRef]
54. López, R.; Gomez, R. Band-Gap Energy Estimation from Diffuse Reflectance Measurements on Sol–Gel and Commercial TiO2: A Comparative Study. *J. Sol-Gel Sci. Technol.* **2011**, *61*, 1–7. [CrossRef]

Article

A Pilot Study Combining Ultrafiltration with Ozonation for the Treatment of Secondary Urban Wastewater: Organic Micropollutants, Microbial Load and Biological Effects

Cátia A. L. Graça [1], Sara Ribeirinho-Soares [2], Joana Abreu-Silva [3], Inês I. Ramos [4], Ana R. Ribeiro [1], Sérgio M. Castro-Silva [5], Marcela A. Segundo [4], Célia M. Manaia [3,*], Olga C. Nunes [2,*] and Adrián M. T. Silva [1,*]

1 Laboratory of Separation and Reaction Engineering-Laboratory of Catalysis and Materials (LSRE-LCM), Faculdade de Engenharia, Universidade do Porto, Rua Dr. Roberto Frias, 4200-465 Porto, Portugal; catiaalgraca@fe.up.pt (C.A.L.G.); ritalado@fe.up.pt (A.R.R.)
2 LEPABE—Laboratory for Process Engineering, Environment, Biotechnology and Energy, Faculdade de Engenharia, Universidade do Porto, Rua Dr. Roberto Frias, 4200-465 Porto, Portugal; saramariasoares@hotmail.com
3 Universidade Católica Portuguesa, CBQF—Centro de Biotecnologia e Química Fina—Laboratório Associado, Escola Superior de Biotecnologia, Rua Diogo Botelho 1327, 4169-005 Porto, Portugal; jsilva@porto.ucp.pt
4 LAQV, REQUIMTE, Departamento de Ciências Químicas, Faculdade de Farmácia, Universidade do Porto, Rua de Jorge Viterbo Ferreira 228, 4050-313 Porto, Portugal; iibmramos@gmail.com (I.I.R.); msegundo@ff.up.pt (M.A.S.)
5 Adventech-Advanced Environmental Technologies, Centro Empresarial e Tecnológico, Rua de Fundões 151, 3700-121 São João da Madeira, Portugal; sergio.silva@adventech.pt
* Correspondence: cmanaia@porto.ucp.pt (C.M.M.); opnunes@fe.up.pt (O.C.N.); adrian@fe.up.pt (A.M.T.S.)

Received: 8 October 2020; Accepted: 6 December 2020; Published: 9 December 2020

Abstract: Ozonation followed by ultrafiltration (O_3 + UF) was employed at pilot scale for the treatment of secondary urban wastewater, envisaging its safe reuse for crop irrigation. Chemical contaminants of emerging concern (CECs) and priority substances (PSs), microbial load, estrogenic activity, cell viability and cellular metabolic activity were measured before and immediately after O_3 + UF treatment. The microbial load was also evaluated after one-week storage of the treated water to assess potential bacteria regrowth. Among the organic micropollutants detected, only citalopram and isoproturon were not removed below the limit of quantification. The treatment was also effective in the reduction in the bacterial loads considering current legislation in water quality for irrigation (i.e., in terms of enterobacteria and nematode eggs). However, after seven days of storage, total heterotrophs regrew to levels close to the initial, with the concomitant increase in the genes 16S rRNA and *intI*1. The assessment of biological effects revealed similar water quality before and after treatment, meaning that O_3 + UF did not produce detectable toxic by-products. Thus, the findings of this study indicate that the wastewater treated with this technology comply with the water quality standards for irrigation, even when stored up to one week, although improvements must be made to minimise microbial overgrowth.

Keywords: advanced oxidation; membrane technology; micropollutants; biological contaminants; cytotoxicity; wastewater reuse

1. Introduction

Urban wastewater reuse is considered an important strategy when addressing water scarcity issues [1]. This is a common practice in some countries, where the treated wastewater is mostly directed for agricultural irrigation [2]; however, urban wastewater often contains a variety of contaminants, such as salts, metals, metalloids, pathogens, and organic micropollutants, such as residual drugs, endocrine-disrupting chemicals, and residues from personal care products, among others [3,4]. Moreover, there is growing evidence that conventional urban wastewater treatment plants (UWWTPs) are not completely effective in eliminating bacteria and chemical micropollutants [5,6], rendering the effluent unsuitable for crops irrigation. Failure to properly treat and manage wastewater can generate adverse health effects, accumulation of heavy metals in crops, and the production of low-quality agricultural goods [3]. A new regulation on minimum preconditions for water reuse for agricultural irrigation has entered into force in the EU, which encompasses coordinated water-quality monitoring requisites for the safe reuse of treated urban wastewater [7]. These new rules will be put into practice in 2023 and are expected to promote water reuse. This regulation also demands an established water reuse risk management plan that should consider the environmental quality standards for priority substances and certain other pollutants, as well as additional requirements, such as heavy metals, pesticides, disinfection by-products, pharmaceuticals, and other substances of emerging concern, including micropollutants and microplastics. It also addressed the identification of some preventive measures that can be taken to limit risks, namely additional disinfection or pollutant removal measures.

Advanced oxidation processes (AOPs) and technologies (AOTs), such as ozonation, have emerged as effective tertiary treatments for the removal of both chemical and biological contaminants in UWWTPs [8,9]. Ozonation is among the few AOTs that have been applied to large-scale water treatment, due to its strong oxidation ability and broad-spectrum disinfection [10]. Ozone can react either by direct oxidation of organic pollutants (mostly at acidic conditions), or via hydroxyl radical formation (mainly produced under alkaline conditions) [10]. Studies employing ozone-based AOTs in UWWTP effluents have yielded remarkable results regarding the simultaneous removal of CECs and the reduction in the microbial load at different ozone doses and contact times [11–16]; however, bacterial regrowth in stored treated wastewater has been observed [14–16], which might be the result of the bacteria's ability to repair injuries, promoting fast regrowth, when stress levels are lowered. This may jeopardize water quality in the long term, thus prompting its immediate reuse rather than storing this water. Additionally, the use of chlorine as the traditional disinfection agent in stored water may not ensure its safety, because injured bacteria can also survive and regrow at low chlorine doses [17]. A suitable approach would be a physical separation step, using membrane-like technology. Although ozone may damage cell components, such as lipids, proteins and DNA, membrane filtration acts via size exclusion and adsorption, retaining microorganisms [18]. Among the available options in the market for full-scale applications, ultrafiltration (UF) membranes are favourable alternatives for bacteria removal due to their small pore size (0.01 to 0.1 μm). Moreover, studies have shown that UF is preferred to other filtration alternatives to avoid the regrowth of antibiotic-resistant bacteria (ARB) [19,20]. For example, Hembach et al., 2019 [18] reported the efficiency of UF in the disinfection of a secondary effluent of a UWWTP, and the results were compared with those obtained with single ozonation. The authors reported that UF (using a membrane pore size of 20 nm) was not able to remove the entire bacterial community, whereas ozonation presented limited effectiveness on the reduction in the same contaminants when using an ozone concentration optimised for micropollutant removal. Thus, these authors suggested further investigations coupling both technologies to achieve both micropollutant removal and bacteria mitigation, which was the target of the present study.

Thus, the present study investigated the potential of using UF in combination with ozonation, operating in continuous mode at a pilot scale, for the treatment of the secondary effluent of a UWWTP. Parameters commonly legislated in different countries were considered when assessing the suitability of treated wastewater for reuse in irrigation (Portuguese laws, US EPA, FAO guidelines and WHO). Moreover, envisaging higher quality criteria, the following parameters were also included in this work: (i) priority substances and CECs identified in Directive 2013/39/EU and Decision 495/2015/EU [21,22], respectively; (ii) load of selected microbial groups; and (iii) potential estrogenic activity, cytotoxicity, and cell viability (biological effects). All these parameters were analysed in both freshly collected and

O_3 + UF treated wastewater to assess treatment efficiency. Biological effects are particularly important to evaluate, due to the possibility of formation of toxic by-products after ozonation. Moreover, microbiological indicators were re-examined after a 7-day storage period to assess potential bacteria regrowth. Regarding other studies coupling O_3 to UF, only a few evaluate the feasibility of this system for urban wastewater reclamation [23–26] and, as far as it is known, none of those comprise the simultaneous evaluation of physico-chemical parameters, removal of priority substances and CECs, microbial inactivation and regrowth, and investigation of biological effects, which are important parameters for safe wastewater reuse, this work bringing a valuable contribution to the knowledge on this field.

2. Materials and Methods

2.1. Chemicals and Materials

All reference and isotopically labelled internal standards for liquid chromatography (>98% purity) were acquired from Sigma-Aldrich (Steinhein, Germany). Ethanol 99.5% (HPLC grade) was obtained from Fisher Scientific U.K. Ltd. (Loughborough, UK). Acetonitrile (MS grade) was purchased from VWR International (Fontenay-sous-Bois, France), whereas formic and sulphuric acid were obtained from Merck (Darmstadt, Germany). Multichannel tubular ceramic membranes with a selective layer of α-Al_2O_3 (nominal pore size of 10 nm) were provided by Rauschert Distribution GmbH, Inopor® (Schesslitz, Germany). Membrane dimensions were 305 mm in length with 15 mm glazed ends. The external diameter was 25 mm, and it contained 19 internal channels of 3.5 mm diameter each.

For microbial culture analyses, water samples were filtered through cellulose nitrate membranes (0.22 μm pore size, 47 mm diameter), provided by Sartorius (Gottingen, Germany). For DNA-based analyses, water samples were filtered through track-etched polycarbonate membranes (0.22 μm pore size, 47 mm diameter) from Whatman® Nuclepore™, provided by VWR (Alfragide, Portugal).

For cell culture experiments, dimethyl sulfoxide (DMSO; ≥99.9%), Triton™ X-100, and thiazolyl blue tetrazolium (MTT) were purchased from Sigma-Aldrich (Steinhein, Germany). Dulbecco's modified Eagle medium (DMEM; ref: 31966-021), heat-inactivated foetal bovine serum (FBS), penicillin-streptomycin (PenStrep), and trypsin-EDTA (1X) were purchased from Gibco® through Life Technologies™ (Warrington, UK). Murine fibroblasts L929 were obtained from the American Type Culture Collection (ATCC, Wesel, Germany). Caco-2 cell line was also purchased from ATCC and used between passage number 35 and 42. LDH Cytotoxicity Detection Kit was acquired from Takara Bio Inc. (Shiga, Japan). The XenoScreen YES/YAS assay kit for estrogenic activity assessment was acquired from Xenometrix® (Allschwil, Switzerland).

The ultrapure water used in the experiments and analytical methods was supplied by a Milli-Q water system (18.2 MΩ cm).

2.2. Secondary Effluent and Treated Samples

The secondary effluent used in the advanced treatment assays was collected at three different dates (between September and October 2019) from a full-scale UWWTP located in northern Portugal. In this UWWTP, the water line treatment includes a preliminary step (trash racking and dredging) followed by decantation, biological treatment with activated sludge, and a final decantation stage before discharging the effluent to the river. In this study, freshly collected samples of this UWWTP secondary effluent were divided into two aliquots, one of which was immediately analysed (WW) and another was directed to the O_3 + UF treatment unit. Details of the analytical methods employed to characterise the UWWTP secondary effluent (WW) are given in Section 2.4, and its chemical and biological characterisation can be found in Tables 1 and 2. Samples collected after O_3 + UF treatment (TWW_0) were also immediately processed for microbiological analyses and DNA extraction. In addition, aliquots of TWW_0 were stored for seven days in sterile glass bottles under dark conditions and at room temperature (herein named as TWW_7) to assess possible bacterial regrowth in a hypothetical storage scenario for wastewater reuse.

Table 1. Characterisation of the urban wastewater treatment plant (UWWTP) secondary effluent, before (WW) and immediately after treatment (TWW_0), and standards of water for irrigation (Decree-Law 236/98) and wastewater reuse in irrigation without restriction, for urban wastewaters which treatment includes a disinfection step (Decree-Law 119/2019) and for wastewater reuse in the Eastern Mediterranean Region—WHO, 2016.

			Decree-Law		WHO
Parameters	**UWWTP Secondary Effluent (WW)**	**After O_3 + UF Treatment (TWW_0)**	**236/98 [27] MVR**	**119/2019 [28] PV**	**2016 [29] MVR**
Al (mg/L)	9.55×10^{-5}	6.10×10^{-5}	5.0	5	5.0
As (mg/L)	1.12×10^{-5}	$<5 \times 10^{-6}$	0.1	n.a	0.1
Ba (mg/L)	4.25×10^{-5}	1.52×10^{-5}	1.0	n.a	n.a
Be (mg/L)	$<5 \times 10^{-6}$	$<5 \times 10^{-6}$	0.5	0.1	0.1
B (mg/L)	1.29×10^{-4}	1.06×10^{-4}	0.3	variable	n.a
Cd (mg/L)	$<5 \times 10^{-6}$	$<5 \times 10^{-6}$	0.01	n.a	0.1
Pb (mg/L)	6.73×10^{-6}	6.55×10^{-6}	5.0	n.a	5.0
Cl^- (mg/L)	80.8	79.5	70	n.a	142 [b]
Co (mg/L)	$<5 \times 10^{-6}$	$<5 \times 10^{-6}$	0.05	0.05	0.05
Cu (mg/L)	1.26×10^{-5}	6.28×10^{-5}	0.2	n.a	0.2
Total Cr (mg/L)	$<5 \times 10^{-6}$	$<5 \times 10^{-6}$	0.1	n.a	0.1
Sn (mg/L)	$<5 \times 10^{-6}$	$<5 \times 10^{-6}$	2.0	n.a	n.a
Fe (mg/L)	1.06×10^{-4}	2.36×10^{-5}	5.0	2.0	5.0
F^- (mg/L)	<DL	<DL	1.0	2.0	1.0
Li (mg/L)	1.98×10^{-5}	1.96×10^{-5}	2.5	2.5	2.5
Mn (mg/L)	4.56×10^{-5}	3.77×10^{-5}	0.2	0.2	0.2
Mo (mg/L)	2.45×10^{-5}	8.60×10^{-5}	0.005	0.01	0.01
Ni (mg/L)	$<5 \times 10^{-6}$	$<5 \times 10^{-6}$	0.5	n.a	0.2
NO_3^- (mg/L)	0.9 ± 0.4	7.70	50	n.a	9.5 [b]
Salinity (µS/cm)	848	782	1000	variable	700 [b]
TDS (mg/L)	335	191	640	n.a	450 [b]
SAR (meq/L)	2.49	1.50	8	variable	3.0
Se (mg/L)	$<5 \times 10^{-6}$	$<5 \times 10^{-6}$	0.02	0.02	0.02
TSS (mg/L)	24.50	0.00	60	≤10	20 [c]
SO_4^{2-} (mg/L)	45.2	50.0	575	n.a	n.a
V (mg/L)	$<5 \times 10^{-6}$	$<5 \times 10^{-6}$	0.1	n.a	0.1
Zn (mg/L)	4.70×10^{-5}	2.61×10^{-5}	2	n.a	2.0
pH	7.0 ± 1.0	8.0 ± 0.2	6.5–8.4	n.a	6.5–8.4
E. coli (log CFU/100 mL)	6.67	<DL	2.0	≤10	2.3 [c]
Intestinal parasite eggs [a]	0.00	0.00	n.a	≤1	n.a

DL stands for detection limit; MVR stands for maximum value recommended; n.a stands for not applicable/available; SAR stands for sodium adsorption ratio; PV stands for parametric value; TDS stands for total dissolved solids; TSS stands for total suspended solids. [a] Analysed by an external laboratory—the maximum value allowed (MVA) for this parameter in the Decree-Law 236/98 is 1. [b] Value up to which there is no restriction to use in irrigation. [c] Permitted limit for greywater reuse in irrigation of vegetables likely to be eaten uncooked.

Table 2. Additional analyses made to the UWWTP secondary effluent, before (WW) and immediately after treatment (TWW_0).

			Becerra et al., 2015 [30]		Decree-Law 119/2019 [28]
Additional Analyses	**UWWTP Secondary Effluent (WW)**	**After O_3 + UF Treatment (TWW_0)**	**MVA**	**MVR**	**PV**
Dissolved organic carbon (DOC, mg/L)	11.0 ± 0.8	9.6 ± 0.8	n.a.	n.a.	n.a.
Biological oxygen demand (BOD_5, mg/L)	15.1 ± 1.1	0	10 [b]	n.a	≤10 [c]
Chemical oxygen demand (COD, mg/L)	22.7 ± 0.7	5.4 ± 0.8	60–200	n.a	n.a.
Turbidity (NTU)	3.25 ± 0.15	0.28 ± 0.02	2	n.a	≤5
NH_4^+	<DL	0.59	n.a	n.a	10
PO_4^{3-}	<DL	<DL	n.a	n.a	n.a

DL stands for detection limit; MVA stands for maximum value allowed; MVR stands for maximum recommended value; PV stands for parametric value; n.a stands for not applicable/available.

2.3. Experimental Setup and Procedure

A scheme of the experimental apparatus is depicted in Figure 1. Ozonation was performed in a packed-bed column (2.2 I.D × 70 cm height) with a useful volume of approximately 0.35 L and containing glass Raschig rings (6 mm I.D × 6 mm height), because the water–ozone mass transfer achieved in the column packed with these Raschig rings was up to 3 times higher than that in a bubble column [31]. Firstly, the reactor was filled with ultrapure water (through a peristaltic pump) to regulate the desired concentration of ozone in the liquid phase. Ozone was produced from pure oxygen in a BMT 802X ozone generator and bubbled at the bottom of the column. The ozone concentration in the gas inlet was regulated by adjusting the oxygen gas flow rate with a mass flow controller and the electric intensity of the ozone generator (BMT 802X). The concentration of ozone in the liquid phase (dissolved ozone) was measured with an ATI model Q45H dissolved ozone analyser placed at the exit of the column. High ozone doses and contact time increase the capital and operating costs, therefore a low ozone dose (0.9 ± 0.1 gO_3/gDOC) and a short hydraulic retention time (HRT: 8 min obtained with a liquid flow rate of 46 mL min^{-1}) were investigated. These experimental conditions were selected in preliminary tests and fixed for all the subsequent experiments.

After a period, ultrapure water in the inlet liquid stream was replaced by the UWWTP effluent to start the ozonation experiments. Samples of ozonised wastewater were only collected after a period of two residence times (~16 min), in order to ensure that the steady state was achieved (i.e., when the outlet wastewater achieved a constant concentration of pollutants in two subsequent measurements). Then, the ozonised effluent was directed to the feed tank of the UF pilot reactor, aiming for the physical removal of microbial cells. Fifteen litres of ozonised effluent was pumped to the UF pilot through a peristaltic pump (Varmec®) and filtered through the 10 nm α-Al_2O_3 membrane operating in cross-flow mode (1 bar of transmembrane pressure). The UF pilot was designed in a way that the liquid flow of ozonised wastewater was automatically regulated to maintain the pressure constant inside the membrane housing compartment. The concentrate was recirculated to the feed tank [32], while a composite sample of the permeate was collected and split for microbiological and chemical analysis (TWW_0 immediately after O_3 + UF treatment and TWW_7 after being stored for seven days). UF was performed after O_3 and not the other way around, because by doing so, the membrane fouling is minimised [33,34]. At the end of the treatment, the membrane was left with H_2O_2 (30% *w/v*) overnight, followed by abundant washing with boiling water and autoclaved before starting another experiment. This cleaning procedure was defined to restore the membrane permeance and sterility.

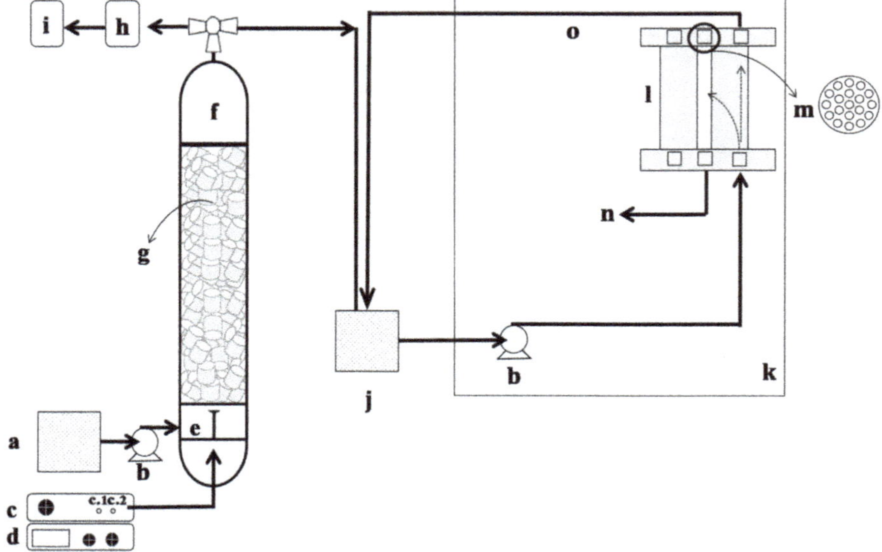

Figure 1. Scheme of the experimental apparatus. (**a**) feed tank containing deionised water or UWWTP effluent; (**b**) peristaltic pump; (**c**) ozone generator (c.1—O_2 entrance; c.2—O_3 exit); (**d**) mass flow controller; (**e**) ozone diffuser; (**f**) packed-bed column; (**g**) Raschig rings; (**h**) ozone analyser; (**i**) ozone destroyer; (**j**) feed to the ultrafiltration (UF) pilot; (**k**) UF pilot system; (**l**) membrane housing; (**m**) 19 channel ceramic membrane (top view); (**n**) permeate stream; (**o**) concentrate stream.

2.4. Chemical Analyses

The anionic and cationic contents (Cl^-, NO_3^-, SO_4^{2-}, Na^+, K^+) in water samples were determined by ion chromatography, as, using a Metrohm 881 Compact IC Pro apparatus equipped with a Metrosep C4 Cationic Exchange Column (250 mm × 4.0 mm) for the quantification of cations and a Metrosep A Supp 7 Anionic Exchange Column (250 mm × 4.0 mm) for quantification of anions. The content of metals was determined by using an inductively coupled plasma-optical emission spectrometer (ICP-OES, thermo scientific, model iCAP 7000 Series). The pH and conductivity of water were measured with pHenomenal® pH 1100L apparatus (VWR, Germany) and a conductivity meter (Crison GLP 31), respectively. Other relevant parameters (referred to as "additional analyses" in Table 2) were considered to assess the quality of water for irrigation: dissolved organic carbon (DOC) determined in a TOC-L analyser (Shimadzu TOC-5000A); turbidity measured with a turbidimeter (Hanna instruments, model HI88703); chemical oxygen demand (COD) determined by the closed reflux method (EPA standard method 5220D); and biochemical oxygen demand measured according to the EPA standard method 5210B (respirometric method) for a 5 day period (BOD_5). These analyses were performed as recommended in the standard methods for the examination of water and wastewater [35].

Moreover, the concentration of target organic micropollutants was determined using ultra-high performance liquid chromatography with tandem mass spectrometry (UHPLC-MS/MS) with Shimadzu Corporation apparatus (Tokyo, Japan) consisting of a triple quadrupole mass spectrometer detector (Ultra-Fast Mass Spectrometry series LCMS-8040) with an ESI (Electrospray Ionisation) source operating in both positive and negative ionisation modes. The mobile phase and operating conditions of the UHPLC-MS/MS system for the detection and quantification of the target pollutants are described elsewhere [16,36]. Prior to UHPLC-MS/MS analysis, WW and TWW_0 samples were pre-concentrated and cleaned up by solid-phase extraction (SPE) using Oasis® HLB (Hydrophilic-Lipophilic-Balanced sorbent, 150 mg, 6 mL) cartridges (Waters, Milford, Massachusetts, USA), according to the methodology described elsewhere [37]. For internal calibration, isotopically labelled internal standards were added to the samples before SPE. The preconcentration procedure was performed in duplicate for all the samples. This methodology allows to determine a total of 14 organic micropollutants.

2.5. Microbial Culture Analyses

Volumes ranging from 100 mL to 1 mL of WW, TWW_0 or TWW_7 samples or of serial 10-fold dilutions thereof were filtered in triplicate and placed onto the appropriate culture media of the target microbial group: Plate Count Agar (PCA, VWR International (Pennsylvania, USA)) (30 °C, 48 h) for culturable heterothrops, m-Faecal Coliform Agar (mFC, Thermo Fisher Scientific, Massachusetts, USA) (37 °C, 24 h) for enterobacteria, Slanetz Bartley Agar (Thermo Fisher Scientific, Massachusetts, USA) (37 °C, 48 h) for enterococci, and Rose Bengal Chloramphenicol Agar (Thermo Fisher Scientific, Massachusetts, U.S.A.) (25 °C, 5 days) for fungi. Results were expressed as colony forming units per 100 mL of sample (CFU/ 100 mL).

2.6. DNA Extraction, 16S rRNA and Inti1 Genes Quantification

Volumes of 100 mL of WW, 2 L of TWW_0, and 800 mL to 1 L of TWW_7 were vacuum-filtrated and processed in three independent samplings as biological replicates. DNA extraction was performed using the DNeasy® PowerWater® Kit (QIAGEN, Hilden, Germany) according to Rocha et al., 2020 [38] and with two additional steps suggested in the manufacturer's troubleshooting guide: after adding the lysis solution, a heating step at 65 °C for 10 min was included in the protocol; and to ensure the removal of residual ethanol before DNA elution, the centrifugation step was conducted in a clean collection tube for an additional minute. DNA samples were stored at −20 °C until quantitative PCR (qPCR) analysis.

The 16S rRNA gene (a marker for total bacteria) and the *intI*1 gene encoding a class 1 integron-integrase (a marker of anthropogenic impact) were quantified based on qPCR to assess the removal efficiency of bacteria after treatment [39,40]. Gene-specific primer sequences are listed in previous studies [41,42] and provided as supplementary information in Table S1. Gene quantification was based on SYBR Green qPCR assays in a StepOnePlus™ Real-Time PCR System (Life Technologies, USA) and interpolation to the standard curve run in each assay, as described elsewhere [39,43].

The data that met the quality criteria described in Rocha et al., 2018 [44] were expressed as the ratio of gene copy number per 100 mL of water sample (WW, TWW_0, and TWW_7). The secondary wastewater effluent (WW) was used as reference to assess the removal efficiency of both 16S rRNA and *intI*1 genes in treated samples, immediately after treatment (TWW_0) and after storage for 7 days (TWW_7). The duration of 7 days was selected to allow enough time for eventual injured cells surviving the treatment to fully recover, as we have verified in previous works with other treatment solutions [14,15].

2.7. Biological Effect Assays

2.7.1. Cell Culture and Incubation with Water Samples

Murine fibroblasts L929 and Caco-2 cells were cultured in Dulbecco's modified Eagle medium (DMEM) with D-glucose (4.5 g L^{-1}), sodium pyruvate (0.11 g L^{-1}), L-alanyl-L-glutamine (0.86 g L^{-1}) and further supplemented with 10% (*v/v*) heat inactivated foetal bovine serum (FBS), and 5% (*v/v*) of PenStrep (37 °C, 5% CO_2 and 95% of humidity). For cell viability and cytotoxicity assessment, the cells were detached from the culture flask as described elsewhere [45]. After cell counting in Neubauer chamber (Boeco, Germany), the suspension was centrifuged at 300 *g* for 5 min, and the cell pellet was suspended in culture medium to a final concentration of 5×10^4 cells per well. Cells were then seeded in a 96-well microplate (100 µL per well) and cultured for 24 h at 37 °C (5% CO_2 and 95% humidity).

2.7.2. Thiazolyl Blue Tetrazolium Reduction (MTT) and Lactate Dehydrogenase (LDH) Assays

Cellular metabolic activity was evaluated as indicator of cytotoxicity by the thiazolyl blue tetrazolium reduction (MTT) assay, whereas cell membrane integrity was evaluated through the lactate dehydrogenase (LDH) assay, providing information about cell viability. Briefly, test water samples were filtered using Corning® syringe filters (Sigma-Aldrich®, St. Louis, MO, USA) with 0.20 μm pore diameter and diluted 1:10 and 1:5 in DMEM. After discarding culture supernatant, 100 μL of diluted samples were added to cell layers and incubated at 37 °C (5% CO_2 and 95% humidity). After 24 h, the supernatant was removed for LDH assay, while the remaining content of the wells was used for MTT assay. For MTT assay, absence of cytotoxicity (100%) was estimated by replacing water test sample by culture medium. For LDH assay, the absence of cell viability (100%) was estimated by replacing water test sample by 1% (*v/v*) Triton X-100 solution prepared in culture medium.

2.7.3. Yeast Estrogen Screen (YES) Assay for Estrogenic Activity Assessment

WW and TWW_0 samples were filtered through 0.21 μm hydrophilic membranes and analysed directly, without any preconcentration. The YES assay and data analysis were performed according to the kit manufacturer's instructions. Calibration was established using standard solutions of the natural estrogen 17-β-estradiol (E2), at concentrations between 10^{-6}–10^{-9} mol L^{-1}. E2 also worked as positive control while ultrapure water was used as negative control. E2 standard solutions were prepared in DMSO (<1% in the assay medium), therefore a solvent blank was also assayed. Samples, standards, and control solutions were transferred to a 96-well microplate, mixed with assay medium, and inoculated with the transformed yeast cells. The mixture was then incubated for 48 h at 31 °C under orbital shaking. Spectrophotometric measurements at 570 nm (β-galactosidase expression) and 690 nm (yeast growth) were carried out using a Cytation3® microplate reader (Bio-Tek Instruments, Winooski, USA). The potential estrogen agonistic activity was estimated through the calculation of the parameters growth factor (G) and induction ratio (IR). The G parameter was calculated as the ratio of absorbance values measured at 690 nm for the sample and for the solvent $(A_{690})_{sample}/(A_{690})_{solvent}$. The IR parameter was calculated as $(1/G) \times ((A_{570} - A_{690})_{sample}/(A_{570} - A_{690})_{solvent})$.

3. Results and Discussion

3.1. Micropollutant Removal, Mineralisation, and Other Physico-Chemical Parameters

Under the regulation on minimum requirements for water reuse in agricultural irrigation, the environmental quality standards for priority substances and certain other pollutants should be targeted [7,21]. Moreover, the same regulation refers to additional requirements for risk assessment, including micropollutants. From the chemical organic micropollutants analysed in fresh (WW) and O_3+UF treated water samples (TWW_0), only 9 out of 14 were detected. The antiplatelet clopidogrel, the herbicide isoproturon, the anti-inflammatory diclofenac, the industrial compound PFOS (perfluorooctanesulfonic acid), and the lipid regulator bezafibrate were detected with a frequency of 100% in WW samples during the sampling campaign (Figure 2). Alachlor was also detected in all WW samples but below the limit of quantification ($LOQ_{alachlor}$ < 25 ng L^{-1}), whereas warfarin, citalopram, and clofibric acid were detected only in some samples. According to the Directive 2013/39/EU and Decision 495/2015/EU [21,22], alachlor, isoproturon and PFOS are considered PSs, whereas the others are considered CECs. After treatment, most micropollutants presented values below LOD—Limit Of Detection. Only alachlor, clopidogrel, citalopram and isoproturon were detected: the first two were below the LOQ—Limit Of Quantification (25 and 5 ng L^{-1}, respectively), whereas the latter two were found at concentrations up to 529 and 10.6 ng L^{-1}, respectively. In fact, isoproturon was the micropollutant with the lowest removal percentage (i.e., 80% of maximum removal). All priority substances (alachlor, isoproturon and PFOS) were below their environmental quality standards defined in the EU Directive 2013/39 [21], complying with the requirements of the EU Regulation 2020/741 [7].

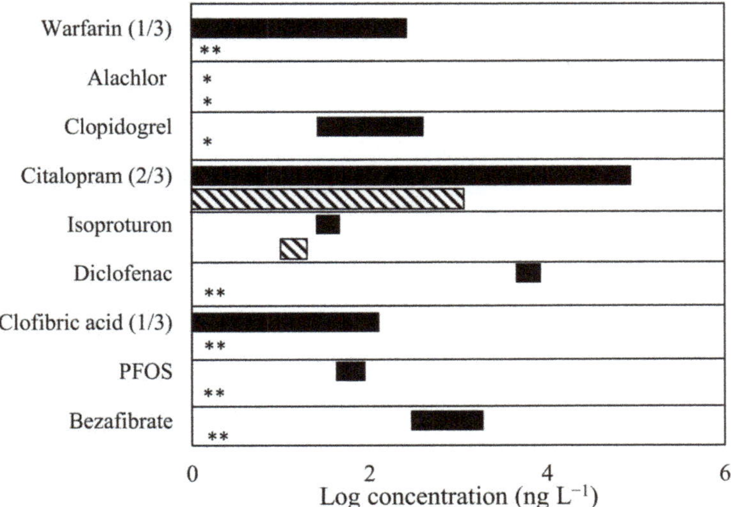

Figure 2. Logarithmic range of concentrations (ng L^{-1}) of the detected micropollutants in WW (black bar) and TWW_0 (striped bar) for samples with concentrations above LOQ. The frequency of occurrence was 100% (3/3) for all compounds, except when indicated in brackets after the compound name. * <LOQ and ** <LOD (compounds with concentrations < LOD before treatment are not shown in this figure for the sake of simplicity).

DOC and pH values did not remarkably vary after treatment (Tables 1 and 2, respectively). Values for DOC are not regulated and both pHs (before and after treatment) comply with the maximum value allowed (MVA). Thus, considering that regulations of water quality for irrigation often do not inform about adequate levels of organic matter, it can be assumed that the achieved values of DOC and micropollutants in treated water do not invalidate its use for irrigation. Moreover, the available literature mentioning the monitoring of DOC in water for irrigation recommends the evaluation of DOC when COD and BOD_5 are at the so-called alarming levels (>60 mgO_2 L^{-1} and >10 mgO_2 L^{-1}, respectively) [28,30,46], which is not the case of TWW_0 (Table 2).

In the combined process, ozonation was expected to be mainly responsible for the removal of micropollutants and dissolved organic matter rather than UF [18]. These results are coherent with other studies performing solely ozonation, in which the authors attributed the low yield of mineralisation to the formation of recalcitrant organic intermediates deriving from the organic micropollutants or, more likely, from the oxidation of dissolved organic matter naturally present in the wastewater [14,16]. For instance, using a similar experimental apparatus for the continuous ozonation of a secondary-UWWTP effluent (without UF), Moreira et al., 2016 [14] reported a DOC removal of ~30% (retention time of 26 min), whereas Iakovides et al., 2019 [16] obtained a DOC removal of ~10% (with similar ozone dosage and retention time).

Regarding other physico-chemical parameters, TWW_0 presents values below the maximum recommended in the Portuguese Laws of (i) water for irrigation [31] and (ii) treated wastewater for reuse [28,30]. The only exception is for the concentration of chloride in Table 1 (ca. 80 mg L^{-1} before and after treatment) which is slightly higher than the maximum value recommended (MVR) of 70 mg L^{-1} in the oldest law [27], which is not included in the newest one [28]. It is worth mentioning that this maximum value recommended for chloride was stipulated considering the sensitivity of tobacco crops; therefore, TWW_0 might not be appropriate for irrigation of this specific crop, but not necessarily inappropriate in the case of crops tolerant to these concentrations of chloride. For instance, some crops of fruits and vegetables are highly tolerant to chloride, such as Rangpur lime and cauliflower, for which the water for irrigation can contain up to 600 and 710 mg L^{-1} of chloride, respectively [47]. In fact, TWW_0 can be applied for irrigation according to the WHO (World Health Organization) and FAO (Food and Agriculture Organization of the United Nations) guidelines of water quality for surface irrigation, where the allowed chloride concentration is up to 142 mg L^{-1} (Table 1) [29], i.e., well above the value determined for the wastewater in this study (ca. 80 mg L^{-1}). The value of salinity (782 µS/cm) is slightly higher than that recommended by FAO and WHO [28] for the use of water for irrigation with no restriction (<700 µS/cm), but this value is not defined in Portuguese guidelines. Another interesting observation is the increase in the nitrate concentration after treatment, although still below the maximum value recommended [27], which can be attributed to the oxidation of nitrogen-containing substances that are likely to be present in the secondary effluent of UWWTPs [48]. Sulphate and copper contents also suffer a slight increase after treatment, which can be due to their release from sediments/soil particles after ozonation [49,50].

Future work must consider the energy demand of these processes [51] and life cycle assessment (LCA) [52–54] for the elimination of micropollutants from urban wastewater—these studies being particularly scarce with data at full scale. For instance, it has been concluded that ozonation has a lower energy demand compared to the use of membranes or UV/H_2O_2 [9]. Conversely, the electrical energy demand of ozonation is higher than those determined for powdered activated carbon (PAC) addition or granular activated carbon (GAC) filtration, but always being a plant-specific issue [51]. Performing LCA, it was suggested that ozonation has a better overall environmental performance than the photo-Fenton process [53], whereas reverse osmosis causes higher environmental burdens than ozonation due to the high energy and material consumption [52]. In these processes, generated impacts result mainly from the production of energy needed (and the respective energy mix) and from the use of some specific reagents [54].

3.2. Microbial Inactivation and Regrowth

As expected, a reduction in the load of the microbiological groups analysed was observed immediately after treatment (Figure 3). Reductions of nearly 3.5 log-units of 16S rRNA gene (indicative of the abundance of total bacteria) and 3.7 log-units of culturable heterotrophs occurred. The abundance of *intI*1 followed a similar trend, with a reduction of ~4.6 log-units immediately after treatment, whereas enterobacteria, enterococci, and fungi, with reductions higher than five log-units, reached values below the detection limit (0.33 CFU per 100 mL). Microbial inactivation can be transient [14,55,56], therefore further assays testing the regrowth capacity after seven days of storage of the treated wastewater were performed (TWW_7 samples). It is known that bacterial reactivation is influenced by factors such as storage conditions, in particular temperature, availability of nutrients, ultraviolet light, and assimilable organic carbon content, among others [57,58]. Therefore, the conditions to perform this assessment were selected to mimic the most common real storage conditions, i.e., room temperature (25 ± 2 °C) and absence of light to minimise DNA repair mechanisms [59]. The abundance of the 16S rRNA and *intI*1 genes, as well as the heterotrophic counts, recovered to values close to those observed in WW samples. The same pattern was observed for fungi, although with a lower regrowth extent (~1.6 log-units). The transient effect of single ozone-based processes for the treatment of UWWTP effluents was reported before [14–16]. In fact, even when operating with close ozone doses (0.75 gO_3/gDOC) and higher HRT (10–60 min) to those used here (0.9 gO_3/gDOC, HRT 8 min), reactivation of all the microbial groups analysed in the current study has been described in the literature [14,15]. In contrast, in the present study, regrowth of faecal indicators (enterobacteria and enterococci) was not observed in TWW_7 samples. Notwithstanding, from a microbiological quality point of view, both TWW_0 and TWW_7 comply with the biological parameters included in the quality standards of water for crops irrigation, both in Portugal [27] and United States [60], or the Portuguese/European Union quality standards of wastewater reuse in irrigation without restriction [28,61]. In fact, faecal coliforms or *Escherichia coli* (enterobacteria) and nematode eggs are the only biological parameters included in these quality standards, for which values were found below the stipulated thresholds (Table 1).

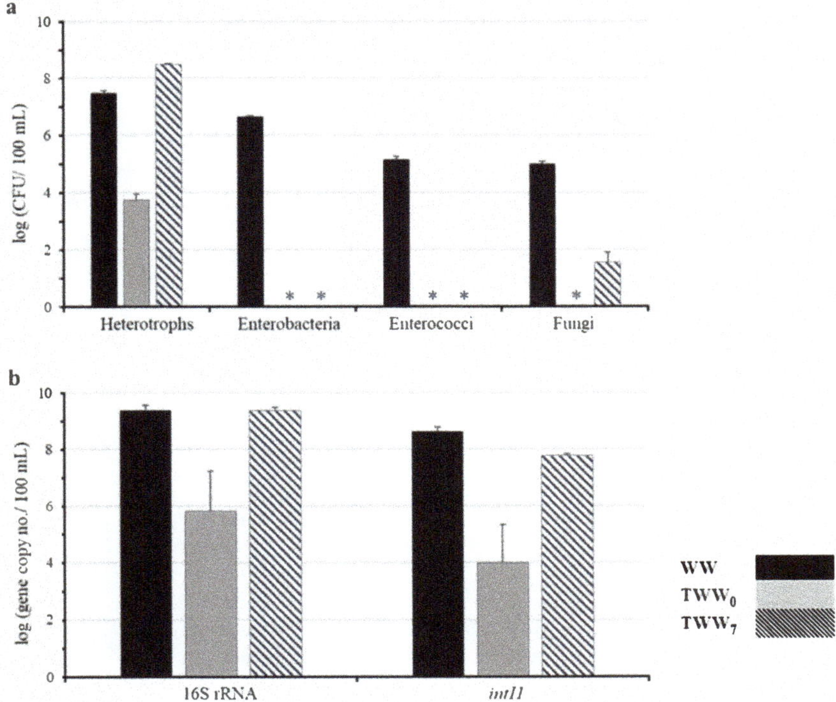

Figure 3. Microbiological water quality. (**a**) Culturable heterotrophs, enterobacteria, enterococci, and fungi, expressed as log (CFU/100 mL of sample); and (**b**) qPCR-based quantification of 16S rRNA and intI1 genes, expressed as log (gene copy number/ 100 mL of sample). * below the detection limit (0.33 CFU/100 mL).

Based on the abundance of enterobacteria in wastewater immediately after ozonation (10^2–10^3 CFU 100 mL $^{-1}$) or after 3 day storage (10^3–10^4 CFU 100 mL^{-1}) reported by Moreira et al. (2016) [14] and Iakovides et al., 2019 [16], the utilisation of ozonation alone would not produce wastewater compatible with its further use in irrigation. In contrast, the combination of UF with O_3 utilised here improved the efficiency of the treatment. The membrane fouling observed during the filtration process, which was evidenced by the permeate flow decrease from ~60 mL min^{-1} to ~16 mL min^{-1}, was most likely derived from bacteria that survived ozonation, cell debris and undissolved (in)organic matter. Nevertheless, the total suspended solid (TSS) value after O_3 was unquantifiable. In spite of the considerable improvements demonstrated in this study, the post-storage increase in total heterotrophs and genes shows that there is still room for additional tuning of the process to prevent the possible contamination of the permeate tank with spores of heterotrophic bacteria or fungi.

3.3. Evaluation of Biological Effects

Cytotoxic and cell viability effects of wastewater collected before (WW) and after treatment with O_3 + UF (TWW_0) were evaluated for skin (L929) and digestive epithelium (Caco-2) cell models by performing complementary MTT and LDH assays (Table 3). Considering that cell viability upon exposure to water samples depends on the final composition of the growth medium [62], test samples were diluted 5 and 10 times in culture medium before incubation with cell layers. Similar cytotoxicity (MTT) and cell viability (LDH) values were obtained for both dilution levels (Table 3). Moreover, cell viability was equivalent to that obtained for cell incubation with a plain culture medium. For both cell lines, no difference in cytotoxicity was observed for water samples collected before and after treatment (Table 3, MTT assay). Cell viability was also maintained after treatment (Table 3, LDH assay), providing similar or even higher values than those obtained for plain culture media or tap water. Additionally, samples analysed right after ozonation (i.e., before UF) rendered percentages of 91 ± 6% and 23 ± 8% in the MTT and LDH assays for L929 cells, respectively, indicating that no cytotoxic compounds were produced during this step.

Table 3. Results (percentage) from MTT [a] and LDH [b] assays obtained for urban wastewater before (WW) and after treatment (TWW_0).

Cell Line	MTT Assay		LDH Assay [c]	
	WW	TWW_0	WW	TWW_0
L929	102 ± 13	112 ± 15	20.7 ± 2.0 (28.3 ± 3.2)	19.7 ± 1.6 (32.6 ± 4.5)
Caco-2	116 ± 8	96 ± 9	58.6 ± 4.4 (59.1 ± 6.8)	53.2 ± 7.7 (59.5 ± 5.6)

[a] Values for culture media were 100% (Relative Standard Deviation—RSD < 20%) and between 1 and 9% for Triton X-100 (total disruption of cells). Samples were diluted 5 times in culture media before incubation with cells. [b] Values for Triton X-100 (total disruption of cells) were 100% (RSD < 10%). Values for tap water were 111 ± 5 for L929 cells and 110 ± 12 for Caco-2 cells. [c] Values between brackets correspond to blank values obtained in culture media only (intact cells). Values for tap water were 26.3 ± 5.6 for L929 cells and 55.9 ± 4.7 for Caco-2 cells.

The presence of estrogenic activity was also evaluated using the YES assay for WW and TWW_0 samples. Yeast growth inhibition was not observed for any of the tested samples. Induction ratios (IR) were 1.02 ± 0.09 for WW, and 0.74 ± 0.02 for TWW_0. These values were below the kit threshold value IR10 (corresponding to 10% of the maximum IR, value of 2.82, obtained for E2 standards), which indicated no estrogenic activity.

Work on toxicity assessment of effluents treated by ozonation has provided contradictory evidence. The biological toxicity of the influent of sewage treatment plants was significantly decreased after applying different advanced treatment processes, including ozone combined with UV, using *Daphnia magna*, zebrafish (*Danio rerio*), and *Vibrio fischeri* [63] as target organisms. However, when ozone and hydrogen peroxide were used together, a slight acute toxicity was perceived for *V. fischeri* while acute toxicity was observed for *D. magna* [64]. Other work, also applying the algae *Desmodesmus quadricauda*, indicated that the toxicity class of treated wastewater may change from completely non-toxic to very high hazard category, with a clear relationship between the time of ozonation and the increase in ecotoxicity [65]. This compound-dependent behaviour was also observed in a study with zebrafish embryos where different pharmaceutical compounds were tested [66]. Therefore, our results with cell lines are in agreement with previous works, where no toxic effect was observed after treatment, particularly when low doses of ozone are applied.

4. Conclusions

The results of this study indicate that UF performed after ozonation can be a suitable approach to allow the safe reuse of urban wastewater for irrigation. The combined process resulted in an effective treatment, especially against micropollutants detected in the UWWTP secondary effluent, and in the reduction in the microbial load. Treated wastewater stored for seven days maintained the quality required for irrigation, with the physico-chemical parameters, and enterobacteria and nematode egg counts below the maximum values recommended in water quality standards. In addition, no harmful biological effects were detected concerning the viability and estrogenicity tests. However, the fact that total bacterial cells, total cultivable heterotrophs as well as the *intI*1 gene reactivated to values close to those observed for untreated wastewater, shows that there is still room for additional improvement of this process.

Supplementary Materials: The following are available online at http://www.mdpi.com/2073-4441/12/12/3458/s1, Table S1: Quantitative PCR conditions used in the present study for absolute gene quantification in all WW, TWW_0, and TWW_7 samples.

Author Contributions: Conceptualisation, A.M.T.S., O.C.N., C.M.M., M.A.S., S.M.C.-S.; methodology, A.M.T.S., O.C.N., C.M.M., M.A.S., S.M.C.-S.; investigation, C.A.L.G., S.R.-S., J.A.-S., I.I.R., A.R.R.; writing—original draft preparation, C.A.L.G., S.R.-S., J.A.-S., I.I.R.; writing—review and editing, A.M.T.S., O.C.N., C.M.M., M.A.S., A.R.R.; supervision, A.M.T.S., O.C.N., C.M.M., M.A.S.; project administration, A.M.T.S., O.C.N., C.M.M., M.A.S., S.M.C.-S.; funding acquisition, A.M.T.S., O.C.N., C.M.M., M.A.S., S.M.C.-S. All authors have read and agreed to the published version of the manuscript.

Funding: This work is a result of the project "DEPCAT—Demonstration of new EquiPment involving integrated CATalytic processes for treatment of organic pollutants and disinfection of water", with the reference NORTE-01-0247-FEDER-033330, co-funded by European Regional Development Fund (ERDF), through the North Portugal Regional Operational Programme (NORTE2020), under the PORTUGAL 2020 Partnership Agreement.

Acknowledgments: We would like to thank the scientific collaboration under Base Funding-UIDB/50020/2020 of the Associate Laboratory LSRE-LCM and Base Funding-UIDB/00511/2020 of the Laboratory for Process Engineering, Environment, Biotechnology and Energy—LEPABE, both funded by national funds through the FCT/MCTES (PIDDAC), and FCT project UID/Multi/50016/2013 (Associate Laboratory CBQF) and UIDB/50006/2020 (LAQV, REQUIMTE).

Conflicts of Interest: The authors declare no conflict of interest.

References

1. UN General Assembly. Transforming Our World: The 2030 Agenda for Sustainable Development, 21 October 2015, A/RES/70/1. Available online: https://www.refworld.org/docid/57b6e3e44.html (accessed on 22 April 2020).
2. Jimenez, B.; Asano, T. *Water Reuse: An International Survey of Current Practice, Issues and Needs (Scientific and Technical Report)*; IWA Publishing: London, UK, 2018.
3. Bixio, D.; De Heyder, B.; Cicurel, H.; Muston, M.; Miska, V.; Joksimovic, D.; Schäfer, A.I.; Ravazzini, A.; Aharoni, A.; Savic, D.; et al. Municipal wastewater reclamation: Where do we stand? An overview of treatment technology and management practice. *Water Sci. Tech. W. Sup.* **2005**, *5*, 77–85. [CrossRef]
4. Pedrero, F.; Kalavrouziotis, I.; Alarcón, J.J.; Koukoulakis, P.; Asano, T. Use of treated municipal wastewater in irrigated agriculture-Review of some practices in Spain and Greece. *Agric. Water Manag.* **2010**, *97*, 1233–1241. [CrossRef]
5. Vaz-Moreira, I.; Nunes, O.C.; Manaia, C.M. Bacterial diversity and antibiotic resistance in water habitats: Searching the links with the human microbiome. *FEMS Microbiol. Rev.* **2014**, *38*, 761–778. [CrossRef] [PubMed]
6. Barbosa, M.; Moreira, N.F.F.; Ribeiro, A.R.; Pereira, M.F.R.; Silva, A.M.T. Occurrence and removal of organic micropollutants: An overview of the watch list of EU Decision 2015/495. *Water Res.* **2016**, *94*, 257–279. [CrossRef] [PubMed]
7. EUR-Lex. Regulation (EU) 2020/741 of the European Parliament and of the Council of 25 May 2020 on minimum requirements for water reuse. *Off. J. Eur. Union L* **2020**, *177*, 32–55.
8. Michael-Kordatou, I.; Karaolia, P.; Fatta-Kassinos, D. The role of operating parameters and oxidative damage mechanisms of advanced chemical oxidation processes in the combat against antibiotic-resistant bacteria and resistance genes present in urban. *Water Res.* **2018**, *129*, 208–230. [CrossRef]
9. Rizzo, L.; Malato, S.; Antakyali, D.; Beretsou, V.G.; Đolić, M.B.; Gernjak, W.; Heath, E.; Ivancev-Tumbas, I.; Karaolia, P.; Ribeiro, A.R.L.; et al. Consolidated vs new advanced treatment methods for the removal of contaminants of emerging concern from urban wastewater. *Sci. Total Environ.* **2019**, *655*, 986–1008. [CrossRef]
10. Wei, C.; Zhang, F.; Hu, Y.; Feng, C.; Wu, H. Ozonation in water treatment: The generation, basic properties of ozone and its practical application. *Rev. Chem. Eng.* **2016**, *33*, 49–90. [CrossRef]
11. Von Gunten, U. Oxidation processes in water treatment: Are we on track? *Environ. Sci. Technol.* **2018**, *52*, 5062–5075. [CrossRef]
12. Von Gunten, U. Ozonation of drinking water: Part, I. Oxidation kinetics and product formation. *Water Res.* **2003**, *37*, 1443–1467. [CrossRef]
13. Ikehata, K.; Jodeiri Naghashkar, N.; Gamal El-Din, M. Degradation of aqueous pharmaceuticals by ozonation and advanced oxidation processes: A review. *Ozone Sci. Eng.* **2006**, *28*, 353–414. [CrossRef]

14. Moreira, N.F.F.; Sousa, J.M.; Macedo, G.; Ribeiro, A.R.; Barreiros, L.; Pedrosa, M.; Faria, J.L.; Pereira, M.F.R.; Castro-Silva, S.; Segundo, M.A.; et al. Photocatalytic ozonation of urban wastewater and surface water using immobilized TiO2 with LEDs: Micropollutants, antibiotic resistance genes and estrogenic activity. *Water Res.* **2016**, *94*, 10–22. [CrossRef] [PubMed]
15. Sousa, J.M.; Macedo, G.; Pedrosa, M.; Becerra-Castro, C.; Castro-silva, S.; Pereira, M.F.R.; Silva, A.M.T.; Nunes, O.C.; Manaia, C.M. Ozonation and UV254 nm radiation for the removal of microorganisms and antibiotic resistance genes from urban wastewater. *J. Hazard. Mater.* **2017**, *323*, 434–441. [CrossRef]
16. Iakovides, I.C.; Michael-Kordatou, I.; Moreira, N.F.F.; Ribeiro, A.R.; Fernandes, T. Continuous ozonation of urban wastewater: Removal of antibiotics, antibiotic-resistant Escherichia coli and antibiotic resistance genes and phytotoxicity. *Water Res.* **2019**, *159*, 333–347. [CrossRef] [PubMed]
17. Rizzo, L.; Manaia, C.; Merlin, C.; Schwartz, T.; Dagot, C.; Ploy, M.C.; Michael, I.; Fatta-Kassinos, D. Urban wastewater treatment plants as hotspots for antibiotic resistant bacteria and genes spread into the environment: A review. *Sci. Total Environ.* **2013**, *447*, 345–360. [CrossRef] [PubMed]
18. Hembach, N.; Alexander, J.; Hiller, C.; Wieland, A.; Schwartz, T. Dissemination prevention of antibiotic resistant and facultative pathogenic bacteria by ultrafiltration and ozone treatment at an urban wastewater treatment plant. *Sci. Rep.* **2019**, *9*, 12843. [CrossRef]
19. Czekalski, N.; Imminger, S.; Salhi, E.; Veljkovic, M.; Kleffel, K.; Drissner, D.; Von Gunten, U. Inactivation of antibiotic resistant bacteria and resistance genes by ozone: From laboratory experiments to full-scale wastewater treatment. *Environ. Sci. Technol.* **2016**, *50*, 11862–11871. [CrossRef] [PubMed]
20. Rizzo, L.; Gernjak, W.; Krzeminski, P.; Malato, S.; McArdell, C.S.; Sanchez Perez, J.A.; Schaar, H.; Fatta-Kassinos, D. Best available technologies and treatment trains to address current challenges in urban wastewater reuse for irrigation of crops in EU countries. *Sci. Tot. Env.* **2020**, *710*, 136312. [CrossRef]
21. EUR-Lex. Directive 2013/39/EU of the European Parliament and of the Council of 12 August 2013 amending directives 2000/60/EC and 2008/105/EC as regards priority substances in the field of water policy. *Off. J. Eur. Union L* **2013**, *226*, 1–17.
22. EU Decision 495/2015. Commission implementing Decision (EU) 2015/495 of 20 March 2015 establishing a watch list of substances for Union-wide monitoring in the field of water policy pursuant to directive 2008/105/EC of the European Parliament and of the Council. *Off. J. Eur. Union L* **2015**, *78*, 40–42.
23. Spencer, P.; Domingos, S.; Edwards, B.; Howes, D.; Shorney-Darby, H.; Scheerman, H.; Milton, G.; . Clement, J. Ozone enhanced ceramic membrane filtration for wastewater recycling. *Water Pract. Technol* **2019**, *14*, 331–340. [CrossRef]
24. Si, X.; Hu, Z.; Huang, S. Combined process of ozone oxidation and ultrafiltration as an effective treatment technology for the removal of endocrine-disrupting chemicals. *Appl. Sci.* **2018**, *8*, 1240. [CrossRef]
25. Wang, H.; Park, M.; Liang, H.; Wu, S.; Lopez, I.J.; Ji, W.; Li, G.; Snyder, S.A. Reducing ultrafiltration membrane fouling during potable water reuse using pre-ozonation. *Water Res.* **2017**, *125*, 42–51. [CrossRef] [PubMed]
26. Acero, J.L.; Benitez, F.J.; Real, F.J.; Rodriguez, E. Elimination of selected emerging contaminants by the combination of membrane filtration and chemical oxidation processes. *Water Air Soil Pollut.* **2015**, *226*, 139. [CrossRef]
27. FAOLEX. *Decree-Law 236/98 Establishing Water Quality Standards*; Republic Diary No. 176/1998, Series I-A of 1998-08-01; Portuguese Presidency of the Council of Ministers: Lisboa, Portugal, 1998; pp. 3676–3716.
28. FAOLEX. *Decree-Law No. 119/2019 Establishing the Legal Scheme of the Production of Water for Reuse*; Republic Diary No. 159/2019, Series I of 2019-08-21; Portuguese Presidency of the Council of Ministers: Lisboa, Portugal, 2019; pp. 21–44.
29. WHO. *A Compendium for Standards for Wastewater Reuse in the Eastern Mediterranean Region*; World Health Organisation (WHO): Cairo, Egypt, 2006.
30. Becerra-Castro, C.; Rita, A.; Vaz-Moreira, I.; Silva, E.F.; Manaia, C.M.; Nunes, O.C. Wastewater reuse in irrigation: A microbiological perspective on implications in soil fertility and human and environmental health. *Environ. Int.* **2015**, *75*, 117–135. [CrossRef]
31. Graça, C.A.L.; Lima, R.B.; Pereira, M.F.R.; Silva, A.M.T.; Ferreira, A. Intensification of the ozone-water mass transfer in an oscillatory flow reactor with innovative design of periodic constrictions: Optimization and application in ozonation water treatment. *Chem. Eng. J.* **2020**, *389*, 124412.
32. Marchese, J.; Ochoa, N.A.; Pagliero, C.; Almandoz, C. Pilot-scale ultrafiltration of an emulsified oil wastewater. *Environ. Sci. Technol.* **2000**, *34*, 2990–2996. [CrossRef]

33. Mansas, C.; Mendret, J.; Brosillon, S.; Ayral, A. Coupling catalytic ozonation and membrane separation: A review. *Sep. Purif. Technol.* **2020**, *236*, 1161221. [CrossRef]
34. You, S.-H.; Tseng, D.-H.; Hsu, W.-C. Effect and mechanism of ultrafiltration membrane fouling removal by ozonation. *Desalination* **2007**, *202*, 224–230. [CrossRef]
35. Rice, E.W.; Baird, R.B.; Eaton, A.D. *Standard Methods for the Examination of Water and Wastewater*; American Public Health Association, American Water Works Association, Water Environment Federation: Washington, DC, USA, 2017.
36. Barbosa, M.O.; Ribeiro, A.R.; Ratola, N.; Hain, E.; Homem, V.; Pereira, M.F.R.; Blaney, L.; Silva, A.M.T. Spatial and seasonal occurrence of micropollutants in four Portuguese rivers and a case study for fluorescence excitation-emission matrices. *Sci. Total Environ.* **2018**, *644*, 1128–1140. [CrossRef]
37. Ribeiro, A.R.; Pedrosa, M.; Moreira, N.F.F.; Pereira, M.F.R.; Silva, A.M.T. Environmental friendly method for urban wastewater monitoring of micropollutants defined in the Directive 2013/39/EU and Decision 2015/495/EU. *J. Chromatogr. A* **2015**, *1418*, 140–149. [CrossRef] [PubMed]
38. Rocha, J.; Manaia, C.M. Cell-based internal standard for qPCR determinations of antibiotic resistance indicators in environmental water samples. *Ecol. Indic.* **2020**, *113*, 106194. [CrossRef]
39. Narciso-da-Rocha, C.; Rocha, J.; Vaz-Moreira, I.; Lira, F.; Tamames, J.; Henriques, J.L.; Manaia, C.M. Bacterial lineages putatively associated with the dissemination of antibiotic resistance genes in a full-scale urban wastewater treatment plant. *Environ. Int.* **2018**, *118*, 179–188. [CrossRef] [PubMed]
40. Pärnänen, K.M.M.; Narciso-da-Rocha, C.; Kneis, D.; Berendonk, T.U.; Cacace, D.; Do, T.T.; Elpers, C.; Fatta-Kassinos, D.; Henriques, I.; Jaeger, T.; et al. Antibiotic resistance in European wastewater treatment plants mirrors the pattern of clinical antibiotic resistance prevalence. *Sci. Adv.* **2019**, *5*, eaau9124.
41. Denman, S.E.; McSweeney, C.S. Development of a real-time PCR assay for monitoring anaerobic fungal and cellulolytic bacterial populations within the rumen. *FEMS Microbiol. Ecol.* **2006**, *58*, 572–582. [CrossRef] [PubMed]
42. Barraud, O.; Baclet, M.C.; Denis, F.; Ploy, M.C. Quantitative multiplex real-time PCR for detecting class 1, 2 and 3 integrons. *J. Antimicrob. Chemother.* **2010**, *65*, 1642–1645. [CrossRef] [PubMed]
43. Brankatschk, R.; Bodenhausen, N.; Zeyer, J.; Bürgmann, H. Simple absolute quantification method correcting for quantitative PCR efficiency variations for microbial community samples. *Appl. Environ. Microbiol.* **2012**, *78*, 4481–4489. [CrossRef]
44. Rocha, J.; Cacace, D.; Kampouris, I.; Guilloteau, H.; Jäger, T.; Marano, R.B.M.; Karaolia, P.; Manaia, C.M.; Merlin, C.; Fatta-Kassinos, D.; et al. Inter-laboratory calibration of quantitative analyses of antibiotic resistance genes. *J. Environ. Chem. Eng.* **2018**, *8*, 102214. [CrossRef]
45. Ferreira, M.; Chaves, L.L.; Lima, S.A.C.; Reis, S. Optimization of nanostructured lipid carriers loaded with methotrexate: A tool for inflammatory and cancer therapy. *Int. J. Pharm.* **2015**, *492*, 65–72. [CrossRef]
46. Dorais, M.; Alsanius, B.W.; Voogt, W.; Pepin, S.; Tüzel, H.; Tüzel, Y.; Möller, K. *Impact of Water Quality and Irrigation Management on Organic Greenhouse Horticulture*; BioGreenhouse COST Action FA1105: Bleiswijk, Netherlands, 2020; ISBN 978-94-6257-538-7.
47. Water Salinity and Plant Irrigation. Available online: https://www.agric.wa.gov.au/water-management/water-salinity-and-plant-irrigation (accessed on 8 October 2020).
48. Khuntia, S.; Majumder, S.K.; Ghosh, P. Removal of Ammonia from Water by Ozone Microbubbles. *Ind. Eng. Chem. Res.* **2013**, *52*, 318–326. [CrossRef]
49. Ballabio, C.; Panagos, P.; Lugato, E.; Huang, J.-H.; Orgiazzi, A.; Jones, A.; Fernández-Ugalde, O.; Borrelli, P.; Montanarella, L. Copper distribution in European topsoils: An assessment based on LUCAS soil survey. *Sci. Total Environ.* **2018**, *636*, 282–298. [CrossRef] [PubMed]
50. Lucheta, A.R.; Lambais, M.R. Sulfur in agriculture. *Rev. Bras. Ciênc. Solo* **2012**, *36*, 1369–1379. [CrossRef]
51. Mousel, D.; Palmowski, L.; Pinnekamp, J. Energy demand for elimination of organic micropollutants in municipal wastewater treatment plants. *Sci. Total Environ.* **2017**, *575*, 1139–1149. [CrossRef] [PubMed]
52. Li, Y.; Zhang, S.; Zhang, W.; Xiong, W.; Ye, Q.; Hou, X.; Wang, C.; Wang, P. Life cycle assessment of advanced wastewater treatment processes: Involving 126 pharmaceuticals and personal care products in life cycle inventory. *J. Environ. Manag.* **2019**, *238*, 442–450. [CrossRef] [PubMed]
53. Arzate, S.; Pfister, S.; Oberschelp, C.; Sánchez-Pérez, J.A. Environmental impacts of an advanced oxidation process as tertiary treatment in a wastewater treatment plant. *Sci. Total Environ.* **2019**, *694*, 133572. [CrossRef] [PubMed]

54. Pesqueira, J.F.J.R.; Pereira, M.F.R.; Silva, A.M.T.S. Environmental impact assessment of advanced urban wastewater treatment technologies for the removal of priority substances and contaminants of emerging concern: A review. *J. Clean. Prod.* **2020**, *261*, 121078. [CrossRef]
55. Spuhler, D.; Andrés Rengifo-Herrera, J.; Pulgarin, C. The effect of Fe^{2+}, Fe^{3+}, H_2O_2 and the photo-Fenton reagent at near neutral pH on the solar disinfection (SODIS) at low temperatures of water containing Escherichia coli K12. *Appl. Catal. B Environ.* **2010**, *96*, 126–141. [CrossRef]
56. Zhao, X.; Hu, H.-Y.; Yu, T.; Su, C.; Jiang, H.; Liu, S. Effect of different molecular weight organic components on the increase of microbial growth potential of secondary effluent by ozonation. *J. Environ. Sci.* **2014**, *26*, 2190–2197. [CrossRef]
57. Giannakis, S.; Merino Gamo, A.I.; Darakas, E.; Escalas-Cañellas, A.; Pulgarin, C. Monitoring the post-irradiation E. coli survival patterns in environmental water matrices: Implications in handling solar disinfected wastewater. *Chem. Eng. J.* **2014**, *253*, 366–376. [CrossRef]
58. Ubomba-Jaswa, E.; Navntoft, C.; Polo-López, M.I.; Fernandez-Ibáñez, P.; McGuigan, K.G. Solar disinfection of drinking water (SODIS): An investigation of the effect of UV-A dose on inactivation efficiency. *Photochem. Photobiol. Sci.* **2009**, *8*, 587–595. [CrossRef]
59. Clancy, S. DNA damage & repair: Mechanisms for maintaining DNA integrity. *Nat. Educ.* **2008**, *1*, 103.
60. EPA. *Guidelines for Water Reuse*; Environmental Protection Agency (EPA): Wasghinton, DC, USA, 2012; (EPA/600/R-12/618).
61. Alcalde-Sanz, L.; Gawlik, B.M. *Minimum Quality Requirements for Water Reuse in Agricultural Irrigation and Aquifer Recharge—Towards A Water Reuse Regulatory Instrument at EU Level, EUR 28962 EN*; Publications Office of the European Union: Luxembourg, 2017.
62. Trintinaglia, L.; Bianchi, E.; Silva, L.; Nascimento, C.; Spilki, F.; Ziulkoski, A. Cytotoxicity assays as tools to assess water quality in the Sinos River basin. *Braz. J. Biol.* **2015**, *75*, 75–80. [CrossRef] [PubMed]
63. Zhang, Y.; Yuan, Y.X.; Wang, Y.F.; Li, C.; Zhu, J.; Li, R.F.; Wu, Y.H. Comprehensive evaluation on the bio-toxicity of three advanced wastewater treatment processes. *Water Air Soil Pollut.* **2020**, *231*, 110. [CrossRef]
64. Nahim-Granados, S.; Rivas-Ibanez, G.; Perez, J.A.S.; Oller, I.; Malato, S.; Polo-Lopez, M.I. Synthetic fresh-cut wastewater disinfection and decontamination by ozonation at pilot scale. *Water Res.* **2020**, *170*, 115304. [CrossRef] [PubMed]
65. Affek, K.; Muszynski, A.; Zaleska-Radziwill, M.; Doskocz, N.; Zietkowska, A.; Widomski, M. Evaluation of ecotoxicity and inactivation of bacteria during ozonation of treated wastewater. *Desalin. Water Treat.* **2020**, *192*, 176–184. [CrossRef]
66. Pohl, J.; Ahrens, L.; Carlsson, G.; Golovko, O.; Norrgren, L.; Weiss, J.; Orn, S. Embryotoxicity of ozonated diclofenac, carbamazepine, and oxazepam in zebrafish (*Danio rerio*). *Chemosphere* **2019**, *225*, 191–199. [CrossRef]

Publisher's Note: MDPI stays neutral with regard to jurisdictional claims in published maps and institutional affiliations.

Article

A MATLAB-Based Application for Modeling and Simulation of Solar Slurry Photocatalytic Reactors for Environmental Applications

Raúl Acosta-Herazo [1,*], Briyith Cañaveral-Velásquez [2], Katrin Pérez-Giraldo [2], Miguel A. Mueses [2], María H. Pinzón-Cárdenas [1] and Fiderman Machuca-Martínez [1]

1 School of Chemical Engineering, Universidad del Valle, Cali A.A. 25360, Colombia; maria.pinzon@correounivalle.edu.co (M.H.P.-C.); fiderman.machuca@correounivalle.edu.co (F.M.-M.)

2 Modeling and Applications of Advanced Oxidation Technologies Research Group, Photocatalysis and Solar Photoreactors Engineering, Department of Chemical Engineering, Universidad de Cartagena, Cartagena A.A. 1382-195, Colombia; briyithcarolina@hotmail.com (B.C.-V.); klpg_15@hotmail.com (K.P.-G.); mmueses@unicartagena.edu.co (M.A.M.)

* Correspondence: raul.acosta@correounivalle.edu.co

Received: 2 July 2020; Accepted: 27 July 2020; Published: 4 August 2020

Abstract: Because of the complexity caused by photochemical reactions and radiation transport, accomplishing photoreactor modeling usually poses a barrier for young researchers or research works that focus on experimental developments, although it may be a crucial tool for reducing experimental efforts and carrying out a more comprehensive analysis of the results. This work presents PHOTOREAC, an open-access application developed in the graphical user interface of Matlab, which allows a user-friendly evaluation of the solar photoreactors operation. The app includes several solar photoreactor configurations and kinetics models as well as two variants of a radiation absorption-scattering model. Moreover, PHOTOREAC incorporates a database of 26 of experimental solar photodegradation datasets with a variety of operational conditions (model pollutants, photocatalyst concentrations, initial pollutant concentrations); additionally, users can introduce their new experimental data. The implementation of PHOTOREAC is presented using three example cases of solar photoreactor operation in which the impact of the operational parameters is explored, kinetic constants are estimated according to experimental data, and comparisons are made between the available models. Finally, the impact of the application on young researchers' projects in photocatalysis at the University of Cartagena was investigated. PHOTOREAC is available upon request from Professor Miguel Mueses.

Keywords: computer-based learning; solar photocatalysis; water contaminants; kinetic modeling; photoreactor design

---

1. Introduction

Heterogeneous photocatalysis is an example of an emerging environmental technology with a variety of promising applications, such as air and water disinfection and decontamination, clean fuel production and green product manufacturing [1–3].

Modeling and computer simulation of photoreactors are crucial for their design, scale-up and technology transfer; since they allow engineers and researchers to understand the role of the design parameters and operational conditions without performing an excessive number of experiments. However, modeling a solar photoreactor is a very complicated task, because it requires a combination of knowledge in applied solar energy, geometric optics, radiative transfer, materials science and photochemical reaction engineering.

The implementation of commercial packages for photoreactor simulations is limited. Simulation packages for chemical plants, such as Aspen HYSYS® (Aspen Technology, Inc., Bedford, MA, USA) or Aspen plus® (Aspen Technology, Inc., Bedford, MA, USA), do not incorporate photocatalytic reactors. On the other hand, modeling and simulation of photoreactors can be carried out in Computational Fluid Dynamics (CFD) packages, such as COMSOL Multiphysics® (COMSOL, Inc., Burlington, MA, USA) and ANSYS® Fluent (ANSYS, Inc., Southpointe, PA, USA). However, they do not have modules dedicated to photoreactor engineering. Therefore, the simulations are performed by adapting the existing simulation modules for the simulation of photocatalytic reactors. This configuration of the CFD modules must be carried out manually by the user, which may result in an approach not intuitive enough for non-experts in photoreactor engineering. Another alternative is to perform the direct coding of the photoreactor model in a programming language. Still, this may result in a challenge for researchers that have not taken advanced courses in programming and numerical methods.

For the above reasons, the direct coding or the use of CFD simulators to implement a photoreactor model could be found inconvenient by non-expert researchers in photoreactors engineering, such as young researchers or those focused on experimental developments. However, implementing a photoreactor model may be a crucial tool for reducing the experimental efforts and carrying out a more comprehensive analysis of the results.

In this work, we present PHOTOREAC, an open-access computational application developed in the graphical user interface of Matlab wholly dedicated to the modeling and simulation of large-scale slurry solar photocatalytic reactors for environmental applications. It is based on the experience gathered by our research groups at Cartagena University (Cartagena de Indias, Colombia) and the Universidad del Valle (Cali, Colombia) during the last twenty years of research in heterogeneous solar photocatalysis, and also on extensive literature research in photoreactor engineering.

The application aims to provide non-expert researchers in photoreactors engineering a user-friendly, dedicated and efficient tool for the modeling and simulation of solar photoreactors, providing them with valuable information without implementing very sophisticated methods.

By employing PHOTOREAC, the users will be able to explore the role of critical parameters of the system on the radiation absorption performance of the photoreactor and the overall kinetic behavior of the photocatalytic process; parameters include the photoreactor geometry, the photoreactor dimensions, the model pollutant, the kinetic expression, the photocatalyst concentration, the photocatalysts optical properties, the initial pollutant concentration, the volume of treated water and the incident radiation. Additionally, PHOTOREAC incorporates a database of experimental information collected in our laboratory regarding the solar photodegradation of a variety of model pollutants under different operational conditions. Therefore, users will have empirical data available to carry out analyses and comparisons with their data.

2. Solar Photoreactors Modeling by PHOTOREAC

PHOTOREAC performs the modeling and simulation of the photoreactors following the general algorithm described in Figure 1. The algorithm considers mathematical simplifications to maintain the approach as rigorously and computationally efficient as possible, and thus it provides the users with valuable information without implementing sophisticated numerical methods that, although they can improve the quantitative results, may not affect the qualitative analysis. These assumptions and simplifications will be described and discussed in the upcoming sections.

The basis of the PHOTOREAC approach is that the radiation field modeling can be carried out independently of the photocatalytic kinetics modeling since the radiation balance in the photoreactor is not a function of the concentration of the chemical species. Therefore, the radiation balance is decoupled from the mass and momentum balances of the system. Besides, the radiation field described by the local volumetric rate of photon absorption (LVRPA) profile inside the photoreactor is considered to be in a steady-state, i.e., it does not vary along the reaction rime its reaction time does not change [4,5]. On the other hand, to carry out a kinetic analysis independent of the radiation absorption effects, i.e., the optimized kinetic parameters are not a function of the irradiation conditions, it is mandatory to know the radiation field in the photoreactor beforehand [6,7].

Thus, PHOTOREAC considers two modules: (i) the photon absorption-scattering module, in which the user will be able to determine the radiation field of the available photoreactor configurations by following the procedure described by the red box in Figure 1; and (ii) the kinetic modeling module, in which the user will be able to estimate the radiation-independent kinetic parameters for the four available kinetics expressions following the procedure described in the blue box in Figure 1.

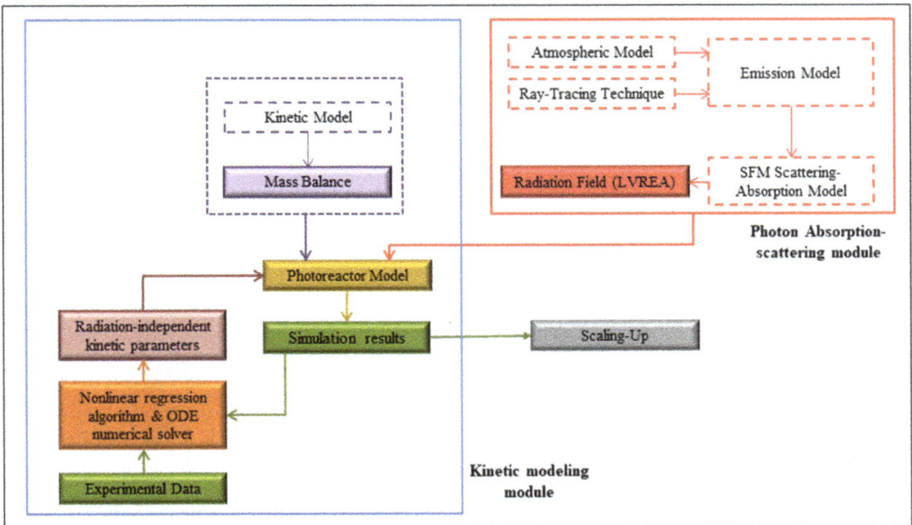

Figure 1. General algorithm for the modeling and simulation of solar slurry photoreactors in PHOTOREAC. ODE: Ordinary Differential Equation; SFM: Six Flux Model.

2.1. The Photoreactors Set-Up in PHOTOREAC

PHOTOREAC includes three configurations of pilot-scale solar photoreactors: a flat plate photoreactor (FPP), a compound parabolic collector photoreactor (CPCP) and a tubular-type photoreactor (TTP). These are the most common configurations for solar-pilot applications of heterogeneous photocatalysis; a detailed description of them can be found in the literature [3,8,9]. For the TTP, a novel prototype is also included, the offset multi-tubular photoreactor (OMTP) [10]. All of the photoreactors operate in recirculation, a flow-through mode with the water passing through an external tank, as shown in Figure 2. The photoreactor is exposed to the sunlight, facing the sun, while the reservoir tank is in the dark. The flow consists of an aqueous suspension of photocatalyst powder and the dissolved contaminant. The Evonik TiO_2 P25 was selected as the model photocatalyst in PHOTOREAC because it is considered the most promising alternative for commercial applications due to its low cost, photochemical stability, and high oxidation power [3]. Therefore, it is widely studied, and its physicochemical and optical properties are well known in the literature [11].

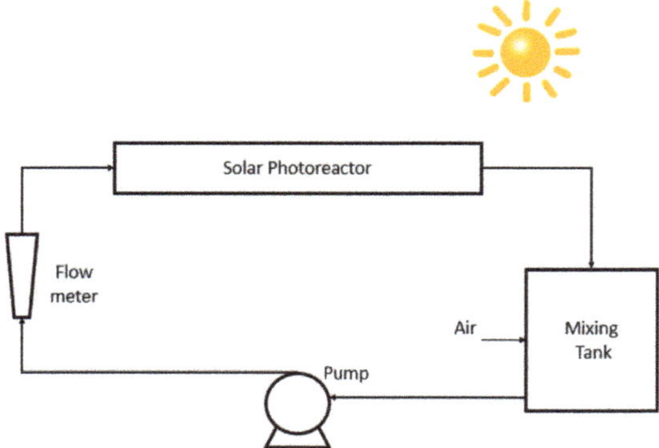

Figure 2. Scheme of a solar pilot photoreactor set-up.

2.2. The Input Data for the Use of PHOTOREAC

The availability and reliability of the input data provided to PHOTOREAC are crucial for good results. Table 1 shows a summary of the input information that is required to simulate the photoreactors in PHOTOREAC. Additionally, it is indicated in which module is the information used.

Table 1. Summary of the input information for the PHOTOREAC modules.

Parameter	Symbol	Units	Belonging to PHOTOREAC Module [a]
Photoreactor radius (CPCP and OMTP)	R	m	PASM
Water film thickness (FFP)	δ	m	PASM
Photoreactor length	L	m	PASM
Solar incident radiation	I_0	W/m^2	PASM
Reaction volume	V_R	L	PASM/KMM
Photocatalyst concentration	C_{cat}	g/L	PASM/KMM
Total volume	V_T	L	KMM
Number of experimental photodegradation data	N	Dimensionless	KMM
Concentration vs. accumulated energy data [b]	C_i vs. ξ_{AE}	ppm vs. J/m^2	KMM
Concentration vs. standard time data [b]	C_i vs. t	ppm vs. min	KMM

[a] PASM: photon absorption-scattering module; KMM: kinetic modeling module. [b] If it is a multicomponent mixture, C_i is replaced by *TOC*.

At the same time, the experimental photodegradation data for kinetic analysis in PHOTOREAC deserves special attention. The effects of the adsorption must be carefully considered in the solar photocatalytic experimental test. The photodegradation data used to feed PHOTOREAC must be reported at the zero-point of photodegradation, where adsorption has already been allowed to homogenize, which is usually achieved by allowing the system to recirculate under darkness for 30 min to establish adsorption–desorption equilibrium conditions before being exposed to solar light. Thus, although the kinetic models in PHOTOREAC do not contemplate the competitive effects of molecular adsorption, the data used will already be corrected with that effect. Therefore, there is no problem with the application of the models [10,12].

During the exposure time to sunlight, the data should be reported as the pollutant concentration C_i against the accumulated ultraviolet (UV) energy ξ_{AE}. The experiments finish when the desired accumulated UV total energy in J/m^2 is reached. Additionally, it is required to record the corresponding standard time for each sample.

2.3. The PHOTOREAC Photon Absorption-Scattering Module

The PHOTOREAC photon absorption-scattering module performs the radiation field modeling of the three available configurations of solar photoreactors: FPP, CPCP and OMTP. It provides the

LVRPA spatial distribution inside the photoreactor and the overall volumetric rate of photon absorption (OVRPA), which corresponds to the LVRPA averaged over the entire volume of the reactor. The latter is a critical magnitude for the kinetic assessment [12].

The PHOTOREAC modeling approach is focused on the six-flux absorption-scattering model (SFM). SFM is an analytical equation in which the leading hypothesis is that scattering only occurs in the six Cartesian directions [5]. Despite being a simplified model, it retains the key aspects of the radiation field modeling in photoreactors and has been implemented successfully at the solar pilot scale [13,14]. Other modeling approaches for solar photoreactors, such as the discrete ordinate method (DOM) or the Monte Carlo simulation, offer a more accurate description of the radiation transport phenomena. However, they are more time-consuming in the computations and their mathematical formulation is of high complexity. The SFM short computation times are ideal for exploring the impact of operational parameters, including the photocatalyst concentration, photoreactor dimensions and incident radiation, in particular for users that are dabbling in photoreactor engineering, to which PHOTOREAC is oriented. Independently of the photoreactor configuration, the central equation of SFM is given by [14]:

$$LVRPA = \frac{I_0}{\lambda_{\omega_{corr}}\omega_{corr}(1-\gamma)}\left[\left(\omega_{corr} - 1 + \sqrt{1-\omega_{corr}^2}\right)e^{-\frac{r_p}{\lambda_{\omega corr}}} + \gamma\left(\omega_{corr} - 1 - \sqrt{1-\omega_{corr}^2}\right)e^{\frac{r_p}{\lambda_{\omega corr}}}\right] \quad (1)$$

where I_0 is the incident solar radiation in W/m^2 and r_p is a spatial coordinate in the reactor domain whose definition depends on the reactor geometry. Finally, the corrected photon path length $\lambda_{\omega corr}$ in m, the dimensionless corrected scattering albedo ω_{corr} and the dimensionless parameter γ are all parameters derived from the SFM formulation. PHOTOREAC also includes a more recent variant of the SFM, the Six Flux Model coupled to the Henyey-Greenstein scattering phase function (SFM-HG). In it, the Henyey–Greenstein (HG) scattering phase function is used to describe the optical properties of the TiO_2 P25 photocatalyst. By contrast, the SFM describes TiO_2 based on a diffuse reflectance scattering phase function [15]. By incorporating both variants of SFM, the users will be able to observe the role of the scattering phase function. The parameters and implementation of Equation (1) are detailed in the literature, and the modeling details for the FFP are given in previous work [16].

On the other hand, for the CPCP and the OMTP, a ray-tracing technique together with Equation (1) must be implemented, since, besides the incident radiation, the direction with which solar rays impact the photoreactor is crucial. A complete description of the SFM implementation for CPCP and OMTP is reported elsewhere [10,13,14].

2.4. The PHOTOREAC Kinetic Modeling Module

The PHOTOREAC kinetic modeling module estimates the kinetic parameters from the photodegradation experimental data provided. Table 2 shows the photocatalytic kinetic models in PHOTOREAC. These models explicitly consider the effect of the radiation absorption on the average reaction rate in $\langle -r_i \rangle_{V_R}$ by including the E_g, and the overall rate of photon absorption (OVRPA) in W/m^3, which corresponds to the LVRPA averaged over the entire volume of the reactor. Additionally, C_i is the concentration of the water contaminant in mol/m^3, $\kappa_P = 2/S_g\ C_{cat}$ is the particle constant in m^3/m^2, S_g is the catalyst specific surface area m^2/kg, C_{Cat} is the photocatalyst concentration kg/m^3, C_{O2} is the oxygen concentration in mol/m^3 and ϕ_g^{eff} is the effective quantum yield in mol/(s watts). Finally, k^{L-H}, K_{kin}, α_1 and α_2 are the kinetic constants of the models, which are independent of the irradiation conditions.

Table 2. Photocatalytic kinetic models in PHOTOREAC.

Kinetic Model	Mathematical Expression	Fitting Parameters	Refs.
Langmuir–Hinshelwood	$\langle -r_i \rangle_{V_R} = -\frac{K_{Kin} k^{L-H} C_i}{1 + k^{L-H} C_i} (E_g)^{0.5}$	k^{L-H} (L/mol), K_{kin} (mol L^{-1} s^{-1} $W^{-0.5}$)	[13]
Zalazar et al.	$\langle -r_i \rangle_{V_R} = -\frac{\phi_g^{eff} E_g}{\frac{1}{2} + \left[\frac{1}{4} + K_{kin} \frac{\phi_g^{eff} E_g}{2 C_{cat}^2 C_i C_{O_2}}\right]^{0.5}}$	ϕ_g^{eff} (mol s^{-1} $watts^{-1}$), K_{kin} (mole s kg^2 m^{-9})	[17]
Ballari et al.	$\langle -r_i \rangle_{V_R} = -2 \frac{\alpha_1}{\kappa_p} \left[-1 + \sqrt{1 + \kappa_P \frac{\alpha_2 E_g}{C_i}} \right] C_i$	α_1 (cm s^{-1}), α_2 (mol $watts^{-1}$ cm^{-1})	[18]
Mueses et al.	$\langle -r_i \rangle_{V_R} = -2 \frac{\alpha_1}{\kappa_P} \left[-1 + \sqrt{1 + \frac{\kappa_P}{\alpha_1} \phi_g^{eff} E_g} \right] \frac{k^{L-H} C_i}{1 + k^{L-H} C_i}$	α_1 (mol m^{-2} s^{-1}), ϕ_g^{eff} (mol s^{-1} $watts^{-1}$), k^{L-H}(m^3 mol^{-1})	[12]

Each of these previous expressions has its features and limitations, from either a phenomenological or a numerical point of view. For instance, the Langmuir–Hinshelwood expression is a semi-empirical model. By contrast, the other models were deduced from a detailed reaction mechanism. Zalazar et al. and Mueses et al.'s kinetic expressions consider the effect of the effective quantum yield ϕ_g^{eff} explicitly, a critical parameter to evaluate photocatalytic reactions. However, the expression proposed by Mueses et al. is the only one with three fitting parameters, unless the effective quantum yield of the system is previously known [12].

To determine the kinetics parameters, it is necessary to follow a rigorous approach to account for the effects of the diffusion and convection in the material balance of the photoreactor. Although the inclusion of these effects will provide more accurate results for the kinetic parameters (such parameters will be independent of the diffusion and convection), it also implies the implementation of more advanced numerical techniques, e.g., finite differences and orthogonal collocation [7,19]. PHOTOREAC considers the photoreactor-tank system as a batch mode reactor; therefore, the effects of the diffusion and convection are lumped in the kinetic parameters, which simplifies the numerical approach.

The following assumptions are established for the mass balance of the system (represented by Figure 2): (i) the system is perfectly mixed; (ii) there are no mass transport limitations; (iii) the conversion per pass in the reactor is differential; and (iv) parallel dark reactions can be neglected. The mass balance in the reservoir tank can then be expressed as follows [7,18]:

$$\frac{dC_i}{dt} = \frac{V_R}{V_T} \langle -r_i \rangle_{V_R} \tag{2}$$

where C_i is the concentration of the water contaminant in mol/m^3 at time t, t is time in s, $\langle -r_i \rangle_{V_R}$ is the average reaction rate in (mol m^3 s^{-1}), and V_R and V_T are the volumes of the photoreactor and the total reaction volume in m^3, respectively. However, for solar photoreactors, the standard time may not be the more appropriate magnitude for following the concentration of the water pollutant due to the fluctuation of the incident solar irradiance because of the atmospheric phenomena and the time of day. Therefore, a change of variable is proposed as follows [10]:

$$\frac{dC_i}{dt} = \left(\frac{dC_i}{d\xi_{AE}}\right)\left(\frac{d\xi_{AE}}{dt}\right) \tag{3}$$

$$\frac{dC_i}{d\xi_{AE}} = \frac{\beta}{\xi_t} \langle -r_i \rangle_{V_R} \tag{4}$$

With the initial condition, C_i ($\xi_{AE} = 0$) = $C_{i,0}$, where C_i is the water contaminant concentration for a given ξ_{AE} is the accumulated energy in J/m^2, $\xi_t = \left(\frac{d\xi_{AE}}{dt}\right)$ in J/m^2s is the slope of the straight line resulting from the experimental data relationship of the accumulative incident solar radiation vs. time for each experimental test, and the dimensionless factor $\beta = V_R/V_T$.

The search for the best values for the kinetic parameters of the model is carried out using a non-linear regression procedure, as is shown in Figure 1. It starts with an initial guess and follows an optimization criterion until the required convergence is reached. The error function is given by the sum of the squared errors of the experimental water contaminant concentration $C_{i,\exp}$ and the value determined from the numerical solution of Equation (4) $C_{i,calc}$:

$$F_{obj} = \sum_{i=1}^{N} \left(C_{i,\exp} - C_{i,calc} \right)^2 \tag{5}$$

where N is the number of experimental data. The Matlab function fminsearch, which uses the Nelder–Mead algorithm, is implemented as the optimization solver together with the Matlab function ode 45 for solving the ordinary differential equation (ODE) given by Equation (4).

For the photodegradation of multicomponent mixtures, the concentration C_i may be replaced by a global concentration parameter such as total organic carbon (*TOC*) [12]. Therefore, Equation (4) is written as:

$$\frac{dTOC}{d\xi_{AE}} = \frac{\beta}{\xi_t} \langle -r_{TOC} \rangle_{V_R} \tag{6}$$

with the initial condition $TOC\ (\xi_{AE} = 0) = TOC_0$, where TOC is the total organic carbon of the mixture mol/m^3 for a given ξ_{AE}, ξ_{AE} is the accumulated energy in J/m^2, TOC_0 is TOC of the mixture measured at the starting point of the experiment, and V_R and V_T are the volumes of the photoreactor and the total reaction volume in m^3, respectively. $\langle -r_{TOC} \rangle_{V_R}$ is the average reaction rate of the TOC of the mixture in (mol m^3 s^{-1}). The mathematical expressions for $\langle -r_{TOC} \rangle_{V_R}$ are the same given in Table 2, replacing $\langle -r_i \rangle_{V_R}$ by $\langle -r_{TOC} \rangle_{V_R}$ and C_i by TOC.

Similarly, Equation (5) is rewritten as:

$$F_{obj} = \sum_{i=1}^{N} \left(TOC_{i,\exp} - TOC_{i,calc} \right)^2 \tag{7}$$

Then, for multicomponent mixtures, the *TOC* of the mixture must be provided to PHOTOREAC as a function of the accumulated energy instead of the concentration of a pure component water contaminant. This approach is particularly useful in real environmental applications because in such cases the most usual situation is that the content of the wastewater is unknown, and it would be tough and resource-consuming to determine it. Therefore, it is easier to establish a global parameter such as the *TOC*, which shows the mineralization of both intermediates and the precursor compounds in the wastewater. By contrast, the monitoring of each initial pure component in the mixture does not consider the formation of intermediates.

The Kinetic Modeling Module Database

In the kinetic modeling module, PHOTOREAC incorporates a database that consists of 26 datasets of the solar photocatalytic degradation of water contaminants using TiO_2 P25 Evonik as a photocatalyst. The information was collected by the Modeling and Applications of Advanced Oxidation Technologies Research Group at Cartagena University (Cartagena de Indias, Colombia) and the Research Group on Advanced Processes for Biological and Chemical Treatments (GAOX) at the Universidad del Valle (Cali, Colombia). Table 3 details the information available in the database: two solar photoreactor configurations (CPCP and OMTP) and five model pollutants at different initial concentrations and photocatalyst concentrations. By selecting the dataset to perform the kinetic analysis, PHOTOREAC loads the information about the experimental test: the pollutant concentration vs. accumulated energy data, the OVRPA and the $\beta = V_R/V_T$ factor.

Table 3. PHOTOREAC database of the solar photodegradation of water contaminants.

Water Contaminant	Photoreactor Configuration	Initial Concentration of the Contaminant, ppm	Photocatalyst Concentration, g/L
Dichloroacetic acid (DCA)	CPCP	30 60 120	0.1, 0.5 0.1, 0.35 0.1, 0.35, 0.5
	OMTP	60 120	0.35 0.35
Phenol (PH)	CPCP	60 120	0.1 0.1
	OMTP	60 120	0.1 0.1
4-chlorophenol (4-CP)	CPCP	60 120	0.5 0.5
	OMTP	60 120	0.5 0.5
Methylene Blue (MB)	CPCP	10	0.25
	OMTP	10	0.2, 0.25, 0.3, 0.35
Amoxicillin (AMX)	CPCP	20	0.3, 0.6, 0.9, 1.0

3. Implementation of PHOTOREAC in Solar Photoreactors

In this section, three example cases to demonstrate the use of the PHOTOREAC application are presented. All of the cases are based on an experimental test already performed in the solar photoreactor platforms of our research groups in Cartagena, Colombia (10°25′25″ N, 75°31′31″ W) and Cali, Colombia (3°27′00″ N, 76°32′00″ W). Further information about the set-up and operation of the experimental solar tests can be found in previous works [10,12].

3.1. Example Case I: Solar Photodegradation of Dichloroacetic Acid (DCA) in a CPCP

This example shows the implementation of PHOTOREAC for an analysis of the solar photocatalytic degradation of DCA in a CPCP. The photoreactor consists of ten borosilicate tubes with radius R = 0.016 m and length L = 1.2 m providing a reaction volume of V_R = 9.7 L. The DCA initial concentration was C_i = 30 ppm using a TiO_2 P25 Evonik concentration of C_{cat} = 0.5 g/L. The main objective of the example case was to determine the radiation-independent kinetic parameters of the system from the experimental data provided to the application using the SFM as the radiative model.

First, the radiation field is determined by the photon absorption-scattering module. Figure 3 shows the main screen of the PHOTOREAC GUI: (1) the photoreactor panel, where the photoreactor configuration was selected; (2) the system properties panel, where the input data were introduced for the simulation; (3) the SFM model panel, where the SFM variant for the simulation is selected; (4) the SFM scattering phase function probabilities are displayed according to the SFM variant that was selected, in this case, the SFM; (5) the resulting LVRPA spatial distribution in the cross-section of the CPCP tube is plotted; (6) the resulting OVRPA of the system is displayed; (7) the options menu. Together with the main screen shown in Figure 3, PHOTOREAC generates a secondary screen with the results of the ray-tracing simulation (Figure 4).

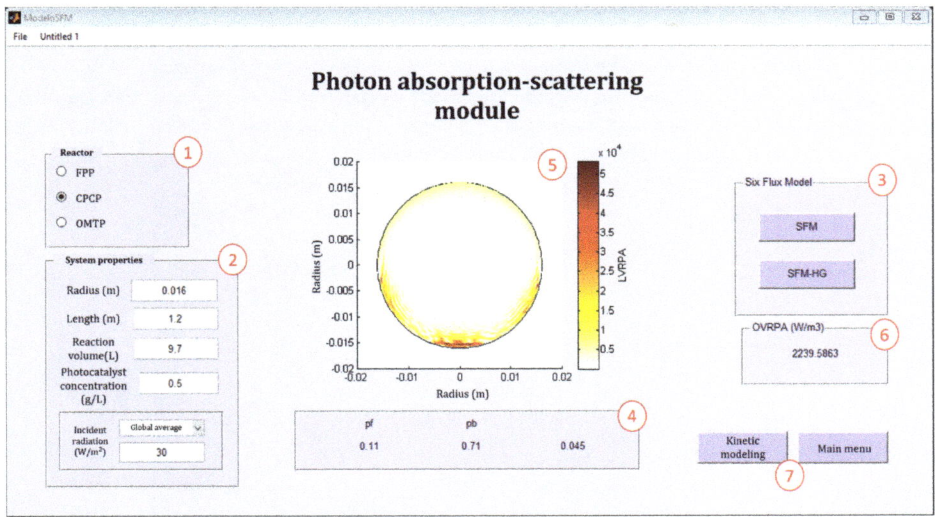

Figure 3. Radiation field simulation for a CPCP in the photon absorption-scattering module. (1) photoreactor panel; (2) system properties panel; (3) the SFM panel; (4) display the corresponding SFM scattering probabilities used in the simulation; (5) LVRPA spatial distribution plot; (6) OVRPA; (7) options menu.

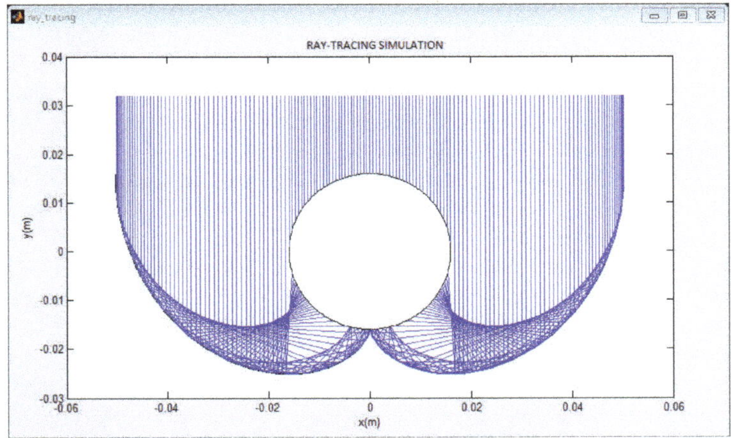

Figure 4. Ray-tracing simulation for the CPCP.

From the results presented by the photon absorption-scattering module, the user will be able to extract essential findings regarding the impact of variables on the photocatalyst concentration. For instance, for this example case, in the LVRPA distribution plot shown in Figure 3, it is observed that the highest values of the LVRPA are around $y = -0.015$ m and $y = -0.005$ m. This result is due to the fact that at these coordinates there is a high concentration of rays that come from the CPCP reflectors, as can be observed in Figure 4. Additionally, it is observed that the LVRPA is concentrated near to the CPCP wall, and the center of the tube shows very low LVRPA values, as a result of the relatively high photocatalyst concentration used in the simulation ($C_{cat} = 0.5$ g/L). This behavior is well-known in the literature: at high concentrations of the photocatalyst, the photons cannot penetrate deeply into the tube and the absorbed energy is concentrated around the boundary wall [14].

Once the radiation field for the CPCP is determined, the application proceeds to the kinetic modeling module. Figure 5 shows the input panel displayed by PHOTOREAC. The application loads the system parameters determined previously, such as the TiO_2 concentration and the OVRPA. The remaining system parameters must be provided manually by the user. Similarly, the photodegradation vs. accumulated energy data should be introduced in the experimental data panel. Finally, the user may proceed to the kinetic modeling module's main screen.

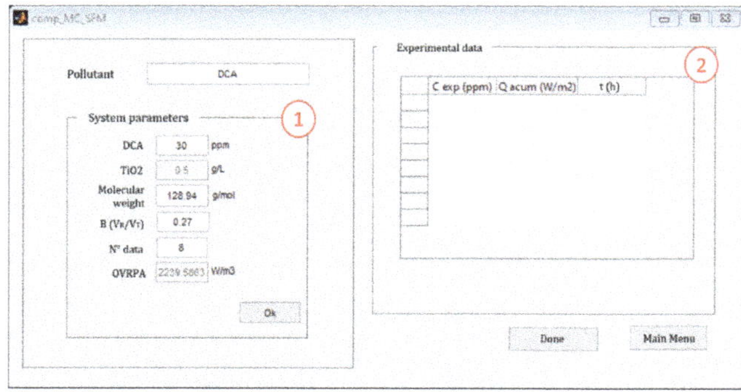

Figure 5. Input panel for the kinetic modeling module. (1) System parameters panel; (2) experimental data panel.

Figure 6 shows the main screen of the PHOTOREAC GUI at the kinetic modeling module: (1) kinetic models panel, where the user can choose the kinetic models to be fitted; (2) experimental data and models simulations plot, where the experimental data and the fitting curves of the models that were previously selected are displayed; (3) fitted kinetic parameters panel, where the values of the fitting parameters of each model chosen are displayed; (4) the x-axis magnitude panel, where the user can determine if the displayed data are presented in accumulated energy or standard time as the x-axis magnitude; (5) correlation coefficient panel, which displays the higher R^2 among the selected kinetic models; (6) correlation coefficient panel, which shows the kinetic model with the highest R^2 value among the chosen ones; (7) export data button, which exports the results of the fitting curve to a Microsoft Excel file; (8) options menu panel.

From the PHOTOREAC kinetic modeling module screen in Figure 6, it is observed that the best fitting is achieved for the Ballari et al. model with $R^2 = 0.97392$. The other models reported $R^2 = 0.97365$ for Mueses et al., $R^2 = 0.63843$ for Langmuir–Hinshelwood and $R^2 = 0.\ 63585$ for Zalasar et al. Due to PHOTOREAC only displaying the model with the higher value for the correlation coefficient R^2, it selected the Ballari et al. model. However, Mueses et al.'s expression showed an almost identical R^2, and it should not be discarded without further analysis. From Table 2, it is observed that the mathematical structure of the Ballari et al. and Mueses et al. expressions are very similar; indeed, the Ballari et al. expression is considered a particular case of the Mueses et al. model for systems with high molecular adsorption [12]. Therefore, it is expected that both models performed similarly, as is the case for the DCA photodegradation. The Langmuir–Hinshelwood and Zalasar et al. expressions may not lead to successful results due to the fact that they do not describe the effects of the absorbed radiation (OVRPA) accurately. On the other hand, Ballari et al. and Mueses et al. may perform better since they include an OVRPA squared root correction factor. the same can be said for Ballari et al. and Mueses et al. regarding the OVRPA squared root correction factor.

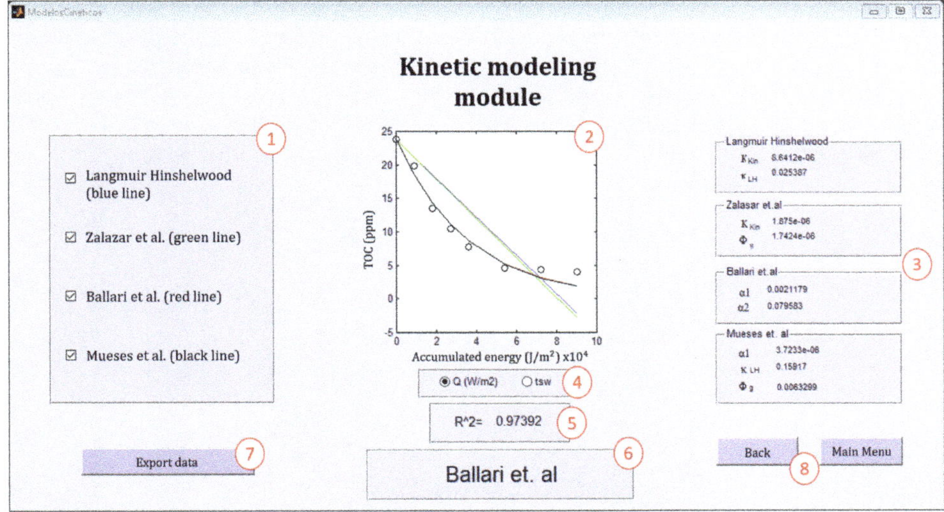

Figure 6. DCA fitting for the available kinetics models in the kinetics modeling module. (1) Kinetic models panel; (2) experimental data and models simulations plot; (3) fitted kinetic parameters panel; (4) *x*-axis magnitude panel; (5) correlation coefficient panel; (6) display of the kinetic model with the highest R^2 value among the ones selected; (7) export data button; (8) options menu panel.

3.2. Example Case II: Solar Photodegradation of Methylene Blue in an OMTP

In the previous example case, the user must provide all the required information to perform the computations. In this example, the use of the database incorporated in the PHOTOREAC kinetic modeling module is shown. Figure 7 shows the PHOTOREAC screen of the kinetic modeling module: (1) the photoreactor configuration panel, for selecting the photoreactor to be studied; (2) the model pollutant panel, for choosing the water contaminants from the five available options in the database; (3) the photocatalyst-pollutant panel, for choosing the photocatalyst concentration-initial pollutant concentration combination from the available options in the database; (4) the experimental data panel, for loading the pollutant concentration vs. accumulated energy (or standard time); (5) the system parameters panel, which displays the OVRPA and the $\beta = V_R/V_T$ factor charged.

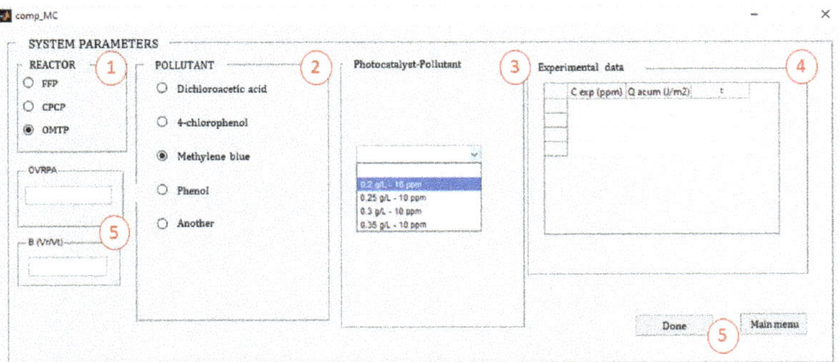

Figure 7. Database in the kinetic modeling module. (1) photoreactor configuration panel; (2) model pollutants panel; (3) photocatalyst-pollutant panel; (4) experimental data panel; (5) system parameters panel.

In this case, an OMTP with methylene blue (MB) was selected as a model pollutant with an initial concentration of C_i = 10 ppm and a photocatalyst concentration of C_{cat} = 0.2 g/L. Figure 8 shows the results obtained by PHOTOREAC. It is observed that the best fitting is achieved for the Mueses et al. and Ballari et al. models with R^2 = 0.99737 for both. The other models reported R^2 = 0.001. As the Ballari et al. model is a particular case of the Mueses et al. model, the first is considered the more appropriate option since it is more specific for this case. These results agree with the discussion presented in the previous section.

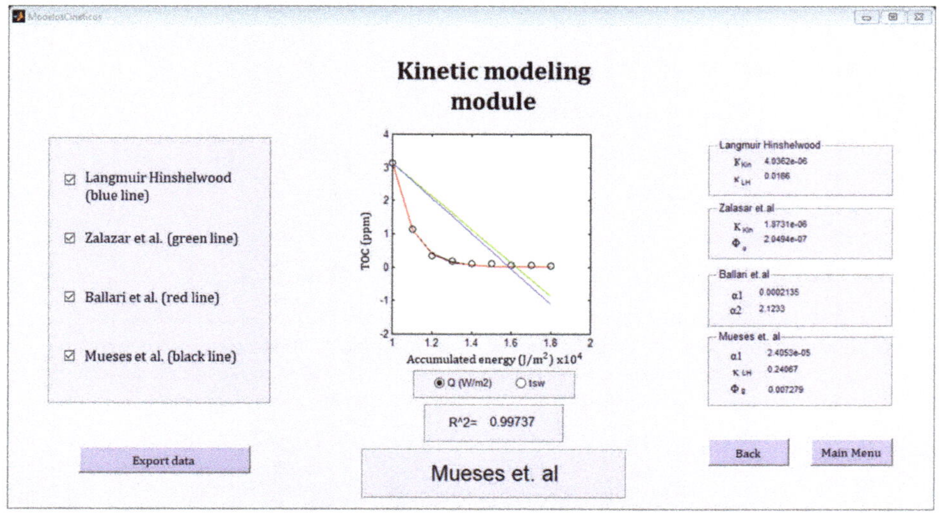

Figure 8. Methylene blue solar photodegradation fitting in an offset multi-tubular photoreactor (OMTP) for the four available kinetics models in the kinetic modeling module.

3.3. *Example Case III: Radiation Field Modeling in a Flat Plate Photoreactor (FPP)*

In this case, the objective was to compare the radiation field simulation for an FPP using SFM and SFM-HG. The photoreactor consists of a titled squared flat plate of length L = 1 m, which is placed facing the sun and uniformly irradiated. A water film of 1 cm thickness flows over its surface, providing a reaction volume of V_R = 10 L. The TiO_2 P25 Evonik concentration is C_{cat} = 0.2 g/L. Figures 9 and S1 show the LVRPA profile in the FPP calculated with the SFM-HG and the SFM, respectively. In both cases, the highest LVRPA values are found near to the surface of the water film (thickness = 0–0.2 cm) because this is the boundary that the solar light irradiates. After 0.2 cm, exponential decay in the LVRPA occurs as a result of the absorption and scattering of photons by the suspended photocatalyst. Due to the photoreactor being considered as uniformly irradiated, the changes in the LVRPA profile are only significant along with the water film thickness.

In Figure S1, which uses the SFM, a shaper exponential LVRPA profile is observed, with higher values near the irradiated boundary (at thickness = 0–0.2 cm) when comparing to values in Figure 9, which uses the SFM-HG. These results are due to the difference in the scattering phase function; the SFM-HG uses a predominantly forward scattering phase function, which causes photons to penetrate deeper into the water film. By contrast, the SFM uses a predominantly backward phase function, which causes that photons to be redirected toward the irradiated boundary and be mostly absorbed in the beginning of the film or escape from the system [15].

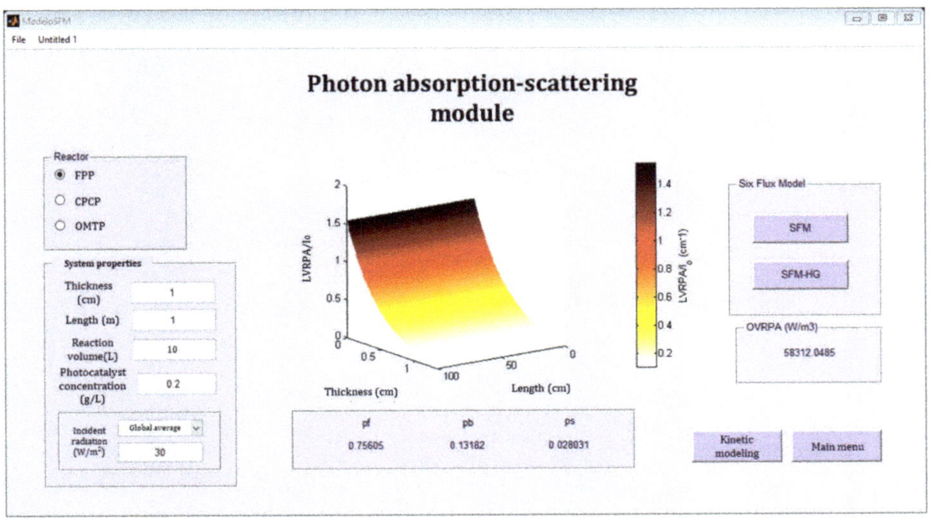

Figure 9. Radiation field simulation of a flat plate photoreactor (FPP) with SFM-HG.

4. PHOTOREAC Implementation in Research Projects in Heterogeneous Photocatalysis and Photoreactor Engineering by Chemical Engineering Undergraduates

Since the year 2015, different versions of PHOTOREAC have supported the final degree projects of chemical engineering students belonging to the Modeling and Applications of Advanced Oxidation Technologies Research Group at Cartagena University. The students developed research on heterogeneous photocatalysis and photoreactor engineering. A survey was done amongst them to determine the perceived impact of PHOTOREAC on their final degree projects. Table 4 shows the results of the survey. Between 2015–2018, ten final degree projects were developed in the research group, with an average impact of 37.5% perceived by the students. The use of PHOTOREAC can be summarized as follows: in four of the degree projects, both modules of PHOTOREAC were implemented since they performed the radiation field simulation and kinetic modeling of model pollutants; in two other projects, the photon-scattering module was used to determine the radiation field in photoreactors; finally, in four projects, the application was used in the learning process for modeling solar photoreactors. As a relevant outcome, two of the degree projects supported publications in high-impact journals. In all of the projects, the authors highlighted the use of PHOTOREAC as a user-friendly tool that allows them to reach the main objective of the projects or to achieve a fast advance in the learning curve, therefore allowing them to focus on more complex research.

Table 4. PHOTOREAC impact on final degree projects in chemical engineering.

Year	Title of the Final Degree Project	Related Publication/Ref.	PHOTOREAC Impact on the Project Perceived by the Students	PHOTOREAC Implementation in the Project
2015	Design and evaluation of a modified compound parabolic collector solar reactor	A Novel Prototype Offset Multi Tubular Photoreactor (OMTP) for solar photocatalytic degradation of water contaminants/ref. [10]	30%	Modeling the radiation field and kinetics of methylene blue for both the CPCP and OMTP
2015	Effect of oxygen transfer from the air on the photocatalytic degradation of dichloroacetic acid using a flat plate reactor	-	50%	Modeling the radiation field and kinetics of dichloroacetic acid in an FFP

Table 4. *Cont.*

Year	Title of the Final Degree Project	Related Publication/Ref.	PHOTOREAC Impact on the Project Perceived by the Students	PHOTOREAC Implementation in the Project
2016	Radiant field modeling in heterogeneous photoreactors implementing Monte Carlo simulation: Modification of the Six Flux Model to new phase functions	Coupling the Six Flux Absorption-Scattering Model to the Henyey–Greenstein scattering phase function: Evaluation and optimization of radiation absorption in solar heterogeneous photoreactors/ref. [15]	60%	Modeling the radiation field for the FFP and a CPCP
2016	Evaluation of the temperature effect on the heterogeneous photocatalytic degradation kinetics	Modeling and experimental evaluation of a non-isothermal photocatalytic solar reactor: temperature effect on the reaction rate kinetics/ref. [20]	20%	The learning process for modeling CPCP
2016	Solar heterogeneous photocatalytic degradation of organic pollutants in a pilot-scale modified tubular collector	A Novel Prototype Offset Multi Tubular Photoreactor (OMTP) for solar photocatalytic degradation of water contaminants/ref. [10]	50%	Modeling the radiation field and kinetics of DCA, PH and 4-CP for both the CPCP and OMTP
2016	Simulation of in series and in parallel arrangements of solar reactors (CPCP) for wastewater treatment	–	30%	The learning process for modeling CPCP
2016	Experimental evaluation and mathematical modeling of the performance of TiO_2-P25 reuse in heterogeneous solar photocatalytic degradation of acetaminophen	–	30%	The learning process for modeling CPCP
2017	Solar photocatalytic ozonation applied to amoxicillin degradation in wastewater at pilot-plant scale	–	30%	The learning process for modeling CPCP
2018	Mathematical modeling and simulation of photocatalytic hydrogen production	–	35%	Radiation field modeling of an FFP
2018	Experimental evaluation and mathematical modeling of the regeneration of commercial TiO_2 by the photocatalytic degradation of glyphosate	–	40%	Modeling the radiation field and the kinetics glyphosate in a CPCP

5. Analysis of the Overall Performance of PHOTOREAC

PHOTOREAC was shown to be a useful tool for modeling and simulation of solar photoreactors, and in particular for a non-expert public. Its user-friendly interface developed in the graphical user interface of Matlab proved to be intuitive enough to be used successfully by chemical engineering undergraduates, which develop research in heterogeneous photocatalysis.

In Section 3, the application was evaluated for disparate operational conditions, showing that it can fit and simulate the photodegradation experimental data provided for the two cases evaluated: CPCP-DCA (example case I) and OMTP-MB (example case II). These example cases were very different from each other, mainly because of the different photoreactor geometries: CPCP can capture more solar radiation per length than the OMTP as a result of being equipped with reflectors. However, OMTP has more volume than the CPCP [10]. Additionally, the employed model pollutant, its initial concentration, and the photocatalyst concentration were different. In both examples, PHOTOREAC performed successfully, allowing the user to evaluate different kinetic expressions and extract relevant findings from it. Finally, example case III focused on the photon absorption-scattering module, in which the impact of the radiation model was evaluated and discussed for an FFP. In this case, PHOTOREAC shows its versatility for researchers with an interest in studying the energy absorption behaviors of photoreactors.

The dedicated interface of PHOTOREAC for photoreaction engineering, together with its numerical algorithm, allowed the evaluation of the performance of large scale solar photoreactors without time-consuming computations and a complex mathematical formulation. The time invested in preparing and launching a simulation in PHOTOREAC is between 5–10 min, and the calculation time does not exceed 45 s. In contrast to commercial CFD simulators in which preparing and starting a first-time simulation may take a couple of hours needed for generating the photoreactor geometry in the system (or it importing it from CAD software), preparing the simulation modules and their models and selecting the proper meshing and numerical algorithms; besides, the computational time for each simulation is, generally, measured in hours [21–23]. Nevertheless, the results obtained by CFD simulators are much more complete and accurate than the results that PHOTOREAC may offer; for instance, CFD simulators provide detailed flow patterns for studying the hydrodynamics in the photoreactor. However, its high computational time may result in a barrier when exploring the impact of numerous parameters on a wide range of values.

Moreover, the most common CFD commercial simulators used in modeling photoreactors are very expensive licensed software. At the same time, PHOTOREAC is an open-access application that is available on-demand, by email to one of the authors of the paper, professor Miguel A. Mueses (mmueses@unicartagena.edu.co).

In conclusion, PHOTOREAC is recommended for the following cases: (i) for an introduction to photoreactor engineering; (ii) when a quantitative margin of error is still acceptable in the calculations; (iii) when qualitative results are the main objective of the work; and (iv) when the parametric space in the study is extensive, i.e., it is required to study the impact of numerous variables in broad ranges. In this case, PHOTOREAC may be employed to reduce the parametric space and then to implement a CFD simulator.

6. Limitations and Future Work

As with every modeling software, PHOTOREAC is limited by the availability and reliability of the input data provided by the users. Additionally, the computational application is limited to Titanium Dioxide P25 Evonik as photocatalyst. Although TiO_2 P25 is the most common photocatalyst, the capability of performing simulations for any photocatalyst will be crucial for the software, since an area of intensive research in heterogeneous photocatalysis is the development and testing of new photoactive materials. On the other hand, expanding the available kinetic models would also be a considerable improvement, because it will allow users to make a more comprehensive analysis by comparing the results of the kinetic models' fitting. Moreover, it is necessary to implement the option that users introduces their own kinetic expression, since some pollutants will require concrete mathematical expression because their kinetic mechanism may not follow the most common postulates. These drawbacks are expected to be overcome in the upcoming version of PHOTOREAC.

In the authors' opinion, some important challenges for PHOTOREAC and, in general, for photoreaction engineering at the pilot-solar scale are that the models account for the variability of the incident radiation on the solar photoreactor caused by fluctuations in atmospheric conditions. This improvement will allow more accurate quantification of the energy absorbed by the suspended photocatalyst, and therefore better quantification of the chemical species produced by the photoactivation of the photocatalyst.

Supplementary Materials: The following are available online at http://www.mdpi.com/2073-4441/12/8/2196/s1, Figure S1: Radiation field simulation of a flat plate photoreactor (FPP) with SFM.

Author Contributions: Conceptualization, B.C.-V., K.P.-G. and M.A.M.; data curation, B.C.-V., and K.P.-G.; formal analysis, R.A.-H.; funding acquisition, M.H.P.-C. and F.M.-M.; investigation, B.C.-V. and K.P.-G.; methodology, B.C.-V. and K.P.-G.; project administration, M.H.P.-C. and F.M.-M.; supervision, R.A.-H. and M.A.M.; validation, B.C.-V. and K.P.-G.; writing—original draft, R.A.-H.; writing—review and editing, R.A.-H., M.H.P.-C., M.A.M. and F.M.-M. All authors have read and agreed to the published version of the manuscript.

Funding: This research was funded by Universidad del Valle, grant number C.I. 21022.

Acknowledgments: The authors gratefully thank Universidad de Cartagena (Cartagena, Colombia) with the project 017-2018: "Plan de Fortalecimiento del Grupo de Investigación Modelado y Aplicación de Procesos Avanzados de Oxidación" and Universidad del Valle (Cali, Colombia) with the Project CI. 21022 "Estudio del efecto hidrodinámico y de transporte de energía radiante en el diseño y optimización de reactores fotocatalíticos heterogéneos solares. CI 21022." for financial support. Acosta-Herazo thanks the CEIBA foundation with the program "Bolivar Gana con Ciencia" for financing his doctoral studies.

Conflicts of Interest: The authors declare no conflict of interest.

References

1. Byrne, C.; Subramanian, G.; Pillai, S.C. Recent advances in photocatalysis for environmental applications. *J. Environ. Chem. Eng.* **2018**, *6*, 3531–3555. [CrossRef]
2. Kumaravel, V.; Mathew, S.; Bartlett, J.; Pillai, S.C. Photocatalytic hydrogen production using metal doped TiO2: A review of recent advances. *Appl. Catal. B Environ.* **2019**, *244*, 1021–1064. [CrossRef]
3. Spasiano, D.; Marotta, R.; Malato, S.; Fernandez-Ibañez, P.; Di Somma, I. Solar photocatalysis: Materials, reactors, some commercial, and pre-industrialized applications. A comprehensive approach. *Appl. Catal. B Environ.* **2015**, *170–171*, 90–123. [CrossRef]
4. Cassano, A.E.; Alfano, O.M. Reaction engineering of suspended solid heterogeneous photocatalytic reactors. *Catal. Today* **2000**, *58*, 167–197. [CrossRef]
5. Li Puma, G.; Brucato, A. Dimensionless analysis of slurry photocatalytic reactors using two-flux and six-flux radiation absorption-scattering models. *Catal. Today* **2007**, *122*, 78–90. [CrossRef]
6. Grčić, I.; Li Puma, G. Photocatalytic degradation of water contaminants in multiple photoreactors and evaluation of reaction kinetic constants independent of photon absorption, irradiance, reactor geometry, and hydrodynamics. *Environ. Sci. Technol.* **2013**, *47*, 13702–13711. [CrossRef]
7. Marugán, J.; van Grieken, R.; Pablos, C.; Satuf, M.L.; Cassano, A.E.; Alfano, O.M. Rigorous kinetic modelling with explicit radiation absorption effects of the photocatalytic inactivation of bacteria in water using suspended titanium dioxide. *Appl. Catal. B Environ.* **2011**, *102*, 404–416. [CrossRef]
8. Colina-Marquez, J.; Castilla-Caballero, D.; Machuca-Martinez, F. Modeling of a falling-film photocatalytic reactor: Fluid dynamics for turbulent regime. *Appl. Math. Model.* **2016**, *40*, 4812–4821. [CrossRef]
9. Malato, S.; Maldonado, M.I.; Fernández-Ibáñez, P.; Oller, I.; Polo, I.; Sánchez-Moreno, R. Decontamination and disinfection of water by solar photocatalysis: The pilot plants of the Plataforma Solar de Almeria. *Mater. Sci. Semicond. Process.* **2016**, *42*, 15–23. [CrossRef]
10. Ochoa-Gutiérrez, K.S.; Tabares-Aguilar, E.; Mueses, M.Á.; Machuca-Martínez, F.; Li Puma, G. A Novel Prototype Offset Multi Tubular Photoreactor (OMTP) for solar photocatalytic degradation of water contaminants. *Chem. Eng. J.* **2018**, *341*, 628–638. [CrossRef]
11. Manassero, A.; Satuf, M.L.; Alfano, O.M. Evaluation of UV and visible light activity of TiO2 catalysts for water remediation. *Chem. Eng. J.* **2013**, *225*, 378–386. [CrossRef]
12. Mueses, M.A.; Machuca-Martinez, F.; Li Puma, G. Effective quantum yield and reaction rate model for evaluation of photocatalytic degradation of water contaminants in heterogeneous pilot-scale solar photoreactors. *Chem. Eng. J.* **2013**, *215–216*, 937–947. [CrossRef]
13. Colina-Márquez, J.; Machuca-Martínez, F.; Li Puma, G. Photocatalytic mineralization of commercial herbicides in a pilot-scale solar CPC reactor: Photoreactor modeling and reaction kinetics constants independent of radiation field. *Environ. Sci. Technol.* **2009**, *43*, 8953–8960. [CrossRef] [PubMed]
14. Colina-Marquez, J.; Machuca-Martínez, F.; Puma, G.L. Radiation absorption and optimization of solar photocatalytic reactors for environmental applications. *Environ. Sci. Technol.* **2010**, *44*, 5112–5120. [CrossRef] [PubMed]

15. Acosta-Herazo, R.; Monterroza-Romero, J.; Mueses, M.A.; Machuca-Martínez, F.; Li Puma, G. Coupling the Six Flux Absorption-Scattering Model to the Henyey-Greenstein scattering phase function: Evaluation and optimization of radiation absorption in solar heterogeneous photoreactors. *Chem. Eng. J.* **2016**, *302*, 86–96. [CrossRef]
16. Otálvaro-Marín, H.L.; Mueses, M.A.; Machuca-Martínez, F. Boundary layer of photon absorption applied to heterogeneous photocatalytic solar flat plate reactor design. *Int. J. Photoenergy* **2014**, *2014*. [CrossRef]
17. Zalazar, C.S.; Romero, R.L.; Martín, C.A.; Cassano, A.E. Photocatalytic intrinsic reaction kinetics I: Mineralization of dichloroacetic acid. *Chem. Eng. Sci.* **2005**, *60*, 5240–5254. [CrossRef]
18. De Los Ballari, M.M.; Alfano, O.O.; Cassano, A.E. Photocatalytic degradation of dichloroacetic acid. A kinetic study with a mechanistically based reaction model. *Ind. Eng. Chem. Res.* **2009**, *48*, 1847–1858. [CrossRef]
19. Otálvaro-Marín, H.L.; González-Caicedo, F.; Arce-Sarria, A.; Mueses, M.A.; Crittenden, J.C.; Machuca-Martinez, F. Scaling-up a heterogeneous H2O2/TiO2/solar-radiation system using the DamkÖhler number. *Chem. Eng. J.* **2019**, *364*, 244–256. [CrossRef]
20. Molano, M.; Mueses, M.A.; Fiderman, M.M. Modelado y evaluación experimental de un reactor solar fotocatalítico no isotérmico: Efecto de la temperatura sobre la cinética de la velocidad de reacción. *Ing. Compet.* **2017**, *19*, 143–154. [CrossRef]
21. Casado, C.; García-Gil, Á.; van Grieken, R.; Marugán, J. Critical role of the light spectrum on the simulation of solar photocatalytic reactors. *Appl. Catal. B Environ.* **2019**, *252*, 1–9. [CrossRef]
22. Casado, C.; Marugán, J.; Timmers, R.; Muñoz, M.; van Grieken, R. Comprehensive multiphysics modeling of photocatalytic processes by computational fluid dynamics based on intrinsic kinetic parameters determined in a differential photoreactor. *Chem. Eng. J.* **2017**, *310*, 368–380. [CrossRef]
23. Boyjoo, Y.; Ang, M.; Pareek, V. Some aspects of photocatalytic reactor modeling using computational fluid dynamics. *Chem. Eng. Sci.* **2013**, *101*, 764–784. [CrossRef]

MDPI
St. Alban-Anlage 66
4052 Basel
Switzerland
Tel. +41 61 683 77 34
Fax +41 61 302 89 18
www.mdpi.com

Water Editorial Office
E-mail: water@mdpi.com
www.mdpi.com/journal/water

www.ingramcontent.com/pod-product-compliance
Lightning Source LLC
LaVergne TN
LVHW071017170726
843515LV00006B/1152

* 9 7 8 3 0 3 6 5 1 1 1 2 2 *